HANDBOOK OF
Transcription Factor NF-kappaB

HANDBOOK OF
Transcription Factor NF-kappaB

Edited by
Sankar Ghosh

CRC Press
Taylor & Francis Group
Boca Raton   London   New York

CRC Press is an imprint of the
Taylor & Francis Group, an informa business

CRC Press
Taylor & Francis Group
6000 Broken Sound Parkway NW, Suite 300
Boca Raton, FL 33487-2742

© 2007 by Taylor & Francis Group, LLC
CRC Press is an imprint of Taylor & Francis Group, an Informa business

No claim to original U.S. Government works
Printed in the United States of America on acid-free paper
10 9 8 7 6 5 4 3 2 1

International Standard Book Number-10: 0-8493-2794-6 (Hardcover)
International Standard Book Number-13: 978-0-8493-2794-0 (Hardcover)

Library of Congress Cataloging-in-Publication Data

Handbook of transcription factor NF-kappaB / [edited by] Sankar Ghosh.
    p. ; cm.
   Includes bibliographical references and index.
   ISBN-13: 978-0-8493-2794-0 (alk. paper)
    1. NF-kappa B (DNA-binding protein)--Handbooks, manuals, etc. 2. Transcription factors--Handbooks, manuals, etc. 3. Gene expression--Handbooks, manuals, etc. I. Ghosh, Sankar, 1959-
    [DNLM: 1. NF-kappa B. QU 58.5 H2368 2006]

QP552.N46H36 2006
611'.018166--dc22   2006013488

Visit the Taylor & Francis Web site at
http://www.taylorandfrancis.com

and the CRC Press Web site at
http://www.crcpress.com

Preface

In 1986, my laboratory was searching for transcription factors that might control the activation of the kappa immunoglobulin light chain when we came across NF-κB. We thought it was specific to B-lymphocytes and never imagined that it would turn out to be among the most protean of transcription factors ever discovered. But that realization came during the next few years as we first found that it could be induced in many cells and then found that it was stored in the cytoplasm, ready to be activated by a wide range of inducing stimuli. When its sequence was first determined, we realized that it had a history before 1986; Howard Temin's laboratory had discovered the Rel oncogene years before and it is a slightly altered member of the NF-κB family of proteins.

This book contains much that is known about the basic biology of NF-κB or at least references to it. Its clearest function is as a transcription factor orchestrating the activation and repression of many genes involved in inflammation, but it also helps activate genes of innate and adaptive immunity in response to pathogens, maintains liver integrity in the developing mouse, plays multiple roles in the nervous system, and is implicated in skin, hair, and tooth formation and much more in the vertebrate body. In insects, homologues of the mammalian proteins are involved in defense against pathogens but also play key roles in development (for example, *dorsal* in *Drosophila* establishes dorsal/ventral polarity in the developing embryo).

Any system in the body that plays so many important roles is likely to also cause pathology when it goes awry, and NF-κB is no exception. Its role as an oncogene showed early on that it could be a cause of cancer, and the linkage of cancer and inflammation, which is so widely discussed today, is another reason for linking NF-κB to cancer induction. NF-κB is implicated in a wide range of other pathologies, including heart disease, muscular dystrophy, and many other conditions involving chronic inflammation.

In 1986, we knew only a few transcription factors and only a few of the intracellular proteins that transduce signals from cell surface receptors. Today, we stare at myriads of components whose interactions make urban subway systems look simple. A remarkable fraction of these intracellular mediators are parts of pathways that lead to or through NF-κB. Thus, what was supposed to be a simple regulator of a particular step in differentiation has turned out to be linked by transduction pathways to many of the cellular signaling pathways. This book then is one of many that should be read together if the full richness of the cell's transduction machinery is to be understood. No one person can today absorb this body of knowledge and no one can keep pace with its continued growth. But each member of the community of interested scientists must absorb what she or he can and then share in the interactions that bring together the joint knowledge that one day will describe all of cellular metabolism. Because new students interested in these systems are continually entering the field, a book such as this is an entry point for them, and it is to these newcomers that I dedicate this limited preface.

David Baltimore

The Editor

Sankar Ghosh, Ph.D., is currently a Professor of Immunobiology and Molecular Biophysics and Biochemistry at Yale University School of Medicine. Dr. Ghosh obtained his undergraduate education in Calcutta, India, before coming to the United States in 1982 to pursue his Ph.D. at the Albert Einstein College of Medicine in New York. After obtaining his Ph.D. in 1998 he joined the laboratory of Dr. David Baltimore at the Whitehead Institute/Massachusetts Institute of Technology to carry out postdoctoral studies in molecular immunology. During his postdoctoral fellowship, Dr. Ghosh made major findings in the NF-κB field including establishing the importance of phosphorylation in the regulation of NF-κB, and cloning the genes encoding NF-κB p50, p65, and IκB genes. Dr. Ghosh then joined the faculty at Yale in 1991, where he has continued to investigate mechanisms responsible for regulating NF-κB activity.

Contributors

Albert S. Baldwin
Lineberger Comprehensive Cancer
   Center
University of North Carolina School of
   Medicine
Chapel Hill, North Carolina

Yinon Ben-Neriah
Lautenberg Center for Immunology
Hadassah Medical School
Hebrew University
Jerusalem, Israel

Lin-Feng Chen
Gladstone Institute of Virology and
   Immunology
University of California, San Francisco
San Francisco, California

Gourisankar Ghosh
University of California, San Diego
La Jolla, California

Warner C. Greene
Gladstone Institute of Virology and
   Immunology
University of California, San Francisco
San Francisco, California

Hans Häcker
Laboratory of Gene Regulation and
   Signal Transduction
School of Medicine
University of California, San Diego
La Jolla, California

Matthew S. Hayden
Yale University School of Medicine
New Haven, Connecticut

Tom Huxford
San Diego State University
San Diego, California

Alain Israël
Institut Pasteur
Paris, France

Michael Karin
School of Medicine
University of California, San Diego
La Jolla, California

Steven C. Ley
Division of Immune Cell Biology
National Institute for Medical Research
London, England

Anu K. Moorthy
University of California, San Diego
La Jolla, California

Vinay Tergaonkar
Laboratory of Genetics
The Salk Institute for Biological Studies
La Jolla, California

Inder M. Verma
Laboratory of Genetics
The Salk Institute for Biological Studies
La Jolla, California

Sebo Withoff
Laboratory of Genetics
The Salk Institute for Biological Studies
La Jolla, California

Contents

1 The NF-κB Pathway: A Paradigm for Inducible Gene Expression

Sankar Ghosh

CONTENTS

1.1   INTRODUCTION

The eukaryotic transcription factor NF-κB was first identified by Sen and Baltimore 20 years ago as a protein that bound to a specific decameric DNA sequence (5′-GGGACTTTCC-3′) within the intronic enhancer of the immunoglobulin kappa light chain gene in mature B and plasma cells but not pre-B cells [1]. Subsequently, these same workers demonstrated induction of NF-κB:DNA binding in other cell types in response to exogenously applied stimuli, such as phorbol esters, and showed that this activation was independent of *de novo* protein synthesis [2]. In the decades following its discovery, NF-κB has been shown to exist in most cell types, and specific NF-κB binding sites termed κB sites have been identified in the promoters/enhancers of a very large number of inducible genes. Similarly, the range of biological factors and environmental conditions known to induce NF-κB activity has been shown to be remarkably large and diverse. Efforts to understand how the many intracellular signals evoked by such diverse stimuli can be integrated at the level of activation of this transcription factor has led to fundamental insights about many aspects of gene regulation and signal transduction. These insights in turn have stimulated discoveries in different biological pathways.

In keeping with the enormous research activity targeted toward understanding the regulation of NF-κB in different biological systems, a large number of reviews have been, and are being, published discussing different features of this transcription factor. However, the breadth of information makes it impossible for any particular review, or even a collection of reviews, to provide an adequate summary of the current knowledge about NF-κB. The purpose of this volume is to collect in one place a number of chapters, written by leading authorities who were involved in making the seminal discoveries that have shaped this field, in a form that emphasizes current knowledge, with enough historical information to make the process of discovery apparent. Significant effort has been invested in making the chapters similar in style, giving this volume the feel of a textbook. It is also anticipated that with input from the community, this handbook will be updated at regular intervals, to ensure that the information included remains current and useful.

We begin this handbook on NF-κB by first providing a summary introduction to the key features of this transcription factor. We believe this information will provide an uninitiated reader easier access to the complexities in the NF-κB system. Subsequent chapters will delve in greater detail into different regulatory features of this transcription factor, culminating with a discussion of current efforts to develop therapies aimed at inhibiting the aberrant activity of NF-κB in different diseases.

1.2 NF-κB/REL PROTEINS

In the majority of cell types NF-κB exists in the cytoplasm as homo- or heterodimers of a family of structurally related proteins. Mammalian cells express five members that contain a conserved N-terminal region called the Rel homology domain (RHD) within which lies the DNA-binding and dimerization domains and the nuclear localization signal (NLS) (Chapter 2). In unstimulated cells, NF-κB complexes are sequestered in an inactive form via interaction with a monomer of an inhibitory protein called IκB, which itself belongs to a structurally and functionally related family of molecules. Signals that induce NF-κB activity cause degradation of IκB, allowing NF-κB dimers to translocate to the nucleus and induce gene expression (Figure 1.1). Although in most cases NF-κB activity must be induced, in certain cell types, e.g., mature B cells, thymocytes, monocytes, macrophages, neurons, corneal keratinocytes, vascular smooth muscle cells, and many tumor cells, NF-κB can also be detected as a constitutively active, nuclear protein.

The first reports of purification of NF-κB DNA-binding subunits to homogeneity identified two proteins with molecular weights of approximately 50 and 65 kDa [3,4], respectively. Purification was performed using κB-site-specific DNA affinity chromatography and the two proteins, referred to as p50 (also called NF-κB1) and p65 (RelA), were subsequently shown to form heterodimers that bind specifically to cognate κB-sites. The functional domains of NF-κB/Rel proteins were determined after molecular cloning of p50, which revealed that the N-terminal 300 amino acids, subsequently termed the Rel-homology domain, was highly homologous to the oncogene *v-Rel*, its corresponding cellular homologue c-Rel and the *Drosophila* protein Dorsal [4,5]. Further studies revealed that p65/RelA, p100/p52 (NF-κB2), and Rel-B were also members of the mammalian Rel-family, thereby bringing the total to five. Almost every combination of NF-κB/Rel proteins as homo- or heterodimers, in either the cytoplasm or nucleus of many different cell types, has been described. For example, p50/p65 heterodimers are found in the cytoplasm of most

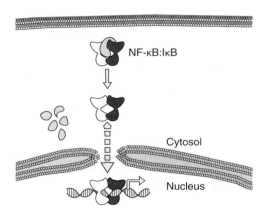

FIGURE 1.1 The basic paradigm of the NF-κB signaling pathway. In uninduced cells, NF-κB remains in the cytosol bound to inhibitory IκB proteins. Stimulation of cells leads to the activation of the IκB kinase complexes that phosphorylate the IκB proteins. The phosphorylated IκB proteins are ubiquitinated and degraded, allowing the released NF-κB to translocate to the nucleus and drive gene expression.

cells, whereas a constitutively active nuclear NF-κB in mature B cells is a dimer of p50/c-Rel. Dimers of p52/c-Rel, p65/c-Rel, p65/p65, p50/p50, p52/p52, and p50/p52 have also been identified. RelB is a notable exception in that it only forms dimers with p50 or p52. These RelB/p52-containing dimers are generated as a result of signaling through the "alternative" NF-κB activation pathway (see Section 1.4 and Chapters 2 and 3) and are believed to be the key dimeric form of NF-κB responsible for expression of genes involved in lymphoid organogenesis.

Not all combinations of NF-κB/Rel proteins are transcriptionally active; and the members of this family can be divided into two groups based upon their transactivating ability (see Chapter 5 for a discussion of additional transcriptional requirements). First, p65, c-Rel, and RelB contain potent transactivation domains within their C-terminal portions. In addition to its C-terminal domain, RelB also contains an N-terminal leucine zipper-like region, and both these domains are required for RelB to be fully active. However, neither p50 nor p52 contains any transactivating regions and, as such, are transcriptionally inactive. Homo- or heterodimers of p50 and p52 have been shown to repress κB-dependent transcription *in vivo*, most likely through recruitment of histone deacetylases to κB-sites. It is believed that p50 homodimers inhibit transcription by constitutively occupying κB sites that might otherwise be bound by transactivating NF-κB/Rel proteins. One interesting property of p50 and p52 homodimers is their ability to specifically associate with two atypical members of the IκB family (see Section 1.3), namely the proto-oncogene BcL-3 and IκBζ. These interactions have been shown to induce κB-dependent transcription by inhibiting DNA-binding of p50 and p52 homodimers, and by providing transactivating potential to these normally inactive dimers.

In contrast to NF-κB/Rel molecules containing C-terminal transactivation domains (p65, c-Rel, and RelB), p50 nor p52 is synthesized as large precursor molecules of 105 kDa (p105) and 100 kDa (p100), respectively. The N-terminal regions of p105 and p100 constitute the RHD of p50 and p52 while the C-termini of each molecule contain multiple copies of ankyrin repeat sequences, which are found in IκB family members (see the next section). Indeed, both p105 and p100 inhibit the nuclear localization and transcriptional activity of NF-κB/Rel proteins with which they dimerize. Generation of p50 from p105 and p52 from p100, although superficially similar, occurs by distinct processing mechanisms (see Chapter 4). Signal-induced processing of p100 to release transcriptionally active p52/RelB dimers occurs following inducible phosphorylation of specific C-terminal serine residues by IKKα, leading to subsequent ubiquitination and incomplete processing by the proteasome. In contrast, the processing of p105 appears to occur constitutively, leading to the generation of a fixed ratio of p105 to p50 in an individual cell. It has been proposed that the constitutive processing of p105 occurs through a cotranslational mechanism, although details about the process and its regulation remain sketchy. The mechanism by which inducible p100 processing occurs is also poorly defined, and it is clear that much more work remains to be done in answering fundamental questions about these important events in the NF-κB activation pathway.

1.3 IκB PROTEINS

During the original purification of NF-κB, it was discovered that DNA-binding of cytoplasmic NF-κB could be induced by treatment of cytosolic extracts with dissociating agents such as sodium deoxycholate or formamide [6,7]. This led to the finding that two distinct proteins associated with NF-κB and inhibited its activity in nonstimulated cells [8–10]. These proteins, termed IκBα and IκBβ, had molecular weights of 37 and 43 kDa respectively, and purified preparations of these proteins could both specifically and reversibly inhibit DNA-binding of NF-κB, and disassociate DNA-bound NF-κB *in vitro*. Functional domains of these inhibitory molecules became apparent after their molecular cloning, which revealed that they contain repeated sequences of 30 to 33 amino acids known as ankyrin repeats [11,12]. As described above, the p50 and p52 precursor molecules p105 and p100 also contain ankyrin repeats in their C-terminal regions and, prior to processing, are capable of inhibiting NF-κB activity. All IκB molecules have subsequently been found to contain between three and seven ankyrin repeats. To date, with the inclusion of IκBζ, eight structurally related members of the mammalian IκB family have been cloned. The IκB family includes IκBα, IκBβ, IκBε, BCL-3, IκBζ, IκBγ, p105, and p100 (Chapter 2). Unlike the other IκB proteins, IκBγ is not the product of a distinct gene but is a 70 kDa protein containing the C-terminal domain of p105 that arises from a unique transcript of p105 and is expressed only in certain murine lymphoid cells. There are two other less well-studied IκB isoforms. The first is IκBR (for IκB-related protein), which is a 52 kDa molecule that appears to be preferentially expressed in epithelial cells; however, the *in vivo* regulatory function of this molecule is not known. An additional IκB-like protein, IκB-NS, was reported to be important in negative selection of thymocytes and also in suppressing inflammatory responses under certain situations.

It is generally believed that different IκB molecules preferentially inhibit distinct subsets of NF-κB/Rel protein dimers (Chapter 2). For example, both IκBα and IκBβ strongly inhibit complexes containing c-Rel and p65, whereas BCL-3 and IκBζ only bind homodimers of p50 or p52. Specific inhibitory property has also been reported for IκBε, which appears to only bind to c-Rel, p65, and their respective homodimers. By comparison, p100 and p105 exhibit little specificity, and they have been reported to form inhibitory complexes with p50, p52, p65, and c-Rel. Interestingly, the IκB molecules most often found associated with RelB are p100, consistent with the observation that processing of p100 in response to the "alternative" pathway leads to release of p52:RelB dimers. Inhibition of NF-κB/Rel proteins by IκB molecules occurs via protein-protein interactions between the ankyrin repeats of IκB and regions of the RHD in NF-κB. With the exception of BCL-3, IκBζ, and an unphosphorylated form of IκB-β, this interaction enables the IκB proteins to mask the NLS and prevent nuclear translocation of NF-κB/Rel protein dimers (Chapter 2).

In addition to the ankyrin repeats, which occur in the interior of IκB proteins, the N- and C-terminal domains of each molecule contain important structural and functional domains. The N-terminal regions contain the signal-responsive sites of phosphorylation in IκBα, IκBβ, and IκBε. Deletion of the C-terminal domain blocks the ability of IκBs to prevent DNA-binding of NF-κB and to dissociate DNA-bound NF-κB dimers. Within their C-termini, IκB proteins contain an acidic and Thr-rich proline, glutamic acid, serine, and threonine (PEST) sequence, which is believed to play a role in stabilization of the molecule. Deletion studies have demonstrated that removal of the PEST domain can protect IκBα from degradation induced during NF-κB activation (see below) and basal phosphorylation of sites within this domain (that is, by casein kinase II) has been implicated in regulating the constitutive turnover of IκB molecules in unstimulated cells.

1.4 IκB KINASE COMPLEX

Degradation of IκB is a tightly regulated event that is initiated upon specific phosphorylation by activated IKK. The IKK activity in cells can be purified as a 700 to 900 kDa complex and has been shown to contain two kinase subunits, IKKα (IKK1) and IKKβ (IKK2), and a regulatory subunit NEMO (NF-κB essential modifier) or IKKγ (see Chapter 3). In the classical NF-κB signaling pathway, IKKβ is both necessary and sufficient for phosphorylation of IκBα on Ser^{32} and Ser^{36}, and IκBβ on Ser^{19} and Ser^{23}. The role of IKKα in the classical pathway is unclear, although recent studies suggest it may regulate gene expression in the nucleus by modifying the phosphorylation status of histones. The alternative pathway, however, depends only on the IKKα subunit, which functions by phosphorylating p100 and causes its inducible processing to p52 [13]. The alternative pathway is activated in response to a subset of NF-κB inducers including LTβ and B cell activating factor (BAFF). Upon phosphorylation by IKKs, IκB proteins are recognized and ubiquitinated by members of the Skp1-Culin-Roc1/Rbx1/Hrt-1-F-box (SCF or SCRF) family of ubiquitin ligases (see Chapter 4). The ubiquitinated IκB proteins are then degraded rapidly by the proteasome. Despite the enormous interest in the IKK complex, there

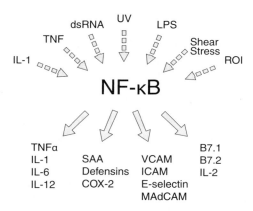

FIGURE 1.2 NF-κB acts as a central coordinator of immune and inflammatory responses. NF-κB is induced by signals that in general represent states of infection or stress. The genes that NF-κB in turn regulates allow the organism to respond effectively to infection and stress. Representative examples of key NF-κB inducers and NF-κB regulated genes are shown.

are still significant gaps in our knowledge of how this protein kinase complex is activated and regulated (Chapter 3).

1.5 NF-κB INDUCERS AND κB-DEPENDENT GENES

As described above, NF-κB/Rel proteins have been shown to play a crucial role in the rapidly induced transcription of a wide variety of genes in response to extracellular stimuli (Figure 1.2). Many of these genes are critical for both innate and adaptive immune responses (Chapters 6 and 7). Many aspects of the inflammatory response are also controlled by NF-κB, including the production of acute phase proteins such as serum amyloid A (SAA), angiotensinogen, and factors of the complement cascade and the induction of proinflammatory cytokines such as IL-1 and TNF-α (Chapter 8). NF-κB also controls later phases of an inflammatory response by mediating induction of genes encoding leukocyte adhesion molecules (e.g., E-selectin, VCAM-1, and ICAM-1) that are expressed by vascular endothelial cells in response to proinflammatory cytokines and bacterial lipopolysaccharide (LPS). NF-κB activation following recognition of microbial products by pattern-recognition receptors, such as Toll-like receptors (TLRs), illustrates its role in responses to infection. There is also a great deal of evidence concerning the role of NF-κB in the lifecycle of a number of viruses, for example, HIV1, CMV, and SV40, that contain κB sites within their promoters and as such have evolved to make use of their host's transcriptional apparatus to direct their own expression and infectivity. Finally, NF-κB has been shown to be involved in many aspects of cell growth, differentiation, and proliferation via the induction of certain growth and transcription factors, as well as by acting as a protective factor against programmed cell death. These functions of NF-κB also help explain the phenotypes observed in genetic diseases such as *incontinetia pigmenti,* where the activity of NF-κB is inhibited [14] (see Chapter 9).

In summary, the NF-κB/Rel proteins are key players of a fascinating system by which the organism responds to changes in its environment or to infection and injury. The subsequent chapters will delve deeper into the various aspects by which members of this transcription factor family are regulated and conclude with a discussion of current efforts to target this transcription factor for treatment of various diseases including arthritis and cancer (Chapter 10).

REFERENCES

[1]   Sen, R. and Baltimore, D., Multiple nuclear factors interact with the immunoglobulin enhancer sequences, *Cell,* 46, 705, 1986.

[2]   Sen, R. and Baltimore, D., Inducibility of kappa immunoglobulin enhancer-binding protein NF-kappaB by a posttranslational mechanism, *Cell,* 47, 921, 1986.

[3]   Baeuerle, P.A. and Baltimore, D., A 65-kappaD subunit of active NF-kappaB is required for inhibition of NF-kappaB by IkappaB, *Genes Dev.,* 3, 1689, 1989.

[4]   Ghosh, S., Gifford, A.M., Riviere, L.R. et al., Cloning of the p50 DNA binding subunit of NF-kappaB: Homology to rel and dorsal, *Cell,* 62, 1019, 1990.

[5]   Kieran, M., Blank, V., Logeat, F. et al., The DNA binding subunit of NF-kappaB is identical to factor KBF1 and homologous to the rel oncogene product, *Cell,* 62, 1007, 1990.

[6]   Baeuerle, P.A. and Baltimore, D., I kappa B: A specific inhibitor of the NF-kappaB transcription factor, *Science,* 242, 540, 1988.

[7]   Baeuerle, P.A. and Baltimore, D., Activation of DNA-binding activity in an apparently cytoplasmic precursor of the NF-kappaB transcription factor, *Cell,* 53, 211, 1988.

[8]   Ghosh, S. and Baltimore, D., Activation *in vitro* of NF-kappaB by phosphorylation of its inhibitor IkappaB, *Nature,* 344, 678, 1990.

[9]   Thompson, J.E., Phillips, R.J., Erdjument-Bromage, H. et al., IkappaB-beta regulates the persistent response in a biphasic activation of NF-kappaB, *Cell,* 80, 573, 1995.

[10] Zabel, U. and Baeuerle, P.A., Purified human IkappaB can rapidly dissociate the complex of the NF-kappaB transcription factor with its cognate DNA, *Cell,* 61, 255, 1990.

[11] Davis, N., Ghosh, S., Simmons, D.L. et al., Rel-associated pp40: An inhibitor of the rel family of transcription factors, *Science,* 253, 1268, 1991.

[12] Haskill, S., Beg, A.A., Tompkins, S.M. et al., Characterization of an immediate-early gene induced in adherent monocytes that encodes IkappaB-like activity, *Cell,* 65, 1281, 1991.

[13] Senftleben, U., Cao, Y., Xiao, G. et al., Activation by IKKalpha of a second, evolutionary conserved, NF-kappaB signaling pathway, *Science,* 293, 1495, 2001.

[14] Smahi, A., Courtois, G., Vabres, P. et al., Genomic rearrangement in NEMO impairs NF-kappaB activation and is a cause of incontinentia pigmenti. The International Incontinentia Pigmenti (IP) Consortium, *Nature,* 405, 466, 2000.

2 Structural Aspects of NF-κB and IκB Proteins

Anu K. Moorthy, Tom Huxford, and Gourisankar Ghosh

CONTENTS

2.1 INTRODUCTION

Expression of eukaryotic genes is regulated by gene-specific enhanceosomes composed of DNA enhancers and multiple activators. Through stimulus-specific interaction with a multitude of other factors on gene enhancers, members of the NF-κB family of transcription factor proteins coordinate enhanceosome assembly and modulate the expression of a large number of genes that, in turn, mediate key cellular functions including immune response, cell proliferation, growth, and death [1,2].

Basic biochemical and structural characterization of various dimeric NF-κB proteins have revealed their molecular architecture and mechanisms of biological activity. X-ray crystal structures of NF-κB in complex with IκB inhibitor proteins, and consensus DNA elements from target gene enhancers, shed light on the mechanisms of NF-κB regulation and suggest testable models of NF-κB-mediated activation of transcription. In this chapter, we review the current state of NF-κB molecular biology from the perspective of protein structure and biochemistry.

2.2. THE NF-κB PROTEINS

The entire mammalian NF-κB family consists of five closely related polypeptide subunits that associate to form transcriptionally active homo- and heterodimers (Figure 2.1a). Each of the subunits shares high sequence homology within a conserved Rel Homology Domain (RHD), a stretch of ~300 residues near the amino-terminus. Contained within the RHD of NF-κB subunits are all of the amino acid

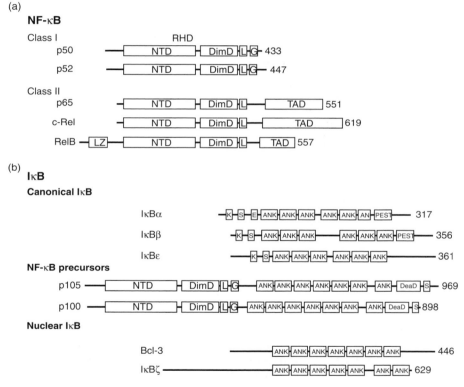

FIGURE 2.1 Domain organization of the mammalian NF-κB and IκB proteins. (a) NF-κB polypeptides are divided into classes I and II based on the absence or presence of a carboxyl-terminal transcriptional activation domain (TAD). Other abbreviations are RHD — Rel homology region; NTD — Amino-terminal domain; DimD — Dimerization domain; L — Nuclear localization sequence; G — Glycine-rich region; LZ — Leucine zipper. The number of amino acids in the full-length human primary sequence is indicated on the right. (b) IκB proteins are categorized into canonical, NF-κB precursors, and nuclear, as in the text. Additional abbreviations used are ANK — Ankyrin repeat; K — Lysine site for signal-dependent ubiquitinylation; S — Serine site of signal-dependent phosphorylation; PEST — region rich in the amino acids proline, glutamic acid, serine, and threonine; DeaD — Death domain.

sequences necessary for subunit dimerization, nuclear localization, DNA binding, and inhibitor binding. Subunit dimerization is mediated entirely by the dimerization domain, a conserved 100 amino acid folded domain located within the carboxyl-terminal third of the RHD. Both the dimerization domain and an amino-terminal domain participate in target DNA binding. In the following section we describe NF-κB subunit dimerization. DNA binding is discussed in Section 2.4.

2.2.1 NF-κB Dimer Formation

As mentioned above, members of the NF-κB family can associate with each other to form various combinations of homo- and heterodimers. Such combinatorial associations can broaden the spectrum of activity and diversify biological function. Although theoretically 15 possible dimeric combinations can be generated, not all of them have been demonstrated to be physiologically important. The p50/p65 heterodimer is by far the most ubiquitous, being seen in almost all cell types. The p50/p50, RelB/p50, and RelB/p52 dimers are also known to be prevalent. However, p65/p65, p50/p50, c-Rel/c-Rel, p65/c-Rel, and p50/c-Rel dimers are only found in a limited subset of cell types. On the other hand there is no convincing evidence for the presence of the p52/p52 or p52/p50 dimers. This is somewhat surprising if one considers the close sequence similarity between p50 and p52 and the relative abundance of the p50/p50 homodimer. Recent studies also point toward the existence of the RelB/p65 and RelB/c-Rel heterodimers, but it is still not clear if the Rel-subunits in these dimers interact through the dimer interface. The RelB/RelB homodimer is yet to be detected in cells.

Cell-specific expression of the different Rel-subunits may help explain the presence of the different dimeric forms. On the other hand, differential affinities may also contribute to the varied distribution of the dimers. Although no direct measurements of binding affinities between the different dimerization domains have been reported, high-resolution crystal structures have provided some clues about how the differential dimers might be generated. Structures of the dimerization domains, and of the NF-κB dimers bound to DNA, show that the dimerization interface remains fairly conserved in both the free and DNA bound forms. All the structures solved to date, with the exception of the RelB dimerization domain, show similar overall features wherein each monomer assumes an independent Ig-fold that packs against each other to form a dimer interface. The variable dimerization specificities may then be explained by amino acid substitutions within the dimerization domain. For example, Y267 in murine p50 is a key residue involved in extensive hydrogen bonding and van der Waals contacts, both of which aid in homodimer formation. This residue is replaced by phenylalanine in p65 and c-Rel, which abolishes hydrogen bonding, thereby weakening subunit association in p65 and c-Rel compared to p50. It is likely that the lack of the Y267 hydrogen bonding network in p65 may explain why the p65 homodimer is weaker than the p50 homodimer. In the p50/p65 heterodimer, however, D254 of p50 forms strong hydrogen bonds with N200 of p65, thereby strengthening the p50-p65 dimer interface and making it a more stable heterodimer compared to the individual homodimers.

2.2.2 THE UNUSUAL RELB DIMER

Although the dimerization domain of RelB is highly homologous to other members of the Rel-family (52% identity and 70% homology to p50 and p52), the architecture of the dimerization interface is markedly different. Unlike other Rel-proteins, RelB exists as a domain swapped dimer where each Ig-like dimerization domain is formed by the complementary halves of two polypeptides. Hence instead of two independently folded monomers, both RelB monomers intertwine with an extra β-sheet connecting the two domains. Close comparison of the RelB dimerization domain structure with other NF-κB structures shows that most of the critical residues in the dimer interface are essentially in a similar structural orientation in p50, p52, p65, c-Rel, and RelB. Y300 (Y267 in p50), H332 (H304 in p50), and Q334 (Q306 in p50) in RelB are the only residues that show a significant change in side chain conformation. Therefore, the reason for the differential folding of RelB is not very apparent. One plausible explanation is the presence of surface polar residues in non-RelB subunits that lead to extensive hydrogen bonding network, resulting in a stable domain structure. In RelB this hydrogen bonding is significantly reduced lowering the folding stability of the individual domains, the effect of which may be alleviated by the extra β-sheet in an intertwined architecture. In all, structural and biochemical studies reveal that smaller number of amino acid variations both within and outside the dimer interface modulate dimerization specificity among the Rel proteins.

2.3 THE IκB PROTEIN FAMILY

The IκB family of transcription factor inhibitor proteins is slightly larger and considerably more diverse than the NF-κB family [3]. Seven IκB proteins have been identified in both the mouse and human genomes. These are IκBα, IκBβ, IκBε, p105, p100, BcL-3, and IκBζ (Figure 2.1b). Based on their domain organization, selectivity toward specific NF-κB dimers, and subcellular localization, the IκB family of proteins can be broadly organized into three separate classes (Figure 2.2). IκBα, IκBβ, and IκBε are prototypical IκB proteins, which inhibit p65 and c-Rel homo- and heterodimers. On the other hand, the p105 and p100 proteins function as IκB proteins as well as precursors of the mature NF-κB subunits p50 and p52, respectively. The third class of IκB proteins functions in the nucleus to modulate the transcriptional potential of some NF-κB dimers. This subgroup, which includes BcL-3 and IκBζ, is referred to as "nuclear IκB." In this section we describe these three types of IκB proteins in terms of their structural and biological functions before returning to NF-κB and its DNA binding properties.

2.3.1 THE CANONICAL IκB PROTEINS

IκB was originally identified as an activity that helped retain NF-κB in the cytoplasm, and inhibited the ability of NF-κB to bind to DNA [4]. Treatment of purified, inactive NF-κB complexes with the detergent deoxycholate (DOC) resulted in the removal of IκB protein and, consequently, the activation of NF-κB DNA binding activity. The original IκB activity was eventually ascribed to two similar proteins known as

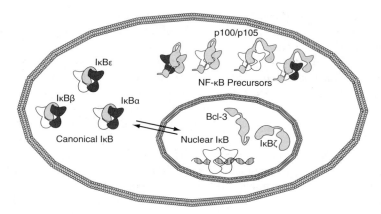

FIGURE 2.2 A schematic diagram of the NF-κB/IκB complexes contained in a resting cell. Classification of IκB proteins based on their NF-κB binding partner specificity and subcellular localization in resting cells results in three groupings referred to in this chapter as canonical, NF-κB precursor, and nuclear IκB. The IκB proteins are depicted as gray with kidney-shaped ankyrin repeat domains that contact NF-κB dimers. The NF-κB subunits with transcription activation potential are filled in black, while those without are filled in white. The canonical IκBα, IκBβ, and IκBε bind and retain to the cytoplasm NF-κB dimers with strong transcription activation potential. The lone exception is IκBα, which has been shown to shuttle rapidly between the nucleus and cytoplasm in resting cells. The NF-κB precursors p105 and p100 are cytoplasmic nonspecific inhibitors of all NF-κB subunits. Nuclear BcL-3 and IκBζ are thought to function as transcriptional coactivators by binding nuclear NF-κB p50 homodimers.

IκBα and IκBβ [5–7]. Both of these proteins exhibit a similar domain arrangement with a centrally located ankyrin repeat (AR)-containing domain (ARD) flanked on its amino-terminal end by a signal responsive region (SRR) and on its carboxyl-terminus by a PEST region (a region rich in the amino acids proline, glutamic acid, serine, and threonine).

Subsequent investigations have shown that the ARD is common to all members of the IκB family. Ankyrin repeats are composed of homologous 33-residue long sequences that adopt similar folds, and in an ARD the folded units stack one upon another to form an ordered protein structure devoid of a central hydrophobic core [8]. The ARD is present in numerous proteins across all phyla and can contain as many as 24 individual ankyrin repeats. The ARD in the IκB family consists of either six or seven repeats. The ARDs typically mediate protein–protein interactions, including the interaction of the centrally positioned ARD of IκB proteins with the DNA binding RHD of NF-κB dimers.

The SRR of IκBα, IκBβ, and IκBε contains a conserved pair of serines within a destruction box motif. Regulated phosphorylation of both serines by the IKK complex is required for degradation of IκB (Chapter 3). The phosphorylated IκB is then ubiquitinylated leading to their proteolysis by the 26 S proteasome in an ATP-dependent manner (Chapter 4) [9].

Carboxyl-terminal to the ARD of IκBα and IκBβ is the PEST region, which is a shared feature of many proteins that undergo rapid turnover in cells. The PEST region of IκBα and IκBβ can be phosphorylated *in vivo* by the protein kinase CK2 (casein kinase II). This region of IκBα has been shown to be required for its NF-κB DNA-inhibitory binding activity [10].

Both IκBα and IκBβ display specificity toward the NF-κB subunits p65 and c-Rel. This is true both *in vivo* and *in vitro*, where equilibrium dissociation binding constants for IκBα and the NF-κB p65 homodimer has been measured at 10 nM, whereas the same inhibitor protein binds NF-κB p50 homodimers in the low micromolar range [11]. The complex of IκBα and the NF-κB p50/p65 heterodimer is the most stable with dissociation constants measured at 1 nM [12]. IκBβ shows similar properties in its NF-κB binding specificity. However, IκBα displays higher binding affinities to both the p65 homodimer and p50/p65 heterodimer than IκBβ [13]. Remarkably, although recombinant IκBα binds to both c-Rel homo- and p50/c-Rel heterodimers, only PEST phosphorylated IκBβ, but not recombinant IκBβ, recognizes these c-Rel dimers with high affinity. These variations in the molecular mechanisms of NF-κB/IκB complex formation point to functional differences exhibited by the IκBα and IκBβ proteins *in vivo*.

In 1997, three laboratories simultaneously announced the cloning and initial characterization of a third canonical IκB member, IκBε [14–16]. Like IκBα and IκBβ, IκBε contains an ARD with six ankyrin repeats. It contains the amino-terminal signal responsive elements, although the amino-terminal domain is longer than either IκBα or IκBβ. The most significant difference in domain arrangement between IκBε and the other canonical IκB proteins is that IκBε lacks a recognizable carboxyl-terminal PEST region. Also, IκBε displays a more limited tissue specific expression pattern compared to either IκBα or IκBβ. However, its specificity toward the NF-κB p65 subunit, and its cytoplasmic localization, firmly places it among the canonical IκB proteins.

X-ray crystal structures of NF-κB/IκB complexes have helped to explain many of the observed binding properties exhibited by canonical IκB members. The x-ray crystal structure of the NF-κB p50/p65 heterodimer bound to the ARD and PEST regions of IκBα reveals an extensive, complex protein–protein interface [17,18]. Both the IκBα ARD and PEST are involved in contacting the NF-κB p65 subunit amino-terminal domain, dimerization domain, and the nuclear localization sequence (NLS) sequence as well as the p50 subunit dimerization domain. In all, 4800 Å^2 of the solvent exposed surface of the NF-κB p50/p65 heterodimer is buried upon binding to IκBα. The NF-κB p50 subunit amino-terminal domain does not interact with IκBα. This was inferred from earlier binding studies and was supported by the observation that the p50 subunit amino-terminal domain was not present in the reported NF-κB/IκBα complex crystal structures [17,18]. However, both structures contain the amino acid sequences encoding the carboxyl-terminal NF-κB p50 subunit NLS. Although present, this region is disordered in both crystals, leading to the hypothesis that NF-κB/IκBα complexes might have inherent nuclear localization potential. This hypothesis has been supported by a number of studies that clearly demonstrate nuclear accumulation of inactive NF-κB/IκBα complexes when the nuclear export receptor Crm1 is rendered inactive by the drug leptomycin B [19,20]. It has been further shown that a nuclear export signal (NES) within the amino-terminal SRR of IκBα functions to transport NF-κB/IκBα complexes back to the cytoplasm of resting cells. Thus, the inactive NF-κB/IκBα complex is not exclusively cytoplasmic in resting cells, but shuttles between the nucleus and cytoplasm (Figure 2.2).

As previously mentioned, IκBβ derives more binding energy through its inter-actions with the C-terminal elements of the NF-κB RHD. This observation was confirmed by the x-ray crystal structure of the NF-κB p65 homodimer in complex with the ARD and PEST of IκBβ [21]. This structure reveals how the NLS polypep-tide of the NF-κB p65 subunit makes additional contacts with a unique hydrophobic pocket on the top of the IκBβ ankyrin repeat stack. Furthermore, the structure suggests how the NLS from the second NF-κB subunit might bind uniquely to IκBβ. Consistent with this structure-based hypothesis, it has since been observed that, in contrast to the shuttling NF-κB/IκBα complexes, NF-κB/IκBβ remains exclusively in the cytoplasm until the inhibitor is removed by proteolysis [13,22]. Finally, the C-terminal PEST of IκBβ is disordered in the complex with NF-κB p65 homodimer. This further strengthens the hypothesis that although IκBα and IκBβ bind the same NF-κB dimers with similar affinities, they do so by unique molecular mechanisms. The functional consequences of these differences continue to be investigated.

2.3.2 THE NF-κB PRECURSORS

A second class of IκB family protein is represented by p105 and p100. These proteins are functional precursors of the mature NF-κB p50 and p52 subunits. The p50 and p52 RHD are contained at the amino-terminal end of the precursor proteins. These are followed by a polypeptide region rich in glycine. Carboxyl-terminal to this glycine-rich region (GRR) is an ARD that contains seven ankyrin repeats. This, in turn, is followed by a death domain (DD). Finally, a consensus destruction box with its pair of serines is encoded by amino acids at the extreme carboxyl-terminus.

The p105 and p100 precursors function as inhibitors of NF-κB. The ankyrin repeats bind with specificity to p50 and p52 subunits while the RHD preferentially interacts with c-Rel and p65 subunits. The architecture of these p105 and p100 bound to NF-κB complexes is not clear. One p105 molecule can bind to an NF-κB subunit such as p50 or p65 through its dimerization domain of amino-terminal RHD, and at the same time the carboxyl-terminal ARD can bind to an intact p50 homodimer or p50 subunit-containing heterodimer. Therefore, p105 homodimers can be formed asymmetrically where one p105 ARD interacts with the N-terminal RHD homodimer portion in *cis* whereas the other ARD can interact with another NF-κB dimer in *trans*. The p105/p65 or p100/p65 complex can be formed similarly to NF-κB/IκB complexes, where p65 or c-Rel form heterodimers with the amino-terminal p50 and p52 regions of p105 and p100, respectively, and the ARD interacts with the dimer platform. Further biochemical studies using purified components will be required to resolve this issue.

Localization studies have revealed that both precursors reside exclusively in the cytoplasm [23]. The observation that p105 and p100 reside primarily in the cell cytoplasm suggests the intriguing hypothesis that these NF-κB precursors might be responsible for maintaining an inactive pool of NF-κB subunits through-out the process of NF-κB induction by proteolysis of canonical IκB. The presence of the IKK-responsive destruction box at the C-termini of p105 and p100 indicate that these proteins can be degraded in response to specific signaling pathways through IKK. Indeed, characterization of B cell activating factor (BAFF) and lymphotoxin-β signaling has revealed that they lead to signal-induced processing of p100 to p52 by a mechanism that is dependent upon the IKKα subunit. It is likely that a significant fraction of cytoplasmic NF-κB is regulated through association with the NF-κB precursors p100 and p105.

2.3.3 THE NUCLEAR IκB PROTEINS

A third class of IκB is the nuclear IκB proteins. This has been recently expanded to include two members, BcL-3 and IκBζ (also known as MAIL and INAP). Although these two proteins differ significantly with respect to their primary sequence and domain organization, they share similarities in their NF-κB subunit binding specificity, subcellular localization, and the curious ability to function as transcriptional coactivators.

BCL-3 was originally discovered as a gene translocation into the immunoglobulin locus in a screen of chronic lymphocytic leukemias [24,25]. As with other IκB proteins, BCL-3 contains a centrally-positioned ARD. The ARD of BCL-3 contains seven consensus ankyrin repeats. Amino-terminal to the BCL-3 is a region of undefined structure. Notably, no consensus IKK phosphorylation or destruction box sequences are contained within the BCL-3 amino-terminal sequence. Carboxyl-terminal to the BCL-3 ARD is a serine-rich segment. In contrast to the canonical IκB proteins and NF-κB precursors, BCL-3 is a nuclear protein that can activate gene transcription. BCL-3 shows preference in binding to NF-κB p50 and p52 subunits leading to the speculation that BCL-3 antagonizes the inhibitory activity of these NF-κB homodimers at NF-κB target gene promoters. Recently, it has been found that the carboxyl-terminus of BCL-3 is phosphorylated constitutively by glycogen synthase kinase 3β (GSK-3β) [26]. This phosphorylation down-regulates BCL-3 function by leading to its ubiquitinylation and eventual proteolysis and, therefore, affects the expression of BCL-3-dependent genes.

The x-ray crystal structure of the seven ankyrin repeat ARD from BCL-3 has been determined [27]. Each of the ankyrin motifs within the BCL-3 ARD exhibits the familiar ankyrin fold. The majority of the amino acids involved in specific contacts between IκBα and NF-κB are conserved in BCL-3. This suggests that BCL-3 contacts NF-κB dimers with a similar orientation to that observed in the NF-κB/IκBα and NF-κB/IκBβ complexes. The principal differences between the BCL-3 ARD and the ankyrin repeats of IκBα and IκBβ are that BCL-3 lacks a recognizable site for binding the NF-κB NLS and that BCL-3 contains a seventh ankyrin repeat. Superposition of the BCL-3 ARD onto the DNA-bound NF-κB structure indicates that the seventh ankyrin repeat of BCL-3 should not disrupt NF-κB/DNA binding. In fact, the bottom surface of the BCL-3 ARD could provide an additional DNA binding surface.

IκBζ was identified as a gene that is upregulated transcriptionally by bacterial lipopolysaccharide, a known strong activator of NF-κB [28–30]. Like BCL-3, IκBζ contains seven ankyrin repeats within its ARD. However, IκBζ contains a long amino-terminal region devoid of sequence homology to any known protein and entirely lacks a carboxyl-terminal sequence. At the amino acid sequence level, IκBζ displays higher homology to other IκB proteins than it does to ankyrin repeat proteins at large. It is, however, somewhat surprising that within the IκB family its closest homolog is IκBε.

Transfection of cells with IκBζ revealed that the exogenously expressed protein localizes to the cell nucleus. A nuclear localization signal within the amino-terminal region of the protein is responsible for this subcellular distribution. IκBζ shares with BcL-3 specificity toward the NF-κB p50 homodimer. Moreover, it was discovered that IκBζ is involved in NF-κB-dependent activation of interleukin-6 (IL-6) gene expression by functioning as a coactivator. Taken together, these data point to a model whereby IκBζ, an NF-κB-regulated gene product, functions as a coactivator to elevate expression of a subset of late NF-κB-dependent genes [31]. It is likely that the absence of a PEST-like sequence at the carboxyl-terminus of IκBζ, and the lack of involvement of the carboxyl-terminal region of BcL-3 in contacting p50, allows these two IκB proteins to function as coactivators rather than inhibitors.

2.4 NF-κB DNA BINDING

Removal of IκB proteins from NF-κB dimers leads to their rapid nuclear accumulation, sequence specific DNA binding, and target gene expression. In the subsequent sections we discuss the mechanisms underlying the specific recognition of binding sites on DNA by NF-κB as revealed by structural studies.

2.4.1 κB DNA RESPONSE ELEMENTS

In response to inducing stimuli, IκB is degraded and the free NF-κB binds DNA to elicit transcriptional response. The NF-κB specific DNA target sites are collectively referred to as κB DNA. Extensive studies over the past decade have helped shed light on rules that discern the selection of closely related target sites by the various NF-κB dimers. Most cellular κB sites seem to be 10 base pairs (bp) in length with the consensus sequence 5'-GGGRN W YYCC-3' (where R denotes a purine base, N denotes any base, W denotes an adenine or thymine base, and Y denotes a pyrimidine base). This κB DNA consensus sequence is highly degenerate and innumerable variations exist in physiological gene promoters.

X-ray crystal structures of several NF-κB/DNA complexes have revealed the preferential target DNA recognition by the different protein subunits. The structure of the p50/p65 heterodimer bound to DNA shows that while the p50 subunit binds to the 5 bp 5'-GGGRN-3' half-site, the p65 subunit prefers the 4 bp 5'-YYCC-3' half-site [32]. Consistent with such preferences, both the p50 and p52 homodimers prefer an 11 bp κB site comprised of two 5 bp half-sites separated by a central A•T bp. Heterodimers containing one of these protein subunits prefer a 10 bp site. On the other hand, p65 and c-Rel dimers bind 9 bp sites containing two 4 bp half-sites. Therefore, κB sites can be broadly classified into two main classes: the class I κB sites, which are 10 bp or 11 bp sites targeted by p50 and p52 containing hetero- or homodimers, and the 9 bp class II κB sites targeted by p65 and c-Rel hetero- and homodimers.

2.4.2 STRUCTURAL BASIS FOR TARGET DNA RECOGNITION

All NF-κB dimer structures solved to date show a remarkably similar global fold wherein the RHD is organized into two independent domains, each of which is a β-sheeted variant of the immunoglobulin (Ig) fold [32–39] (Figure 2.3). The N-terminal domain (NTD) makes both base specific and backbone nonspecific contacts with the DNA. The C-terminal domain (CTD) is primarily responsible for subunit dimerization and nonspecific DNA contacts.

Like its protein counterpart, the conformation of the κB DNA also exhibits high structural similarity in all these structures. The DNA is slightly bent toward the major groove. The extent of bending can be directly correlated to the length of the central tract of A•T base pairs. The AT-rich sequence is known to possess a higher propensity for bending, with longer stretches being more prone to bending. This region of the DNA also makes far fewer direct contacts with the protein, and the bending seems to facilitate NF-κB binding at the flanking recognition sequences. The minor groove is significantly narrow in all these structures.

A common theme among all NF-κB dimers is the use of protein loops, which connect secondary structure elements within the Ig domains, to contact the target DNA. The most significant contribution is made by loop L1, which is located near the start of the NTD. This loop encompasses five base-contacting residues in murine p50 (R54, R56, Y57, E60, and H64) and four base-contacting residues in murine p65 (R33, R35, Y36, and E39) and c-Rel (R21, R23, Y24, and E27). Unlike conventional disordered loops, loop L1 of NF-κBs is a highly structured region, which fits snugly in the DNA major groove contacting the distal end of both the DNA half-sites. The two arginines make bidendate contacts with the conserved guanines at positions −4 and −3. This interaction is further stabilized by hydrogen bonds from a glutamate that binds the cytosine in the opposite strand. These amino acid residues also interact with each other, thereby positioning themselves in an optimal spatial orientation for DNA binding. The conserved tyrosine residue in loop L1, in addition to making van der Waals contacts with two central pyrimidines (preferably thymines), also interacts with the DNA phosphate backbone. The conserved histidine in loop L1 of both p50 and p52 directly contacts the first guanine in the κB site, which accounts for the requirement for an extra G•C bp at the −5 position in both p50 and p52 half-sites. The replacement of this histidine residue by an alanine in p65 and c-Rel prevents this interaction and seems to negatively affect the binding of these subunits. This changes the preference from a 5 bp κB site for p50 to a 4 bp κB site for both p65 and c-Rel. In addition to these base-specific interactions, the loops also provide extensive non-specific contacts with the DNA phosphate backbone that contributes to the high overall DNA binding affinity of these NF-κB dimers. Although the RelB dimerization domain sequence appears somewhat different, the recently solved structure of RelB/p50 het-erodimer with DNA reveals that it too conforms to the other NF-κB/DNA structures.

2.4.3 DNA BINDING AFFINITY AND SPECIFICITY

Structural studies have thus far not revealed any DNA sequence-NF-κB dimer specific interactions. While *in vitro* studies have shown that each NF-κB dimer

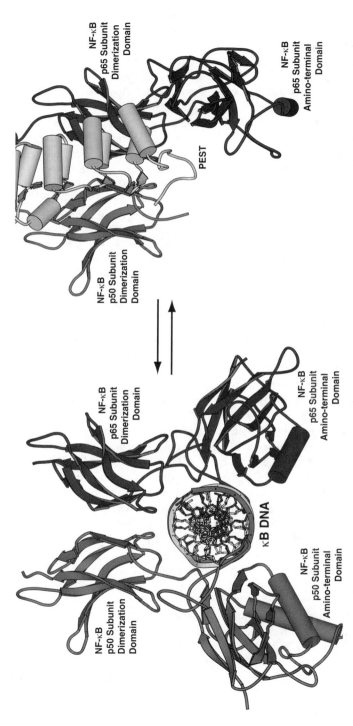

FIGURE 2.3 Ribbon diagrams of NF-κB p50/p65 heterodimer bound to κB DNA and IκBα. These schematic representations of x-ray crystal structures illustrate the two domain structures of the NF-κB RHD as well and the helical ankyrin repeat motif. The beta-strand secondary structure elements are depicted as arrows and helices are represented as cylinders. The arrows symbolize that binding to κB DNA or IκBα is mutually exclusive. In cells, removal of IκBα requires signal-dependent phosphorylation, ubiquitinylation, and proteolysis of the inhibitor. In contrast, resynthesis of IκBα is all that is needed for removal of NF-κB from κB DNA targets.

has some preference for a specific set of κB sites, all NF-κB dimers can bind to most κB sites with reasonably high affinities. The affinities however vary 2.5-fold, which in a more physiological context may be significant. Earlier studies using transiently transfected reporter systems, in conjunction with overexpressed NF-κB proteins, also led to a similar conclusion: certain NF-κB dimers preferred certain κB sites [40]. Reporter assays in NF-κB knockout cells, however, exhibited no such preference [41].

Although the apparent consensus sequence of κB sites is broad and should by and large be tolerant to changes, the κB sites within a gene show high evolutionary stability. This further underlines an element of specificity that is not solely based on the one-to-one binding affinity between the NF-κB dimer and the DNA.

Taking into account all the above observations, the physiological requirement for so many different NF-κB dimers and κB sites, and the signal specific-binding of a dimer on a κB site is indeed intriguing. This can be partially explained by the fact that the relative distribution of various NF-κB dimers may be governed by tissue specific expression. For example, although p65 is present in almost all cell types, c-Rel is expressed primarily in mature monocytic and lymphocytic lineages. However, in a cell type containing numerous dimers and κB sites, the choice of which dimer will bind at a given time may be dictated by several other factors. First, the cellular affinity of the dimer for the κB site may be governed by additional modifications, such as phosphorylation and acetylation (Chapter 5). Previous studies have shown that phosphorylation of Ser337 in p50 by protein kinase A increases the affinity of the p50/DNA complex [42]. Similarly, differential phosphorylation of p65 modulates its transcriptional activity in a promoter specific manner with distinct phosphorylated forms targeted to a particular subset of genes [43]. Signal-induced acetylation has also emerged as an important mechanism to regulate both p50 and p65 DNA binding and transcriptional activity [44]. Second, the nuclear concentration of various binding-proficient dimers may be under signal-dependent temporal control [45]. Hence, the type of signal would determine the type of dimer that translocates to the nucleus and binds a κB site. Third, and perhaps most compellingly, binding may be directly correlated to the ability to recruit and interact with other protein factors on the promoter. There is evidence to suggest that alteration of even a single base on the consensus DNA binding site can dictate the type of coactivator that binds [46]. Some of the recent knowledge pertaining to this mechanism is discussed further.

2.4.4 ALLOSTERIC EFFECTS OF DNA ON TRANSCRIPTION REGULATION

Eukaryotic transcription generally requires the assembly of an enhanceasome containing multiple transcription factors, which bind tandem DNA sites to synergistically activate transcription [47]. It was always believed that the cooperativity between the bound transcription factors mainly arose from extensive protein-protein interactions. Recent structural and biochemical studies however show that this is not entirely true. The transcription factor Oct-1 has a homeodomain and a POU specific domain separated by a flexible linker. These domains bind opposite faces of DNA without extensive protein-protein contacts. This interaction remains cooperative even after removal of the linker, suggesting that conformational changes induced in DNA may govern binding [48]. Similarly, crystal structure of the DNA-binding domains of ATF-2/c-Jun and two IRF-3 molecules in a complex with 31 base pairs of the IFN-β enhancer show that association of the four proteins with DNA creates a continuous surface for the recognition of 24 bp on DNA [49]. Protein–protein interactions are not critical for cooperative binding. Instead, cooperativity arises mainly through nucleotide sequence-dependent structural changes in the DNA that allow formation of complementary DNA conformations.

Structural and biochemical studies have shown that NF-κB dimers are highly flexible, and each dimer can undergo conformational alterations to effectively interact with myriad κB sites. This inherent flexibility also allows for different NF-κB dimers to bind the same κB site with comparable affinities. The bimodular architecture of the RHD provides a unique structural framework for NF-κB dimers. Each dimer has three structural units: the dimerization domain, and two N-terminal domains. These three units are free to move with respect to each other, allowing each dimer to assume different conformations upon binding to various related, but distinct, κB sites. For example, structure of the p50/p65 heterodimer bound to the Ig/HIV-κB DNA is different from that bound to the uPA-κB site [50]. The binding affinities, however, remain comparable. Sequence-dependent structural changes in DNA lead to alteration in the mode of NF-κB binding and conformation. Thus, a specific NF-κB/DNA complex can influence the cooperativity of interactions with nearby bound proteins, including coactivators and corepressors, leading to the formation of a distinct enhanceasome. Hence, DNA can function as an allosteric regulator of transcriptional activity and NF-κB serves as a perfect transmitter of allostery to other components in a transcription complex. The biological role of κB sites in gene specific transcription activation has been demonstrated in two elegant studies. Leonard and colleagues showed that the Ig-κB (5′-GGGACTTTCC-3′) site cannot be swapped with IL-2R-κB site (κB site present in the promoter of the α-subunit of the interleukin 2 receptor gene). This observation was reconfirmed and extended further by Baltimore and colleagues, who showed that a single nucleotide alteration between incontenia pigmenti-10 (IP-10) and monocyte chemoattractant protein-1 (MCP-1) κB sites abolishes promoter activity, although the *in vitro* NF-κB binding activity of these two sites is similar and NF-κB is recruited to the mutated site [46]. They have further shown that coactivator recruitment is κB-site dependent. It is likely that the specific conformation of various κB-sites lies at the heart of the assembly of transcription complex.

2.5 CONCLUSION

Transcriptional regulation is a complex process that requires the coordinated interplay of several distinct DNA and protein–protein binding factors. The IκB proteins may be considered one of the primary means of regulation of the NF-κB signaling pathway. The diversity exhibited by members of the IκB family of inhibitor proteins indicates the active role that these proteins play in integrating diverse signaling messages into a coherent NF-κB response. Greater understanding of the structures of different IκB family proteins and their complexes with NF-κB subunits as well as upstream regulators will aid in determining if this is indeed the case. The NF-κB family of proteins has been shown to function both as a transcriptional activator and repressor, a property that is primarily determined by its interaction with other coactivator or corepressor proteins. Additional levels of regulation may be achieved by covalent modifications, which would affect binding affinities and local chromatin structure. This, in turn, may influence binding site accessibility. Newly emerging evidence suggests that the κB DNA sequence itself may serve as an allosteric regulator of transcription. Earlier experiments and theoretical studies led to the unified theory that DNA was intrinsically inflexible and any sharp bending observed *in vivo* was achieved by protein binding. Newer studies, however, show that DNA itself has an inherent propensity to bend [51]. The sequence of the κB site has been shown to affect NF-κB dimer binding specificity. But rather than influence the binding itself, the sequence seems to dictate which coactivators will form effective interactions with the dimers. Future structural studies on multimeric complexes of NF-κB bound to DNA and coactivators/corepressors will help improve our under-standing of how transcriptional regulation is fine-tuned at the level of protein–protein and protein–DNA binding.

REFERENCES

[1]   Ghosh, S., May, M.J., and Kopp, E.B., NF-kappaB and rel proteins: Evolutionarily conserved mediators of immune responses, *Annu. Rev. Immunol.,* 16, 225, 1998.

[2]   Karin, M., Cao, Y., Greten, F.R. et al., NF-kappaB in cancer: From innocent bystander to major culprit, *Nat. Rev. Cancer,* 2, 301, 2002.

[3]   Whiteside, S.T. and Israel, A., IkappaB proteins: Structure, function and regulation, *Semin. Cancer Biol.,* 8, 75, 1997.

[4]   Baeuerle, P.A. and Baltimore, D., IkappaB: A specific inhibitor of the NF-kappaB transcription factor, *Science,* 242, 540, 1988.

[5]   Zabel, U. and Baeuerle, P.A., Purified human IkappaB can rapidly dissociate the com-plex of the NF-kappaB transcription factor with its cognate DNA, *Cell,* 61, 255, 1990.

[6]   Thompson, J.E., Phillips, R.J., Erdjument-Bromage, H. et al., IkappaB-beta regulates the persistent response in a biphasic activation of NF-kappaB, *Cell,* 80, 573, 1995.

[7]   Davis, N., Ghosh, S., Simmons, D.L. et al., Rel-associated p40: An inhibitor of the Rel family of transcription factors, *Science,* 253, 1268, 1991.

[8]   Sedgwick, S.G. and Smerdon, S.J., The ankyrin repeat: A diversity of interactions on a common structural framework, *Trends Biochem. Sci.,* 24, 311, 1999.

[9]   Yaron, A., Hatzubai, A., Davis, M. et al., Identification of the receptor component of the IkappaBalpha-ubiquitin ligase, *Nature,* 396, 590, 1998.

[10] Ernst, M.K., Dunn, L.L., and Rice, N.R., The PEST-like sequence of IkappaBalpha is responsible for inhibition of DNA binding but not for cytoplasmic retention of c-Rel or RelA homodimers, *Mol. Cell Biol.,* 15, 872, 1995.

[11] Huxford, T., Mishler, D., Phelps, C.B. et al., Solvent exposed non-contacting amino acids play a critical role in NF-kappaB/IkappaBalpha complex formation, *J. Mol. Biol.,* 324, 587, 2002.

[12] Phelps, C.B., Sengchanthalangsy, L.L., Huxford, T. et al., Mechanism of IkappaB alpha binding to NF-kappaB dimers, *J. Biol. Chem.,* 275, 29840, 2000.

[13]   Malek, S., Chen, Y., Huxford, T. et al., IkappaBbeta, but not IkappaBalpha, functions as a classical cytoplasmic inhibitor of NF-kappaB dimers by masking both NF-kappaB nuclear localization sequences in resting cells, *J. Biol. Chem.,* 276, 45225, 2001.

[14]   Li, Z. and Nabel, G.J., A new member of the IkappaB protein family, IkappaBepsilon, inhibits RelA (p65)-mediated NF-kappaB transcription, *Mol. Cell Biol.,* 17, 6184, 1997.

[15]   Simeonidis, S., Liang, S., Chen, G. et al., Cloning and functional characterization of mouse IkappaBepsilon, *Proc. Natl. Acad. Sci. USA,* 94, 14372, 1997.

[16]   Whiteside, S.T., Epinat, J.C., Rice, N.R. et al., IkappaB epsilon, a novel member of the I kappa B family, controls RelA and cRel NF-kappaB activity, *Embo. J.,* 16, 1413, 1997.

[17]   Jacobs, M.D. and Harrison, S.C., Structure of an IkappaBalpha/NF-kappaB complex, *Cell,* 95, 749, 1998.

[18]   Huxford, T., Huang, D.B., Malek, S. et al., The crystal structure of the IkappaBalpha/ NF-kappaB complex reveals mechanisms of NF-kappaB inactivation, *Cell,* 95, 759, 1998.

[19]   Johnson, C., Van Antwerp, D., and Hope, T.J., An N-terminal nuclear export signal is required for the nucleocytoplasmic shuttling of IkappaBalpha, *Embo. J.,* 18, 6682, 1999.

[20]   Huang, T.T., Kudo, N., Yoshida, M. et al., A nuclear export signal in the N-terminal regulatory domain of IkappaBalpha controls cytoplasmic localization of inactive NF-kappaB/IkappaBalpha complexes, *Proc. Natl. Acad. Sci. USA,* 97, 1014, 2000.

[21]   Malek, S., Huang, D.B., Huxford, T. et al., X-ray crystal structure of an IkappaBbeta × NF-kappaB p65 homodimer complex, *J. Biol. Chem.,* 278, 23094, 2003.

[22]   Tam, W.F. and Sen, R., IkappaB family members function by different mechanisms, *J. Biol. Chem.,* 276, 7701, 2001.

[23]   Moorthy, A.K. and Ghosh, G., p105. IkappaBgamma and prototypical IkappaBs use a similar mechanism to bind but a different mechanism to regulate the subcellular localization of NF-kappaB, *J. Biol. Chem.,* 278, 556, 2003.

[24]   McKeithan, T.W., Rowley, J.D., Shows, T.B. et al., Cloning of the chromosome translocation breakpoint junction of the t(14;19) in chronic lymphocytic leukemia, *Proc. Natl. Acad. Sci. USA,* 84, 9257, 1987.

[25]   Ohno, H., Takimoto, G., and McKeithan, T.W., The candidate proto-oncogene BcL-3 is related to genes implicated in cell lineage determination and cell cycle control, *Cell,* 60, 991, 1990.

[26]   Viatour, P., Dejardin, E., Warnier, M. et al., GSK3-mediated BCL-3 phosphorylation modulates its degradation and its oncogenicity, *Mol. Cell,* 16, 35, 2004.

[27]   Michel, F., Soler-Lopez, M., Petosa, C. et al., Crystal structure of the ankyrin repeat domain of BcL-3: A unique member of the IkappaB protein family, *Embo. J.,* 20, 6180, 2001.

[28]   Kitamura, H., Kanehira, K., Okita, K. et al., MAIL, a novel nuclear IkappaB protein that potentiates LPS-induced IL-6 production, *FEBS Lett.,* 485, 53, 2000.

[29]   Haruta, H., Kato, A., and Todokoro, K., Isolation of a novel interleukin-1-inducible nuclear protein bearing ankyrin-repeat motifs, *J. Biol. Chem.,* 276, 12485, 2001.

[30]   Yamazaki, S., Muta, T., and Takeshige, K., A novel IkappaB protein, IkappaBzeta, induced by proinflammatory stimuli, negatively regulates nuclear factor-kappaB in the nuclei, *J. Biol. Chem.,* 276, 27657, 2001.

[31]   Yamamoto, M., Yamazaki, S., Uematsu, S. et al., Regulation of Toll/IL-1-receptor-mediated gene expression by the inducible nuclear protein IkappaBzeta, *Nature,* 430, 218, 2004.

[32] Chen, F.E., Huang, D.B., Chen, Y.Q. et al., Crystal structure of p50/p65 heterodimer of transcription factor NF-kappaB bound to DNA, *Nature,* 391, 410, 1998.

[33] Ghosh, G., van Duyne, G., Ghosh, S. et al., Structure of NF-kappaB p50 homodimer bound to a kappaB site, *Nature,* 373, 303, 1995.

[34] Muller, C.W., Rey, F.A., Sodeoka, M. et al., Structure of the NF-kappaB p50 homodimer bound to DNA, *Nature,* 373, 311, 1995.

[35] Cramer, P., Larson, C.J., Verdine, G.L. et al., Structure of the human NF-kappaB p52 homodimer-DNA complex at 2.1 A resolution, *Embo. J.,* 16, 7078, 1997.

[36] Chen, Y.Q., Ghosh, S., and Ghosh, G., A novel DNA recognition mode by the NF-kappaB p65 homodimer, *Nat. Struct. Biol.,* 5, 67, 1998.

[37] Huang, D.B., Chen, Y.Q., Ruetsche, M. et al., X-ray crystal structure of proto-oncogene product c-Rel bound to the CD28 response element of IL-2, *Structure (Camb.),* 9, 669, 2001.

[38] Escalante, C.R., Shen, L., Thanos, D. et al., Structure of NF-kappaB p50/p65 heterodimer bound to the PRDII DNA element from the interferon-beta promoter, *Structure (Camb.),* 10, 383, 2002.

[39] Berkowitz, B., Huang, D.B., Chen-Park, F.E. et al., The x-ray crystal structure of the NF-kappaB p50.p65 heterodimer bound to the interferon beta-kappaB site, *J. Biol. Chem.,* 277, 24694, 2002.

[40] Kunsch, C., Ruben, S.M., and Rosen, C.A., Selection of optimal kappa B/Rel DNA-binding motifs: Interaction of both subunits of NF-kappaB with DNA is required for transcriptional activation, *Mol. Cell Biol.,* 12, 4412, 1992.

[41] Hoffmann, A., Leung, T.H., and Baltimore, D., Genetic analysis of NF-kappaB/Rel transcription factors defines functional specificities, *Embo. J.,* 22, 5530, 2003.

[42] Guan, H., Hou, S., and Ricciardi, R.P., DNA binding of repressor nuclear factor-kappaB p50/p50 depends on phosphorylation of Ser337 by the protein kinase A catalytic subunit, *J. Biol. Chem.,* 280, 9957, 2005.

[43] Jang, M.K., Goo, Y.H., Sohn, Y.C. et al., Ca2+/calmodulin-dependent protein kinase IV stimulates nuclear factor-kappaB transactivation via phosphorylation of the p65 subunit, *J. Biol. Chem.,* 276, 20005, 2001.

[44] Quivy, V. and Van Lint, C., Regulation at multiple levels of NF-kappaB-mediated transactivation by protein acetylation, *Biochem. Pharmacol.,* 68, 1221, 2004.

[45] Saccani, S., Pantano, S., and Natoli, G., Modulation of NF-kappaB activity by exchange of dimers, *Mol. Cell,* 11, 1563, 2003.

[46] Leung, T.H., Hoffmann, A., and Baltimore, D., One nucleotide in a kappaB site can determine cofactor specificity for NF-kappaB dimers, *Cell,* 118, 453, 2004.

[47] Carey, M., The enhanceosome and transcriptional synergy, *Cell,* 92, 5, 1998.

[48] Klemm, J.D. and Pabo, C.O., Oct-1 POU domain-DNA interactions: Cooperative binding of isolated subdomains and effects of covalent linkage, *Genes Dev.,* 10, 27, 1996.

[49] Panne, D., Maniatis, T., and Harrison, S.C., Crystal structure of ATF-2/c-Jun and IRF-3 bound to the interferon-beta enhancer, *Embo. J.,* 23, 4384, 2004.

[50] Chen-Park, F.E., Huang, D.B., Noro, B. et al., The kappaB DNA sequence from the HIV long terminal repeat functions as an allosteric regulator of HIV transcription, *J. Biol. Chem.,* 277, 24701, 2002.

[51] Cloutier, T.E. and Widom, J., Spontaneous sharp bending of double-stranded DNA, *Mol. Cell,* 14, 355, 2004.

3 Regulation of IKK

Hans Häcker and Michael Karin

CONTENTS

3.1 INTRODUCTION

NF-κB proteins are dimeric transcription factors whose activation is regulated by protein phosphorylation. In their inactive state, NF-κB dimers are sequestered in the cytoplasm, bound to inhibitory IκB proteins (IκBs). Cell stimulation leads to phosphorylation, ubiquitination, and degradation of IκB proteins, which allows NF-κB to translocate to the nucleus and activate target genes. NF-κB activity can be further modulated through direct phosphorylation, although IκB phosphorylation represents the primary mode of regulation. Phosphorylation of IκB proteins is accomplished by several protein kinases, the most important of which are IκB kinase α (IKKα) and IκB kinase β (IKKβ), which together represent the major point of regulation of the NF-κB activation pathway. Although similar, gene knockout studies clearly demonstrate that IKKα and IKKβ are functionally nonredundant — perhaps due to different substrate specificities. While IKKβ is essential for NF-κB activation by proinflammatory factors, such as TNFα, IL-1, and LPS, IKKα is primarily involved in NF-κB activation in response to a subset of TNF-R family members, such as B cell activating factor-receptor (BAFF-R), CD40, and the lymphotoxin β receptor

(LTβ-R). IKKα is also involved in keratinocyte differentiation, but this function is independent of its catalytic activity (see Chapter 6). The signaling pathway relying on IKKβ activity is commonly referred to as the classic NF-κB pathway, whereas IKKα is important for activation of the alternative pathway, which mostly affects NF-κB dimers composed of p52 and RelB. Nonetheless, IKKα does contribute to the classic pathway and most IKKα is present as a heterodimer with IKKβ in a large complex that also contains the regulatory NF-κB essential modifier (NEMO) (IKKγ) subunit. Since its identification nearly 10 years ago, a large body of evidence has accumulated regarding the mechanism of IKK complex activation and its involvement in important physiological and pathophysiological functions including innate and adaptive immunity (Chapters 6 and 7), inflammation, cell survival, and cancer (Chapter 8). In this chapter we will focus on the role of IKK as a key component of the NF-κB pathway and the molecular mechanisms involved in its regulation.

3.2 BACKGROUND

Transcription factor NF-κB consists of a heterogeneous collection of dimers composed of members of the NF-κB/Rel family maintained in an inactive state through interaction with inhibitory IκB molecules (Chapter 2). Site-specific phosphorylation of inhibitory IκB proteins (IκBα, β, and ε) by the IKK complex, followed by their ubiquitination and degradation, is the most critical regulatory event in NF-κB activation. Most of our knowledge about NF-κB activation is based on studies on IκBα regulation. Phosphorylation of conserved serine residues in the N-terminus of IκBα (S32 and S36) leads to its recognition by SCF$^{\beta TrCP}$, resulting in polyubiquitination and degradation of IκBα by the 26S proteasome [1]. Degradation of IκBα leads to nuclear translocation and DNA binding of NF-κB, p50:p65 or p50:c-Rel dimers, initiating gene transcription. IκBβ and IκBε are also regulated by N-terminal phosphorylation, but the kinetics of their phosphorylation and degradation are significantly slower than IκBα and may reflect substrate proclivities of the IKK complex [2]. While activated IKKβ directly phosphorylates IκBα, β, and ε, the p100 precursor is inducibly phosphorylated by IKKα, also leading to recruitment of SCF$^{\beta TrCP}$ and polyubiquitination [3]. In contrast to the classical IκBs, proteolytic degradation of p100 is limited to its C-terminal IκB-like portion, resulting in generation of mature p52, which (primarily) enters the nucleus in a complex with RelB [4, 5]. In contrast to p100, p105 seems primarily processed cotranslationally to generate p50 [6] although IKKβ-dependent phosphorylation and degradation of its C-terminal IκB-like portion were also described [7]. The physiological relevance of IKKβ-dependent inducible p105 processing is not clear yet (Chapter 4).

3.3 THE IKK COMPLEX AND ITS MECHANISM
OF ACTIVATION

Two protein kinases, IKKα and IKKβ, directly regulate NF-κB activation. While IKKβ is most important for the rapid degradation of NF-κB-bound classical IκBs, IKKα-induced p100 processing is substantially slower [8]. Nevertheless, both

IKKα and IKKβ were initially identified by biochemical means as constituents of a 700 to 900kD protein complex that exhibits TNFα-induced IκBα-specific kinase activity [9]. IKKα was also isolated through a two-hybrid screen as a protein that interacted with the NF-κB inducing kinase (NIK) [10]. IKKα and IKKβ are similar in structure (50% sequence identity, 70% protein similarity) and contain N-terminal protein kinase domains and more C-terminally located leucine zipper (LZ) and helix-loop-helix (HLH) motifs (Figure 3.1a). Both proteins exhibit kinase activity toward IκBα. The third component of this complex is a 48kD regulatory subunit, named IKKγ [11], NEMO [12], IKKAP1 [13], or FIP-3 [14]. NEMO was identified biochemically as a component of the IKKα/β-containing complex and, in parallel, by genetic complementation of a cell line nonresponsive to NF-κB activating stimuli [11,12]. The precise functional role of NEMO remains enigmatic. Based on studies using gene-deficient cells or mice, it is clear that NEMO is essential for activation of the classic NF-κB pathway and is required for formation of the large IKK complex, described above [11,12]. Based on structure prediction algorithms, NEMO is primarily a helical protein with large stretches of coiled-coil structure, including a LZ at its C-terminus (Figure 3.1a). The lack of structural data is available on any of the components if the IKK complex remains a major barrier to progress in understanding the mechanism(s) of IKK activation. As such, most of our knowledge is based on mutational analyses, reconstitution experiments, and gene targeting studies.

As discussed, when isolated from cells, the IKK complex has an apparent molecular weight of 700 to 900kD and contains as major stoichiometric components only IKKα, IKKβ, and NEMO [11]. The sum of the molecular weight of the individual IKK components adds up to only 220kD, indicating either that other components are present in the large complex or that the large IKK complex is not a trimer. There is ample evidence that IKKα and IKKβ constitutively form dimers, preferentially heterodimers but also homodimers, via their LZ motif [15,16]. When overexpressed in HEK293 cells or analyzed in IKKγ-deficient cells, the IKKα/β complex elutes with an apparent size of 300kD, implying that it is a dimer of dimers, which is further assembled to the higher-molecular-weight complex by NEMO [12]. Overexpression of IKKα and IKKβ together with IKKγ, both in mammalian cells and yeast cells, leads to formation of a complex that is comparable in size to the large IKK complex and contains catalytic activity [17]. Even when the purified components, IKKα or IKKβ, are added together with NEMO *in vitro*, complex formation occurs, although catalytic activity toward IκB proteins is not restored [17] (or has not been investigated [18]), suggesting that formation of a functional IKK complex may require additional factors or posttranslational modifications — for example, phosphorylation. Consistent with the orchestration of a higher-order complex by NEMO, NEMO itself can form multimers, either trimers [19], or tetramers [18]. This depends on its C-terminal CC and LZ motifs. Binding of IKKβ (and probably IKKα) is conferred by residues located in the N-terminal CC motif of NEMO and requires the very C-terminal 10 amino acids of IKKβ and IKKα, accordingly named the NEMO-binding domain (NBD) [11,20]. Deletion of the NBD interferes with the ability of the affected subunit to interact with NEMO and become activated in response to cell stimulation [21]. Accordingly, a cell permeable peptide

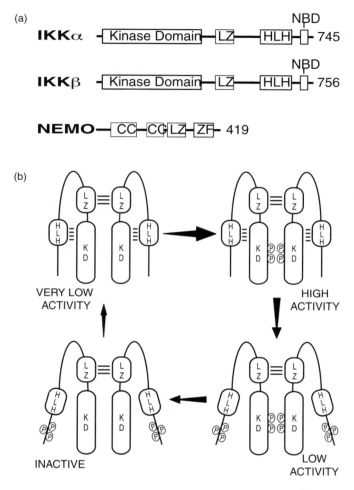

FIGURE 3.1 IKK protein secondary structural motifs and regulation by phosphorylation. Schematic diagram showing the major known subunits of IKK and their structural and functional motifs. (a) LZ — leucine zipper; HLH — helix-loop-helix; CC — coiled coil; ZF — zinc finger; NBD — NEMO binding domain; and amino acid number as indicated at right. Schematic representation of a model of IKK regulation by phosphorylation. (b) The two catalytic subunits of IKK are dimerized through their LZ motifs. The HLH motif interacts with the kinase domain. Phosphorylation of two characterized serines in the activation loop of IKK initiate kinase activity. The non- or low-phosphorylated C-terminus supports kinase activation. Active IKK not only phosphorylates its substrates, the IκBs, but also its own C-terminus by autophosphorylation, thereby inducing a conformational change, which influences the interaction between HLH motif and kinase domain, resulting in a decrease in kinase activity. In this state, IKK is more susceptible to further dephosphorylation and inactivation by phosphatase action.

covering the NBD of IKKβ, which has higher affinity to NEMO than the corresponding part of IKKα, prevents IKK activation by tumor nuclear factor α (TNFα) [21]. Likewise, deletion of NEMO prevents IKK and NF-κB activation by multiple stimuli [22,23]. These data illustrate that activation of IKK by upstream pathways require NEMO. However, our understanding regarding the molecular mechanism of NEMO function and relation to upstream activators is far from being complete. Importantly, some data clearly demonstrate that different stimuli activate NEMO by different molecular mechanisms. For example, the zinc finger motif at the very C-terminus of NEMO is essential for IKK activation by ionizing irradiation but seems to be dispensable for LPS-induced IKK activation [22,24,25]. The mechanisms of IKK activation by distinct stimuli will be discussed in Section 3.4.

Although formation of a catalytically active protein complex of about the size of the endogenous IKK complex with recombinant IKKα/β and NEMO suggests that the core-IKK complex is indeed made up of these three subunits, other proteins have been described to associate with IKK *in vivo*, for instance, Hsp90 [26]. However, Hsp90 is a chaperon protein that interacts with many protein kinases [27] and likely is not specific to IKK.

IKKα and β contain several features that distinguish them from other kinases, such as LZ and HLH motifs. The LZ domain mediates dimerization of the kinases and is essential for kinase activity [15,16,28]. The kinase domain of IKKα/β is similar to that of other serine-threonine kinases, containing a highly conserved adenosine triphosphate (ATP)-binding site. Mutations of the conserved lysine in this region (K44) generate catalytically inactive mutants of IKKα and IKKβ. *In vitro* kinase assays using reconstituted complexes showed that homodimers containing two defective subunits were catalytically inactive, while heterodimers that contain a single active subunit still exhibit kinase activity [16]. Accordingly, transfection of limiting amounts of epitope-tagged IKKα/β (K44)-mutants can be used to immunopurify catalytically active endogenous IKKα or IKKβ from cells, but overexpression of catalytically inactive IKKα or IKKβ can inhibit IKK and NF-κB activation by different stimuli, probably by competition with endogenous active forms of the kinases [15,28,29].

There is considerable evidence that both IKKα and IKKβ need to be phosphorylated to become activated. Purified IKK is inactivated upon incubation with protein phosphatase 2A (PP2A), whereas treatment of cells with the PP2A inhibitor ocadaic acid results in IKK activation [9]. Like other protein kinases, the kinase domains of IKKα and IKKβ contain an activation loop, which is subject to phosphorylation at two serine residues that induces a conformational change leading to kinase activation (Figure 3.1b) [29–31]. Replacement of those two serines in IKKβ (S177/S181) with alanines (IKKβ-AA) prevents kinase activity, while replacement with phosphomimetic glutamates results in a constitutively active kinase [30]. Both of these serines are phosphorylated *in vivo* in response to proinflammatory stimuli, such as TNFα and IL-1 [30]. Substitution of wt IKKβ with IKKβ-AA also prevents IKK and NF-κB activation in response to TNFα or IL-1, similar to the situation in IKKβ-deficient cells [28,30]. The activation loop of IKKα is identical in sequence to IKKβ and also becomes phosphorylated at the corresponding serines during cell stimulation [30]. However, IKKα phosphorylation is not critical for activation of the IKK complex, phosphorylation of IκB proteins and NF-κB activation by most proinflammatory stimuli [32,33]. In contrast, an intact activation loop in IKKα is required for activation of the alternative NF-κB pathway, which leads to processing to p100 (see Section 3.4.1.2). The HLH motif has also been reported to be involved in regulation of IKK activity. Mutations within this motif decreased IKK activity, both in transiently transfected cells and with purified proteins, [15,16] although HLH-deficient mutants still dimerize via their LZ domains and bind to NEMO. Deletion of a C-terminal fragment of IKKβ encompassing the HLH-motif and the NBD destroy kinase activity [30]. Interestingly, the activity of such a C-terminal deletion mutant could be restored by coexpression of separate C-terminal HLH- and NBD-containing fragments [30]. As such, the HLH-motif physically interacts with the kinase domain and seems to serve as an endogenous activator of IKK in a manner similar to the function of the cyclin subunits of cyclin-dependent kinases (CDKs) [34]. C-terminal of the HLH motif and prior to the NBD, IKKα/β contain a stretch of serines, which are heavily phosphorylated during IKK activation [30]. Phosphorylation at these sites requires IKK activity, implicating autophosphorylation as underlying mechanism. In contrast to phosphorylation of the activation loop however, phosphorylation of the C-terminal cluster has a negative autoregulatory function [30]. Replacement of 10 of the C-terminal serines with alanines significantly prolongs TNFα-induced IKKβ activity, whereas replacement of these serines with (phosphomimetic) glutamate residues reduces IKK activity [30]. Given the transient nature of IKK activation by most stimuli investigated so far, this autophosphorylation may be a primary mechanism for negative regulation of IKK activity, followed by phosphatase-dependent dephosphorylation of the activation loop, reverting IKK activity to its baseline levels (Figure 3.1b). Protein phosphatase 2Cβ has been found to associate with IKKβ and reduce its phosphorylation and activity when overexpressed in HEK293 cells [35]. In accordance with these findings, small interfering RNA (siRNA)-based knockdown of PP2Cβ leads to prolonged TNFα-induced IKK activity [35]. Whether PP2Cβ specifically dephosphorylates the serine residues in the activation loop of IKKβ needs to be investigated.

By definition, the alternative NF-κB pathway does not depend on IKKβ and NEMO. Only a subset of TNF receptor family members, such as BAFF-R and CD40 on B cells and LTβR in splenic stromal cells, activate this pathway [8]. There is still much less information about the molecular events involved in IKKα-dependent processing of p100 compared to IKKβ-dependent IκB phosphorylation. Indeed, most of our knowledge on the function of this pathway is based on genetic rather than biochemical analysis. IKKα-deficient cells or cells expressing an IKKα-AA mutant are deficient in p100 processing [36]. Overexpression of a constitutively active form of IKKα, but not catalytically inactive IKKα, induces p100 processing in HEK293 cells [36]. A C-terminal serine residue in p100, which functionally seems to correspond to serines 32 and 36 in IκBα, is required for IKKα-dependent processing and is phosphorylated by purified IKKα *in vitro* [36]. Surprisingly, stimulation-dependent activation of IKKα dimers has not been clearly demonstrated. Whether this is due to a low-level IKKα activation, reflected by the slow kinetics of p100 processing or due to other factors, such as controlled compartmentalization of the kinase and its substrate, is unknown. The inability to show regulated activation of IKKα dimers is also the reason why it is still not clear whether, under physiological circumstances, it is this form of IKKα that phosphorylates p100. Nevertheless, it is clear that the alternative IKKα-dependent pathway is fully operational in the absence NEMO or IKKβ [37], demonstrating that formation of the classical trimolecular IKK complex is not essential for p100 processing.

Notably, however, there are at least two examples, where catalytically active IKKα, as part of the classical NEMO-containing complex, seems to be required for inducible IκBα phosphorylation and degradation. The first one is IKK activation in mammary gland epithelial cells, which proliferate in response to receptor activator of NF-κB (RANK) ligand (RANK-L) in a NF-κB-, cyclin D1-dependent manner [38]. In wild-type (wt) cells, but not in IKKα-AA expressing cells, RANK-L induces IKK activity, which can be isolated with NEMO-specific antibodies. This IKKα-dependent activity regulates a classic NF-κB activation pathway that drives cyclin D1 expression. This pathway is inhibited by expression of a nonphosphorylatable form of IκBα (so-called superrepressor) [38]. Curiously, IKKβ does not seem to be required for proliferation of mammary epithelial cells (unpublished observation). These data indicate that IKKα, in a complex with NEMO can regulate classic "IκB-sensitive" NF-κB activity in a cell type specific manner. The second example of IKKα being required for the classic NF-κB pathway is signaling by the TNF receptor family members cluster of differentiation 27 (CD27) and CD40 [39]. Based on experiments with a human B cell line, both receptors used the upstream kinase NF-κB inducing kinase (NIK) to activate the classic and alternative NF-κB pathways [39]. RNAi-based knock-down experiments indicated that both of these pathways are also IKKα-dependent, consistent with NIK being an IKKα kinase. These data, however, need to be confirmed in primary, gene deficient B cells. Nevertheless, it appears that at least for some pathways leading to IKK activation, the IKK-activating signal is "channeled" through NIK and IKKα into the IKK complex to regulate "classic" NF-κB activity. Other stimuli channel their activity into this complex via IKKβ. It is currently unknown, what dictates the subunit specificity of such stimuli.

3.4 SIGNALING PATHWAYS LEADING TO IKK ACTIVATION

As discussed above, it is likely that activation of IKK at some point involves trans-autophosphorylation, which is needed for amplification of kinase activity and efficient substrate phosphorylation (Figure 3.1b). Different molecular mechanisms were proposed as the initiating event that triggers this autophosphorylation. Notably, for none of the stimuli that lead to IKK activation is the molecular mechanism of kinase activation unequivocally clear. Generally, three mechanisms can be envisioned, which are not mutually exclusive: (i) Direct phosphorylation of one of the IKK subunits, thereby inducing a conformational change in the kinase domains of the catalytic subunits that triggers their activity. In this scenario, akin to other kinase activation pathways, the most likely, but not exclusive phosphorylation site would be the activation loop of the catalytic subunits IKKα or IKKβ. While this may depend on an upstream kinase (IKK-K), it should be noted that phosphorylation of the activation loop also occurs during autophosphorylation of IKKα/β themselves and that so far the most potent IKK-K is IKK itself (D. Rothwarf, M. Delhase, and M. Karin, unpublished observation). As such, phosphorylation of the activation loop during kinase activation does not necessarily mean that this was the original activating event. (ii) Multimerization of IKK, thereby inducing autophosphorylation and kinase activation by close proximity of catalytically active IKKα/β dimers. In this model, multimerization of IKK might be driven by other upstream molecules binding to one of the IKK components, for example, NEMO. However, as discussed above, the IKK complex is already multimeric and therefore the activating event may induce a conformational change that brings the catalytic subunits closer to each other. (iii) Direct modification of one of the IKK components other than phosphorylation, e.g., ubiquitination. Such a modification of NEMO has been reported, at least for some pathways leading to NF-κB activation [40,41]. This modification may alter the conformation of the complex leading to its autoactivation. As mentioned above, based on experiments with zinc finger mutants of NEMO [24,25], it is probable that different mechanisms of IKK activation exist, which are engaged by different NF-κB activating stimuli. Therefore, the different mechanisms proposed to explain IKK activation will be discussed in the context of the major signal transduction pathways that lead to NF-κB activation.

3.4.1 ACTIVATION OF THE NF-κB PATHWAY BY MEMBERS OF THE TNF RECEPTOR (TNFR) FAMILY

As mentioned earlier, based on the different IKK substrates involved during NF-κB activation, a classic and an alternative NF-κB pathway can be discerned. Based on genetic evidence, IKKβ is usually essential for activation of the classic NF-κB pathway (although in some cases IKKα is required), while IKKα is critical for activation of the alternative NF-κB pathway. As NEMO is not required for the alternative pathway, indicating differences in the molecular mechanisms of activation, we will discuss regulation of the two pathways separately.

3.4.1.1 Activation of the Classic NF-κB Pathway by TNFR Family Members

The pathway that leads to NF-κB from members of the TNFR family is one of the best-characterized and first studied NF-κB signaling pathways. Among different TNFR family members, most of our knowledge relies on data obtained from the major TNFα receptor TNF-RI, the study of which led to the biochemical identification of the IKK complex [11]. We will therefore focus this discussion primarily on TNFRI signaling. A hallmark of the receptors of the TNFR family is that they do not contain enzymatic activity, but depend on recruitment of intracellular adaptors and signaling molecules to initiate signal transduction. Several important intracellular molecules have been identified, including members of the TRAF (TNFR associated factor) family, that are critically involved in NF-κB activation [42]. Based on their eponymous C-terminal TRAF-domain, six members of this family have been identified. With the exception of TRAF1, all TRAFs contain N-terminal Zinc-binding motifs including a RING finger, which might be considered as TRAF effector domains, and all contain a C-terminal TRAF domain that can further be divided in an N-terminal coiled-coil (CC) region and the TRAF-C domain, the latter conferring binding to upstream molecules [43].

In some cases, TRAFs bind directly to the intracellular domain of the respective TNFR, e.g., CD40, while in others additional adaptors are required, such as TRADD in case of TNFRI [44]. Tumor necrosis factor receptor associated death domain (TRADD) not only recruits TRAFs, specifically TRAF2 and TRAF5, but also another molecule named RIP1 (receptor interacting protein) [45]. RIP1 contains different structural motifs including a death domain, controlling homotypic interaction with TRADD, a so-called RHIM (RIP homotypic interaction motif) and a kinase domain. Based on sequence similarities in the kinase domain, three other RIP family members have been cloned and three more were found by database homology searches [46]. However, thus far, only RIP1 is known to be required for TNFR-dependent NF-κB activation [47,48]. When overexpressed, all these molecules, TRADD, RIP1, and TRAF2/5 activate NF-κB. A role in NF-κB activation was confirmed genetically for some of these molecules. RIP1 knockout cells fail to activate IKK (and JNK1/2) in response to TNFα and TRAF2-deficient cells show reduced levels of IKK and almost no jun N-terminal kinase (JNK)1/2-activation [48,49]. Residual NF-κB activation in TRAF2 knockout cells might be due to compensation through other TRAF members, as TRAF2/TRAF5 double knockout cells show negligible TNFα-dependent IKK activity [50]. Notably, the kinase activity of RIP1 is dispensable for NF-κB (and JNK1/2) activation, as reconstitution of RIP1-deficient cells with a kinase dead RIP1 mutant confers full responsiveness [47]. Therefore, to some extent all of the above mentioned proteins only postpone the problem of the lack of enzymatic activity of the TNFR family members, when trying to understand how IKK is activated. Furthermore, the mechanisms by which these molecules act are only partially defined. TRAF2 has been demonstrated to recruit the IKK complex to the activated TNFRI via interaction with IKKα/β leucine zipper motifs [51]. This initial recruitment of IKK seems not to depend on RIP1. However, it triggers RIP1-dependent binding to NEMO and activation of the IKK complex [52]. As RIP1 is known to be ubiquitinated during TNFR1 activation and TRAF2 has been found to be critically involved in this process [53], it is possible that TRAF2-dependent ubiquitination of RIP1 is involved in formation of a stable supramolecular complex between TRAF2, RIP1, and IKK.

Although induced proximity of IKK catalytic subunits within this signaling complex might explain IKK activation, there are two other molecules, mitogen activated protein (MAP) extracellular signal regulated protein kinase (ERK) 3 (MEKK3) and transforming growth factor-β (TGF-β)-activated kinase 1 (TAK1), that have been demonstrated to play a role in IKK activation.

When overexpressed, MEKK3 and TAK1 induce NF-κB activation, the response to TAK1 being dependent on coexpression of its associated adaptor proteins, that is, TAB1, TAB2, or TAB3 [54–57]. Data from gene deficient cells have been published for MEKK3 and for TAK1 [58–61]. Mice deficient in either gene die during embryogenesis, MEKK3-/- embryos die around embryonic (E) day 11, exhibiting defects in blood vessel development [62], while TAK1-/- embryos die around E10, exhibiting developmental defects of head fold and neural tube [60]. Unfortunately, these distinct phenotypes do not contribute to our understanding of the roles of the two kinases in NF-κB signaling, but suggest additional functions in other pathways.

Nonetheless, MEKK3-deficient fibroblasts exhibit defects in IL-1 and TNFα-induced IKK (and JNK1/2) activation, a phenotype similar to RIP1-deficient cells [47,48,58]. MEKK3 has been shown to interact with RIP1 and overexpressed MEKK3 can still activate NF-κB and JNK in RIP- and TRAF2-deficient cells, indicating that MEKK3 cooperates in the RIP1-dependent pathway downstream of TRAF2 and RIP1 [58]. Interestingly, when RIP-deficient elongation factor (EF) cells were reconstituted with a MEKK3 fusion protein containing MEKK3 and the death domain (DD) of RIP1, TNFα-induced NF-κB activation was restored, indicating that RIP1 might serve as an adaptor protein that recruits MEKK3 into a TNFRI-induced signaling complex [63]. The NF-κB inducing activity of the MEKK3-RIP1-fusion protein depended on an intact ATP-binding pocket of MEKK3, implying that MEKK3 catalytic activity is required for IKK activation [63]. Although it is formerly possible that binding and oligomerization of the DD of RIP linked to MEKK3 creates a nonphysiological situation, thereby activating MEKK3, an attractive scenario would be that TRADD oligomerization induces recruitment of TRAF2 and RIP1 through direct protein–protein interactions. TRAF2 (together with TRAF5) recruits the IKK complex through binding of its IKKα/β subunits, which may be further supported via interaction between RIP1 and NEMO. RIP1 serves as adaptor that brings MEKK3 into close proximity with the IKK complex, allowing IKK activation, possibly through direct phosphorylation of its catalytic subunits (Figure 3.2a). It needs to be stressed, however, that direct phosphorylation of IKKα/β by MEKK3 has not been demonstrated; therefore, it remains to be established whether MEKK3 acts at all as an IKK-K *in vivo*. TAK1 was originally identified as a kinase involved in TGFβ signaling [64]. Later, it was found to be activated in response to other stimuli, such as TNFα and IL-1 [65,66]. It was also identified as component of an IKK activating complex, coeluting with IKK-inducing activity in an *in vitro*, cell-free reconstitution system [55]. In contrast to MEKK3, TAK1 does not activate NF-κB when overexpressed alone, but depends on coexpression of other molecules, that is, TAB1, TAB2, or TAB3 [55–57,67,68].

However, the analysis of TAB1 and TAB2-knockout mice failed to reveal a role for these molecules in IKK activation. Neither single knockout exhibited defects in TNFα- or IL-1 induced NF-κB activation [60,68]. The results indicate instead that TAB1 is involved in the TGFβ-pathway [69]. However, RNAi knockdown experiments have shown that knockdown of TAB2 or TAB3 alone had no effect on NF-κB activation, but the simultaneous knockdown of both molecules resulted in a significant signaling defect [67]. Given the structural and functional similarities between these proteins, it is possible that TAB2 and TAB3 compensate for each other. Both proteins were found to be recruited to TAK1 during TNFα and IL-1 signaling, and also to interact with TRAF2 and TRAF6, which are involved in TNFα and IL-1 signal transduction, respectively [67]. Interestingly, the C-termini of TAB2/3, which were demonstrated to be required for their function contain a ZnF motif that is typical for ubiquitin-binding proteins [71]. Notably, the TRAF proteins contain RING-finger domains, which may be involved in assembly of K63-linked polyubiquitin chains [72]. Moreover, activation of cells by TNFα and IL-1 induces TRAF2-dependent ubiquitination of RIP1 [52,73]. Further support for atypical (K63-mediated) ubiquitination as an important signaling event comes from the observation that

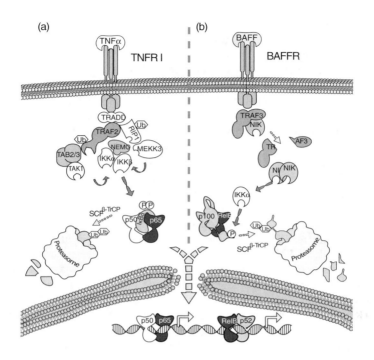

FIGURE 3.2 Schematic representation of the TNFα-induced classic and BAFF-induced alternative NF-κB pathways. TNFRI engagement by TNFα leads to recruitment of various adaptor molecules, including TRADD, TRAF2 (and TRAF5), and RIP1. (a) TRAF2 controls recruitment of the IKK complex through interaction with its catalytic subunits. TRAF2 is also required for ubiquitination of RIP1 and possibly TRAF2 itself. RIP1 binds to NEMO and supports complex formation, possibly also changing the conformation of the IKK complex. Whether RIP1 action depends on ubiquitination is not clear. TAB2 and TAB3 interact with TRAF2 and TAK1, possibly inducing oligomerization and activation of TAK1. This interaction also brings TAK1 near the other components of the signaling complex, including IKKβ, which may be directly phosphorylated by TAK1. How MEKK3 is brought into proximity of the signaling complex is less clear, but may involve interaction with RIP1, rather than TAB2/3 and TRAF2/5. Active IKK phosphorylates IκBs on conserved serines leading to its recognition by the E3 ubiquitin ligase SCF^{βTrCP}, polyubiquitination and degradation by the 26S proteasome. Liberated NF-κB dimers are transported to the nucleus where they drive gene transcription. In nonactivated cells, the kinase NIK is constitutively bound to TRAF3, which controls NIK stability and expression levels. (b) Receptor stimulation leads to recruitment of TRAF3, which is followed by its degradation by an uncharacterized mechanism. Degradation of TRAF3 results in increased stability and expression levels of NIK, possibly also initiating kinase activation by autophosphorylation. NIK itself binds to p100 and supports recruitment of IKKα. NIK directly phosphorylates IKKα at serines in its activation loop, resulting in kinase activation, followed by IKKα-dependent phosphorylation of p100. Phosphorylated p100 is recognized by the E3 ubiquitin ligase SCFβ^{TrCP}, resulting in polyubiquitination and partial proteolytic degradation by the 26S proteasome. Free p52:RelB dimers translocate to the nucleus and drive gene transcription.

an E2 ligase complex containing ubiquitin conjugating enzyme 13 (UBC13) and the UBC-like protein ubiquitin conjugating enzyme E2 variant 1 isoform A (UEV1A) were found to activate IKK in cytoplasmic extracts and that dominant negative mutants or knockdown of UBC13 interfere with IKK activation in cells [40,71]. Putting these data together, an attractive scenario would be that TNFRI activation leads via TRADD recruitment to TRAF2/TRAF5-dependent ubiquitination events (Figure 3.2a). Newly formed K63-mediated polyubiquitin chains on either RIP1 or TRAF proteins themselves may then serve as docking sites for TAB2/3, which recruit TAK1 to the TNFRI signaling complex via protein–protein interaction. Binding and possibly oligomerization of TAK1 through TAB2/3 might also serve to trigger its catalytic activity [71,74]. Concomitant recruitment of the IKK complex via TRAF2 and RIP1, possibly also ubiquitination-dependent, would bring TAK1 and IKK complex into close proximity, leading to IKKα/β phosphorylation and activation (Figure 3.2a). This model is also consistent with other data, showing that TAK1, immuno-purified from HeLa cells and coincubated *in vitro* with different proteins including UBC13/ UEV1A, TRAF6, and ubiquitin, led to specific phosphorylation of IKKβ at its activation loop, supporting a role of TAK1 as an IKK-K [55]. Moreover, ubiquitin-dependent signal transduction may provide a unifying concept for TNFRI and IL-1R signaling, as TRAF6, implicated as an E3 ubiquitin ligase in the IL-1R pathway, may serve a similar function to TRAF2/5 in the TNFRI pathway (see Section 3.4.2).

It should be noted, however, that so far there are many aspects of this model that are not supported by solid experimental data. Importantly, specific ubiquitination sites of proposed TRAF-substrates, such as RIP1, have not been identified and therefore, mutant proteins lacking these sites have not been used to reconstitute knockout cells to demonstrate unequivocally the importance of these ubiquitination events. Furthermore, formation of K63-mediated polyubiquitin chains on any protein involved in NF-κB signaling has not been demonstrated by either mass spectrometry or specific antibodies and their signaling function is mostly a matter of conjecture that rests on the use of different ubiquitin mutants. Moreover, reconstitution of TRAF6-deficient cells with a TRAF6 mutant, lacking the signature motif of E3 RING-finger ligases, that is, the RING finger itself, completely restored IL-1-induced NF-κB and JNK1/2 activation *in vivo* [75]. At the same time, more complex osteo-clast effector functions were found to depend on an intact TRAF6 RING-finger [75], implying a more restricted and specific function of this domain. It is therefore difficult to say how important the proposed E3 ligase activity of TRAF proteins is *in vivo* and whether it is required for some pathways and not for others. Strictly, only *in vivo* mapping of ubiquitination and phosphorylation sites, detailed analysis of reconstitution experiments with appropriate mutant proteins and determination of binding affinities and interaction between endogenous proteins will resolve these issues (see Chapter 4 for further discussion of ubiquitin and NF-κB). As such, it is currently impossible to say whether MEKK3 and TAK1 act in a parallel or serial manner, although, based on the incomplete defects in signaling observed in the absence of either protein, one can suggest that TAK1 and MEKK3 may act in parallel. Obviously, the role of these molecules in NF-κB signaling is not as clear as the role of IKK itself.

3.4.1.2 Activation of the Alternative NF-κB Pathway

As mentioned above, the alternative pathway is activated by a rather limited number of TNFR family members, including BAFF-R and CD40 (on B cells) and LTβR (on stromal cells) [37,76–78]. The hallmark of the alternative NF-κB pathway is inducible NF-κB2/p100 processing, leading to liberation of the mature transcription factor p52. Like phosphorylated IκBα, phosphorylated p100 is recognized by the E3 ligase SCF^{βTrCP} and is targeted for proteasome-dependent proteolysis [3]. The exact reason why the proteolysis of p100, in contrast to IκB proteolysis, is partial, is not fully clear, but may have to do with the presence of stop signals that impede proteasome action [79]. The alternative NF-κB pathway was first discovered and described based on the observation that B cells from IKKα-deficient or IKKα-AA mutant mice have a defect in the constitutive processing of p100 *in vivo*, suggesting that B cells are exposed to a physiological ligand that induces p100 processing in an IKKα-dependent pathway [36]. This ligand and receptor pair was found to be BAFF and BAFF-R [76]. Two kinases are critically involved in activation of the alternative NF-κB pathway: NIK and IKKα [36,80]. Accordingly, mice carrying an inactivating mutation in the NIK gene, the so-called alymphoplasia (aly) mice or NIK knockout mice have defects in BAFF- and LTβR-induced processing of p100 that are similar to those of IKKα-AA mice [37,77]. As mentioned, both IKKβ and NEMO are dispensable for this pathway [37]. There is ample evidence that catalytic activity of both kinases, NIK and IKKα is required for inducible p100 processing. Overexpression of NIK or a constitutive active form of IKKα induces processing of p100 via specific phosphorylation of p100 C-terminal serine residues [36,80]. The response to NIK depends on the presence of catalytically active IKKα and *in vitro*, NIK is a potent IKKα-activating kinase [81]. The IKKα-AA mutant, containing alanine substitutions in its activation loop, is no longer phosphorylated by NIK and fails to induce LTβR- or BAFF-R-dependent processing of p100 [37,82]. Moreover, IKKα has been cloned as a NIK-binding protein, demonstrating direct interaction between the two proteins [10].

Thus, the alternative pathway seems to function as a typical phosphorylation dependent kinase cascade (Figure 3.2b); however, there remain issues in this pathway as well. One is the remarkably slow kinetics of p100 processing, requiring several hours in contrast to the classic NF-κB pathway, in which IκBα degradation occurs within minutes. Also, it is not clear why it is so difficult to demonstrate inducible IKKα kinase activity in response either to BAFF-R, CD40 or LTβR engagement (unpublished observation). It is possible that these problems are related and reflect a low affinity of IKKα for p100. In this context it is important to note that by itself IKKα binds only weakly to p100 and the interaction between the two is significantly enhanced in the presence of NIK, implicating NIK as an adaptor for IKKα docking to its substrate [83,84]. It is well established that protein kinases need to physically dock onto their substrates [84] and in the case of the large IKK complex, such a function may be mediated by the ELKS subunit [85]. Interestingly, regulation of NIK activity has been demonstrated to depend on TRAF proteins, particularly on TRAF3, and suggested to involve ubiquitination [86]. NIK seems to be constitutively bound by TRAF3, resulting in constitutive ubiquitination and degradation of NIK [86]. Receptor-induced degradation of TRAF3 precedes p100 processing,

presumably leading to increased levels of NIK. Accordingly, decreased TRAF3 levels
lead to constitutive p100 processing ([86] and personal observation). Notably,
TRAF2-deficient B cells also exhibit constitutive p100 processing, indicating a
similar role for TRAF2 [87]. Whether TRAF2-dependent activation of the alternative
NF-κB pathway also depends on NIK has not been published so far. Furthermore,
whether NIK activation itself requires other upstream kinases is unclear. Moderate
overexpression of NIK is sufficient for signaling, suggesting it is readily activated
by autophosphorylation. As such, it is possible that reduced degradation of NIK in
the absence of TRAF3 is sufficient to induce NIK activity, which leads to IKKα
activation and, together with NIK's property to support physical interaction of IKKα
with p100, ultimately results in NF-κB activation (Figure 3.2b).

3.4.2 ACTIVATION OF THE NF-κB PATHWAY BY MEMBERS OF THE IL-1R/TLR FAMILY

Members of the IL-1R family, such as IL-1R and IL-18R and members of the TLR
receptor family, such as TLR4, whose activation is triggered by lipopolysaccharide
(LPS), are also potent activators of the classic NF-κB pathway. Both IL-1R- and
TLR-family members share a common structural motif, the so-called TLR/IL-1R
(TIR) homology domain at their cytoplasmic portion [88]. Comparable to TNFR
family members, TIR-containing receptors do not have catalytic activity, but recruit
intracellular adaptors and signal transducing molecules to activate various effector
pathways [89]. Homotypic TIR–TIR interactions between receptor domains and a
limited set of TIR-containing adaptors explain why more than 15 different receptors
trigger only a small number of signaling pathways [89].

Based on genetic evidence and limited biochemical analyses, two adaptor molecules, MyD88 and TRIF/TICAM1, define IL-1R/ TLR signaling. Some receptors,
like TLR3 or TLR9 seem to signal exclusively through either TRIF or MyD88,
respectively, while TLR4 uses both [89]. Most receptors in these groups, including
IL-1R, IL18R, TLR1,2,5,6,7,8,9, seem to signal primarily through MyD88 (Figure
3.3). MyD88-dependent NF-κB activation also depends on a TRAF molecule, in
this case TRAF6, which is also involved in CD40 signaling [90,91]. MyD88-depen-
dent signaling also involves additional molecules that belong to the IRAK (IL-1R
associated kinase) family [88]. Two of the four IRAKs, IRAK1, and IRAK4 were
shown to be involved in IL-1R/TLR induced NF-κB activation [92–95]. Based on
data from knockout fibroblasts, both, IRAK1 and IRAK4 are required for IL-1
dependent NF-κB (and JNK) activation [93,95]. Interestingly, however, kinase activity is not required for IRAK1 and only partially required for IRAK4 signaling at
least in human cells [96,97]. The IRAKs are required to induce recruitment of
TRAF6 to MyD88, possibly through engagement of other molecules, such as TIFA
[98]. Although it cannot be excluded that IRAKs have functions other than TRAF6
recruitment, it is interesting to note that direct, IRAK-independent recruitment of
TRAF6 to CD40 or artificial dimerization/oligomerization of TRAF6 are sufficient
to initiate NF-κB activation [43,91]. Moreover, dimerization of MyD88, but not
TRAF6 leads to activation of IRAK1 [99], further supporting the notion that TRAF6
acts downstream of IRAK1/4, at least in respect to NF-κB activation.

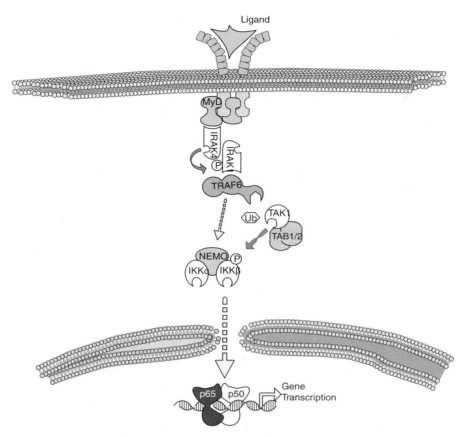

FIGURE 3.3 Schematic representation of the MyD88-dependent NF-κB pathway activated by members of the IL-1R/TLR family. Receptor activation leads to MyD88-mediated recruitment and activation of IRAK4 and IRAK1, which are required to recruit TRAF6. Oligomerization of TRAF6 and possibly ubiquitination initiates recruitment of the adaptor molecules TAB2 and TAB3, which in turn recruit and, possibly, activate TAK1 by oligomerization. Activated TAK1 might then directly phosphorylate IKKβ to activate the IKK complex, ultimately resulting in NF-κB activation (see Figure 3.2a).

TRAF6 is structurally similar to other TRAF proteins, especially TRAF2/5 and thus may function in a similar manner [42]. Indeed, TRAF6 was demonstrated to interact with TAB1/2/3 [55,57,67], similar to TRAF2 and both TAK1- and MEKK3-deficient fibroblasts exhibit significantly reduced NF-κB activation in response to IL-1 [58,61]. However, RIP1 (and Rip2/3) deficient cells have no defect in IL-1 induced NF-κB activation [99]. Whether IRAKs serve a similar function to RIP1 is not clear, although both IRAKs and RIP1 contain similar structural domains, that is, death and kinase domains [46]. Also, ubiquitination (and degradation) of IRAK1 follows its activation, although so far this has only been implicated in regulation of IRAK1 protein levels, rather than being an intrinsic mechanism regulating signal transduction [101]. In support of a role for TAK1 in IL-1R/MyD88 signaling, TAK1-

deficient B cells exhibit defects in NF-κB activation in response to engagement of TLR9, which signals via MyD88 [61,102]. Interestingly, TAK1-deficient B cells or IRAK4-deficient macrophages have diminished NF-κB activation when stimulated with LPS or poly I:C-signaling, which depend either partially (LPS) or completely (poly I:C) on Toll/IL resistance domain containing protein inducing interferon beta (TRIF) [61,95]. However, there is one important difference between MyD88- and TRIF-dependent NF-κB activation: TRAF6 is only essential for MyD88-signaling — at least in bone marrow derived macrophages [103,104].

So far, no data conclusively implicate other known TRAF proteins in TIR-mediated NF-κB activation. Although TRAF3, which also contains a RING-finger motif is recruited into MyD88- and TRIF-dependent signaling complexes, experiments performed with TRAF3 knockout macrophages show that it is not involved in IKK-activation [104]. Instead, TRAF3 is required for recruitment of the IKK-related kinase TBK1/NAK (and probably IKKi/IKKε) into the TIR signaling complex, thereby controlling the TLR-mediated type I interferon and IL-10 response [104]. Notably, the C-terminus of TRIF has been found to bind RIP1 via a so-called RHIM domain and RIP1-deficient fibroblasts exhibit reduced NF-κB activation when incubated with poly I:C [105,106]. However, at least in fibroblasts, TRIF-dependent activation of NF-κB was found to require autocrine production of TNFα [107], which may also explain the involvement of RIP1. It is clear that more detailed experimental data from primary gene deficient cells will be required to resolve these questions. Taken together, although the last few years have witnessed a substantial increase in our understanding of microbial recognition by innate immune receptors of the TLR family and what molecules are involved in TIR-mediated signaling, little information has been gathered about the molecular details of the mechanism of IKK activation beyond what has been learned from studies of TNFR signaling.

3.4.3 ACTIVATION OF THE NF-κB PATHWAY BY T CELL AND B CELL RECEPTORS

During recent years, our understanding of T cell receptor (TCR) and B cell receptor (BCR) dependent cell activation has progressed remarkably through the identification of several novel molecules that play critical roles in NF-κB activation [108]. In contrast to the receptors mentioned above, TCR and BCR are directly associated with cytoplasmic protein kinases and initiate cell activation through a sequence of tyrosine-phosphorylation events. This cascade ultimately leads to activation and recruitment of protein kinase C isozymes (PKCθ for TCR and PKCβ for BCR) to the membrane in the vicinity of the activated antigen receptors. IKK activation requires several molecules, that is, caspase recruitment domain (CARD)-membrane associated guanylate kinase (CARMA1)/CARD11, BCL10, and mucosal associated lymphoid tissue 1 (MALT1), whose roles in NF-κB activation have been substantiated by gene disruption and even forward genetics experiments [108] (Figure 3.4). Based on the use of mutant cells as well as transient transfection experiments it appears that CARMA1, which is constitutively localized at the cell membrane, is required for recruitment of BCL10 and MALT1 to the membrane in response to receptor activation [109,110]. Interestingly, a recent paper showed that TCR-

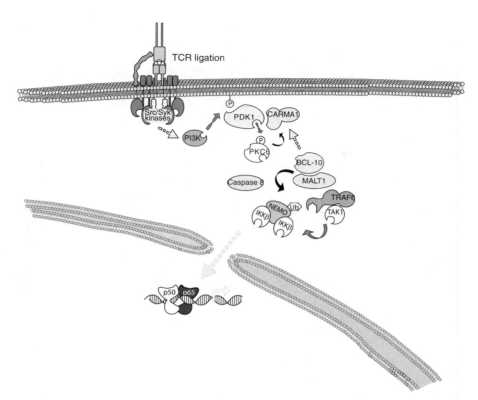

FIGURE 3.4 Schematic representation of the TCR-induced NF-κB pathway. TCR triggering leads to engagement of receptor associated tyrosine kinases of the Src and Syk families, controlling activation of PI3K. PI3K-dependent generation of membrane associated phospholipids leads to recruitment of PDK1, which may directly phosphorylate and activate PKCθ and control further recruitment of CARMA1. Assembly of these molecules in lipid rafts initiates recruitment of BCL10 and MALT1, and possibly TRAF6/TAK1. The molecular mechanism of IKK activation is unclear. It may involve direct ubiquitination of NEMO, or TRAF6-dependent ubiquitination and activation of TAK1, comparable to the mechanism proposed for IL-1R/TLR signaling (see Figure 3.3). The molecular function of caspase 8 is unclear. Activated IKK regulates NF-kB activation as illustrated in Figure 3.2a. A similar overall model depicted here for TCR signaling can also be applied to BCR signaling (see Chapter 7).

dependent PKCθ activation depends on PDK1 (3-phosphoinositide-dependent kinase-1), which may directly phosphorylate PKCθ at its activation loop [111]. PDK1 was also found to interact with CARMA1 by coimmunoprecipitation and to be required for IKK recruitment [111]. As such, PDK1 seems to serve a central role in TCR signaling, initiating the recruitment of several critical signaling molecules and IKK into close proximity [112]. The paracaspase MALT1 directly interacts with BCL10 and activates NF-κB, when coexpressed with BCL10 [113]. The mechanism of IKK activation is still not entirely clear. The role of BCL10 may be the recruitment and possibly oligomerization of MALT1, which seems to be an important intermediate in IKK activation [113]. One mechanism proposed for MALT1-dependent IKK activation is K63-mediated NEMO ubiquitination at a defined lysine residue (K399) within its zinc finger [40]. Phorbol ester/ ionomycin stimulation of the Jurkat T cell line, but also incubation of NEMO with purified recombinant MALT1 protein were found to result in NEMO ubiquitination. Replacement of K399 with arginine (K399R) greatly diminished MALT-dependent NEMO ubiquitination [40]. However, when NEMO-deficient cells were reconstituted with the NEMO (K399) mutant, phorbol ester/ionomycin- and BCL10-dependent NF-κB activation was only partially reduced [40], suggesting that either other ubiquitination sites can compensate for the loss of K399 or that NEMO ubiquitination is not essential for IKK activation. In this context it is also interesting to note that the RING-finger containing proteins TRAF6 and TRAF2, whose role was considered to be limited to IL-1R- and TNFR-signaling may also be involved in TCR-mediated IKK activation [114]. TRAF6 knockdown by RNAi reduced TCR-induced NF-κB activity and purified MALT1 was found to bind and oligomerize TRAF6, resulting in increased E3 ligase activity of TRAF6 towards the zinc finger of NEMO [114]. At the same time, TRAF6-autoubiquitination was suggested to lead to activation of TAK1, which then may phosphorylate IKKβ. Again, many steps in these scenarios need to be recapitulated *in vivo* and backed up by reconstitution experiments using proteins that have been mutated at sites implicated in phosphorylation or ubiquitination. Importantly, neither TRAF2- nor TRAF6-deficient T cells have been reported to have defects in TCR-dependent cell activation, including NF-κB activation. Although this may be explained the redundant functions of these structurally similar proteins, it is difficult to explain why TAK1-deficient B cells have no defect in BCR-dependent NF-κB activation [61], despite the severe signaling defects that result from loss of other components of this pathway suggested to act upstream of IKK: that is, CARMA1, BCL10, and MALT1 [108].

Another twist in TCR- and BCR-induced NF-κB activation came from the observation that cells with a loss of function mutation in the caspase 8 gene or a caspase 8-deficiency exhibit substantial defects in TCR- and BCR-induced NF-κB activation, but show normal responses to TNFα [115,116]. This was very surprising as so far, Caspase 8 had only been implicated in Fas associated death domain-containing protein (FADD)-dependent regulation of cell death [117] and NF-κB activation leads to inhibition of caspase 8 activation through the induction of c-flice-like inhibitory protein (FLIP) [118]. Stimulation of T cell lines with a TCR agonist induced IKKα/β recruitment to BCL10 and MALT1 in a caspase 8-dependent manner [116]. Other pathways, such as mitogen activated protein kinase (MAPK) pathways (ERK1/2, JNK1/2, p38) were not affected by the loss of caspase 8, implicating a rather specific role for caspase 8 in IKK activation. The divergence of these pathways from the CARMA1-BCL10-MALT1 pathway is, however, not entirely clear. Therefore, it is difficult to place caspase 8 into this signaling cascade. The observation that caspase 8 activity is required for IKK activation is even more puzzling as no obligatory proteolytic step upstream of IKK has been described so far. However, it should be noted that one of the NF-κB activation pathways in *Drosophila*, the so-called immune deficiency protein (IMD) pathway, also depends on a caspase, that is, death related ced-3/Nedd2-like protein (DREDD) for activation

[119,120]. Unfortunately, the molecular function of DREDD is, likewise, not clear. Given the observation that caspase activity is required for NF-κB activation, it will be very interesting to see whether it acts on one of the already defined molecules. Taken all together, it is clear that our understanding of the molecular events leading to IKK activation is far from being complete, despite the large number of publications on this topic.

3.4.4 ACTIVATION OF THE NF-κB PATHWAY BY DNA-DAMAGE

Stimuli that induce DNA damage, such as ionizing irradiation (IR) or topoisomerase inhibitors, for example, etoposide (VP16) are known to activate the classic NF-κB pathway in a NEMO-dependent manner [25,121]. Based on genetic evidence it is clear that the kinase ATM (ataxia telangiectasia mutated) is required for NF-κB activation, although the mechanism of its action has been enigmatic [122,123]. A recent paper has uncovered a surprising sequence of events, ultimately leading to NF-κB activation in response to DNA strand breaks (DSBs) [41]. The authors found that NEMO, probably independent of IKKα/β, is constantly modified by attachment of SUMO, a small ubiquitin like polypeptide. This leads to nuclear translocation of a small fraction of NEMO (Figure 3.5). Although sumoylation requires the C-terminal zinc finger domain of NEMO, the sumoylation sites that were identified by mutational analysis (K277 and K309) are not located in this region. Nonetheless,

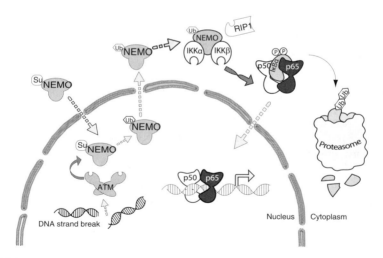

FIGURE 3.5 Schematic representation of NF-κB activation by DNA strand breaks (DSB). NEMO, independent of IKKαβ, is modified by attachment of small ubiquitin-like modifier (SUMO), which targets it to the nucleus. DSB leads to activation of ATM, which controls ubiquitination of nuclear NEMO, possibly through phosphorylation of NEMO, resulting in replacement of SUMO by ubiquitin chains. Ubiquitinated NEMO is transported to the cytoplasm, where it assembles with the catalytic subunits IKKα and IKKβ into the classic large IKK complex. The molecular function of RIP1 is not clear, but it might be required for IKK complex orchestration and activation of ubiquitinated NEMO. Activated IKK regulates NF-κB activation as illustrated in Figure 3.2a.

NEMO sumoylation is required for its nuclear translocation and IKK activation in response to DSB, but not LPS-induced NF-κB activation. The authors also noticed DSB-dependent ubiquitination of NEMO, which was ATM-dependent and followed its sumoylation, most likely involving replacement of NEMO-attached SUMO with polyubiquitin chains. Based on these experiments it has been proposed that SUMO-dependent nuclear translocation juxtaposes NEMO and ATM, which, when activated by DSBs phosphorylate NEMO, triggering its ubiquitination and nuclear export, ultimately resulting in binding of modified NEMO to IKKα/β and IKK activation (Figure 3.5). Certainly, many aspects of this model need to be further confirmed and characterized. The ATM-phosphorylation sites in NEMO, the nuclear export mechanism, and whether ubiquitination of NEMO is required for direct activation of IKK or nuclear export remain to be discovered.

Interestingly, DSB-mediated NF-κB activation also depends directly on RIP1 [124]. Moreover, cell treatment with adriamycin induced interaction of RIP1 with NEMO, which was dependent on ATM, implying that RIP1 acts downstream of ATM. Comparable to TNFα, kinase activity of RIP1 was found to be dispensable for DSB-mediated NF-κB activation. Based on these observations and the above described modifications of NEMO, it is tempting to speculate that NEMO ubiquitination controls interaction of NEMO with RIP1, which might serve as a scaffold protein, possibly leading to oligomerization and activation of IKK (Figure 3.5). Certainly, several aspects of this model are speculation and the exact mode of RIP1 and NEMO interaction and their relation to the active large IKK complex containing IKKα/β and NEMO need to be investigated. Nevertheless, this example of an apparently different mechanism of IKK activation may also shed light on other pathways regulating NF-κB activity.

3.5 SUMMARY

Since IKK was cloned in 1997, substantial progress has been made in understanding activation of IKK and NF-κB. It is now clear that IKK activation is the critical step in all major pathways leading to NF-κB activation. Many new molecules that are involved in different pathways leading to IKK activation have been identified and their biological roles have been confirmed in gene deficient mice or even human disease (Chapter 9). Identification of new molecules in these pathways and characterization of their biological function has further substantiated the overall significance of NF-κB in physiological and pathophysiological conditions, including immune responses, inflammation, and cancer. Yet, despite the undeniable increase in knowledge about IKK activation, the exact molecular mechanism and interactions between IKK and upstream and downstream molecules is still surprisingly unclear. Therefore, it is currently not possible to say whether all major molecules involved in IKK activation have been described and just need to be assembled in the correct way, or, whether we still are missing critical molecules, whose identification is required to allow us to put the pieces of the puzzle together. The characterization of the biochemical events leading to IKK activation is still a major challenge and yet, it will probably be the understanding of the molecular details involved that will allow us to influence this pathway in human disease.

REFERENCES

[1]   Karin, M. and Ben-Neriah, Y., Phosphorylation meets ubiquitination: The control of NF-[kappa]B activity, *Annu. Rev. Immunol.,* 18, 621, 2000.

[2]   Hoffmann, A., Levchenko, A., Scott, M.L. et al., The IkappaB-NF-kappaB signaling module: temporal control and selective gene activation, *Science,* 298, 1241, 2002.

[3]   Amir, R.E., Häcker, H., Karin, M. et al., Mechanism of processing of the NF-kappaB2 p100 precursor: identification of the specific polyubiquitin chain-anchoring lysine residue and analysis of the role of NEDD8-modification on the SCF(beta-TrCP) ubiquitin ligase, *Oncogene,* 23, 2540, 2004.

[4]   Betts, J.C. and Nabel, G.J., Differential regulation of NF-kappaB2(p100) processing and control by amino-terminal sequences, *Mol. Cell Biol.,* 16, 6363, 1996.

[5]   Solan, N.J., Miyoshi, H., Carmona, E.M. et al., RelB cellular regulation and transcriptional activity are regulated by p100, *J. Biol. Chem.,* 277, 1405, 2002.

[6]   Lin, L., DeMartino, G.N., and Greene, W.C., Cotranslational biogenesis of NF-kappaB p50 by the 26S proteasome, *Cell,* 92, 819, 1998.

[7]   Heissmeyer, V., Krappmann, D., Hatada, E.N. et al., Shared pathways of IkappaB kinase-induced SCF(betaTrCP)-mediated ubiquitination and degradation for the NF-kappaB precursor p105 and IkappaBalpha, *Mol. Cell Biol.,* 21, 1024, 2001.

[8]   Bonizzi, G. and Karin, M., The two NF-kappaB activation pathways and their role in innate and adaptive immunity, *Trends Immunol.,* 25, 280, 2004.

[9]   DiDonato, J.A., Hayakawa, M., Rothwarf, D.M. et al., A cytokine-responsive IkappaB kinase that activates the transcription factor NF-kappaB [see comments], *Nature,* 388, 548, 1997.

[10] Regnier, C.H., Song, H.Y., Gao, X. et al., Identification and characterization of an IkappaB kinase, *Cell,* 90, 373, 1997.

[11] Rothwarf, D.M., Zandi, E., Natoli, G. et al., IKK-gamma is an essential regulatory subunit of the IkappaB kinase complex, *Nature,* 395, 297, 1998.

[12] Yamaoka, S., Courtois, G., Bessia, C. et al., Complementation cloning of NEMO, a component of the IkappaB kinase complex essential for NF-kappaB activation, *Cell,* 93, 1231, 1998.

[13] Mercurio, F., Murray, B.W., Shevchenko, A. et al., IkappaB kinase (IKK)-associated protein 1, a common component of the heterogeneous IKK complex, *Mol. Cell Biol.,* 19, 1526, 1999.

[14] Li, Y., Kang, J., Friedman, J. et al., Identification of a cell protein (FIP-3) as a modulator of NF-kappaB activity and as a target of an adenovirus inhibitor of tumor necrosis factor alpha-induced apoptosis, *Proc. Natl. Acad. Sci. USA,* 96, 1042, 1999.

[15] Zandi, E., Rothwarf, D.M., Delhase, M. et al., The IkappaB kinase complex (IKK) contains two kinase subunits, IKKalpha and IKKbeta, necessary for IkappaB phosphorylation and NF-kappaB activation, *Cell,* 91, 243, 1997.

[16] Zandi, E., Chen, Y., and Karin, M., Direct phosphorylation of IkappaB by IKKalpha and IKKbeta: discrimination between free and NF-kappaB-bound substrate, *Science,* 281, 1360, 1998.

[17] Miller, B.S. and Zandi, E., Complete reconstitution of human IkappaB kinase (IKK) complex in yeast. Assessment of its stoichiometry and the role of IKKgamma on the complex activity in the absence of stimulation, *J. Biol. Chem.,* 276, 36320, 2001.

[18] Tegethoff, S., Behlke, J., and Scheidereit, C., Tetrameric oligomerization of IkappaB kinase gamma (IKKgamma) is obligatory for IKK complex activity and NF-kappaB activation, *Mol. Cell Biol.,* 23, 2029, 2003.

[19]   Agou, F., Ye, F., Goffinont, S. et al., NEMO trimerizes through its coiled-coil C-terminal domain, *J. Biol. Chem.*, 277, 17464, 2002.

[20]   May, M.J., Marienfeld, R.B., and Ghosh, S., Characterization of the IkappaB-kinase NEMO binding domain, *J. Biol. Chem.*, 277, 45992, 2002.

[21]   May, M.J., D'Acquisto, F., Madge, L.A et al., Selective inhibition of NF-kappaB activation by a peptide that blocks the interaction of NEMO with the IkappaB kinase complex, *Science*, 289, 1550, 2000.

[22]   Makris, C., Godfrey, V.L., Krahn-Senftleben, G. et al., Female mice heterozygous for IKKgamma/NEMO deficiencies develop a dermatopathy similar to the human X-linked disorder incontinentia pigmenti, *Mol. Cell*, 5, 969, 2000.

[23]   Rudolph, D., Yeh, W.C., Wakeham, A. et al., Severe liver degeneration and lack of NF-kappaB activation in NEMO/IKKgamma-deficient mice, *Genes Dev.*, 14, 854, 2000.

[24]   Makris, C., Roberts, J.L., and Karin, M., The carboxyl-terminal region of IkappaB kinase gamma (IKKgamma) is required for full IKK activation, *Mol. Cell Biol.*, 22, 6573, 2002.

[25]   Huang, T.T., Feinberg, S.L., Suryanarayanan, S. et al., The zinc finger domain of NEMO is selectively required for NF-kappaB activation by UV radiation and topoisomerase inhibitors, *Mol. Cell Biol.*, 22, 5813, 2002.

[26]   Chen, G., Cao, P., and Goeddel, D.V., TNF-induced recruitment and activation of the IKK complex require Cdc37 and Hsp90, *Mol. Cell*, 9, 401, 2002.

[27]   Richter, K. and Buchner, J., Hsp90: Chaperoning signal transduction, *J. Cell. Physiol.*, 188, 281, 2001.

[28]   Woronicz, J.D., Gao, X., Cao, Z. et al., IkappaB kinase-beta: NF-kappaB activation and complex formation with IkappaB kinase-alpha and NIK, *Science*, 278, 866, 1997.

[29]   Mercurio, F., Zhu, H., Murray, B.W. et al., IKK-1 and IKK-2: cytokine-activated IkappaB kinases essential for NF-kappaB activation, *Science*, 278, 860, 1997.

[30]   Delhase, M., Hayakawa, M., Chen, Y. et al., Positive and negative regulation of IkappaB kinase activity through IKKbeta subunit phosphorylation [see comments], *Science*, 284, 309, 1999.

[31]   Johnson, L.N., Noble, M.E., and Owen, D.J., Active and inactive protein kinases: structural basis for regulation, *Cell*, 85, 149, 1996.

[32]   Hu, Y., Baud, V., Delhase, M. et al., Abnormal morphogenesis but intact IKK activation in mice lacking the IKKalpha subunit of IkappaB kinase [see comments], *Science*, 284, 316, 1999.

[33]   Takeda, K., Takeuchi, O., Tsujimura, T. et al., Limb and skin abnormalities in mice lacking IKKalpha, *Science*, 284, 313, 1999.

[34]   Morgan, D.O., Principles of CDK regulation, *Nature*, 374, 131, 1995.

[35]   Prajapati, S., Verma, U., Yamamoto, Y. et al., Protein phosphatase 2Cbeta association with the IkappaB kinase complex is involved in regulating NF-kappaB activity, *J. Biol. Chem.*, 279, 1739, 2004.

[36]   Senftleben, U., Cao, Y., Xiao, G. et al., Activation by IKKalpha of a second, evolutionary conserved, NF-kappaB signaling pathway, *Science*, 293, 1495, 2001.

[37]   Dejardin, E., Droin, N.M., Delhase, M. et al., The lymphotoxin-beta receptor induces different patterns of gene expression via two NF-kappaB pathways, *Immunity*, 17, 525, 2002.

[38]   Cao, Y., Bonizzi, G., Seagroves, T.N. et al., IKKalpha provides an essential link between RANK signaling and cyclin D1 expression during mammary gland development, *Cell*, 107, 763, 2001.

[39] Ramakrishnan, P., Wang, W., and Wallach, D., Receptor-specific signaling for both the alternative and the canonical NF-kappaB activation pathways by NF-kappaB-inducing kinase, *Immunity,* 21, 477, 2004.

[40] Zhou, H., Wertz, I., O'Rourke, K et al., Bcl10 activates the NF-kappaB pathway through ubiquitination of NEMO, *Nature,* 427, 167, 2004.

[41] Huang, T.T., Wuerzberger-Davis, S.M., Wu, Z.H. et al., Sequential modification of NEMO/IKKgamma by SUMO-1 and ubiquitin mediates NF-kappaB activation by genotoxic stress, *Cell,* 115, 565, 2003.

[42] Dempsey, P.W., Doyle, S.E., He, J.Q. et al., The signaling adaptors and pathways activated by TNF superfamily, *Cytokine Growth Factor Rev.,* 14, 193, 2003.

[43] Baud, V., Liu, Z.G., Bennett, B. et al., Signaling by proinflammatory cytokines: oligomerization of TRAF2 and TRAF6 is sufficient for JNK and IKK activation and target gene induction via an amino-terminal effector domain, *Genes Dev.,* 13, 1297, 1999.

[44] Rothe, M., Wong, S.C., Henzel, W.J. et al., A novel family of putative signal transducers associated with the cytoplasmic domain of the 75 kDa tumor necrosis factor receptor, *Cell,* 78, 681, 1994.

[45] Hsu, H., Huang, J., Shu, H.B. et al., TNF-dependent recruitment of the protein kinase RIP to the TNF receptor-1 signaling complex, *Immunity,* 4, 387, 1996.

[46] Meylan, E. and Tschopp, J., The RIP kinases: crucial integrators of cellular stress, *Trends Biochem. Sci.,* 30, 151, 2005.

[47] Ting, A.T., Pimentel-Muinos, F.X., and Seed, B., RIP mediates tumor necrosis factor receptor 1 activation of NF-kappaB but not Fas/APO-1-initiated apoptosis, *Embo. J.,* 15, 6189, 1996.

[48] Kelliher, M.A., Grimm, S., Ishida, Y. et al., The death domain kinase RIP mediates the TNF-induced NF-kappaB signal, *Immunity,* 8, 297, 1998.

[49] Yeh, W.C., Shahinian, A., Speiser, D. et al., Early lethality, functional NF-kappaB activation, and increased sensitivity to TNF-induced cell death in TRAF2-deficient mice, *Immunity,* 7, 715, 1997.

[50] Tada, K., Okazaki, T., Sakon, S. et al., Critical roles of TRAF2 and TRAF5 in tumor necrosis factor-induced NF-kappaB activation and protection from cell death, *J. Biol. Chem.,* 276, 36530, 2001.

[51] Devin, A., Lin, Y., Yamaoka, S. et al., The alpha and beta subunits of IkappaB kinase (IKK) mediate TRAF2-dependent IKK recruitment to tumor necrosis factor (TNF) receptor 1 in response to TNF, *Mol. Cell Biol.,* 21, 3986, 2001.

[52] Zhang, S.Q., Kovalenko, A., Cantarella, G. et al., Recruitment of the IKK signalosome to the p55 TNF receptor: RIP and A20 bind to NEMO (IKKgamma) upon receptor stimulation, *Immunity,* 12, 301, 2000.

[53] Lee, T.H., Shank, J., Cusson, N. et al., The kinase activity of Rip1 is not required for tumor necrosis factor-alpha-induced IkappaB kinase or p38 MAP kinase activation or for the ubiquitination of Rip1 by Traf2, *J. Biol. Chem.,* 279, 33185, 2004.

[54] Zhao, Q. and Lee, F.S., Mitogen-activated protein kinase/ERK kinase kinases 2 and 3 activate nuclear factor-kappaB through IkappaB kinase-alpha and IkappaB kinase-beta, *J. Biol. Chem.,* 274, 8355, 1999.

[55] Wang, C., Deng, L., Hong, M. et al., TAK1 is a ubiquitin-dependent kinase of MKK and IKK, *Nature,* 412, 346, 2001.

[56] Shibuya, H., Yamaguchi, K., Shirakabe, K. et al., TAB1: an activator of the TAK1 MAPKKK in TGF-beta signal transduction, *Science,* 272, 1179, 1996.

[57] Takaesu, G., Kishida, S., Hiyama, A. et al., TAB2, a novel adaptor protein, mediates activation of TAK1 MAPKKK by linking TAK1 to TRAF6 in the IL-1 signal transduction pathway, *Mol. Cell,* 5, 649, 2000.

[58] Yang, J., Lin, Y., Guo, Z. et al., The essential role of MEKK3 in TNF-induced NF-kappaB activation, *Nat. Immunol.*, 2, 620, 2001.
[59] Huang, Q., Yang, J., Lin, Y. et al., Differential regulation of interleukin 1 receptor and Toll-like receptor signaling by MEKK3, *Nat. Immunol.*, 5, 98, 2004.
[60] Shim, J.H., Xiao, C., Paschal, A.E. et al., TAK1 but not TAB1 or TAB2, plays an essential role in multiple signaling pathways *in vivo*, *Genes Dev.*, 19, 2668, 2005.
[61] Sato, S., Sanjo, H., Takeda, K. et al., Essential function for the kinase TAK1 in innate and adaptive immune responses, *Nat. Immunol.*, 2005.
[62] Yang, J., Boerm, M., McCarty, M. et al., MEKK3 is essential for early embryonic cardiovascular development, *Nat. Genet.*, 24, 309, 2000.
[63] Blonska, M., You, Y., Geleziunas, R. et al., Restoration of NF-kappaB activation by tumor necrosis factor alpha receptor complex-targeted MEKK3 in receptor-interacting protein-deficient cells, *Mol. Cell Biol.*, 24, 10757, 2004.
[64] Yamaguchi, K., Shirakabe, K., Shibuya, H. et al., Identification of a member of the MAPKKK family as a potential mediator of TGF-beta signal transduction, *Science*, 270, 2008, 1995.
[65] Sakurai, H., Shigemori, N., Hasegawa, K. et al., TGF-beta-activated kinase 1 stimulates NF-kappa B activation by an NF-kappaB-inducing kinase-independent mechanism, *Biochem. Biophys. Res. Commun.*, 243, 545, 1998.
[66] Ninomiya-Tsuji, J., Kishimoto, K., Hiyama, A. et al., The kinase TAK1 can activate the NIK-I kappaB as well as the MAP kinase cascade in the IL-1 signalling pathway, *Nature*, 398, 252, 1999.
[67] Ishitani, T., Takaesu, G., Ninomiya-Tsuji, J. et al., Role of the TAB2-related protein TAB3 in IL-1 and TNF signaling, *Embo. J.*, 22, 6277, 2003.
[68] Cheung, P.C., Nebreda, A.R., and Cohen, P., TAB3, a new binding partner of the protein kinase TAK1, *Biochem. J.*, 378, 27, 2004.
[69] Sanjo, H., Takeda, K., Tsujimura, T. et al., TAB2 is essential for prevention of apoptosis in fetal liver but not for interleukin-1 signaling, *Mol. Cell Biol.*, 23, 1231, 2003.
[70] Komatsu, Y., Shibuya, H., Takeda, N. et al., Targeted disruption of the Tab1 gene causes embryonic lethality and defects in cardiovascular and lung morphogenesis, *Mech. Dev.*, 119, 239, 2002.
[71] Kanayama, A., Seth, R.B., Sun, L. et al., TAB2 and TAB3 activate the NF-kappaB pathway through binding to polyubiquitin chains, *Mol. Cell*, 15, 535, 2004.
[72] Deng, L., Wang, C., Spencer, E. et al., Activation of the IkappaB kinase complex by TRAF6 requires a dimeric ubiquitin-conjugating enzyme complex and a unique polyubiquitin chain, *Cell*, 103, 351, 2000.
[73] Legler, D.F., Micheau, O., Doucey, M.A. et al., Recruitment of TNF receptor 1 to lipid rafts is essential for TNFalpha-mediated NF-kappaB activation, *Immunity*, 18, 655, 2003.
[74] Chen, Z.J., Ubiquitin signalling in the NF-kappaB pathway, *Nat. Cell Biol.*, 7, 758, 2005.
[75] Kobayashi, N., Kadono, Y., Naito, A. et al., Segregation of TRAF6-mediated signaling pathways clarifies its role in osteoclastogenesis, *Embo. J.*, 20, 1271, 2001.
[76] Kayagaki, N., Yan, M., Seshasayee, D. et al., BAFF/BLyS receptor 3 binds the B cell survival factor BAFF ligand through a discrete surface loop and promotes processing of NF-kappaB2, *Immunity*, 17, 515, 2002.
[77] Claudio, E., Brown, K., Park, S. et al., BAFF-induced NEMO-independent processing of NF-kappaB2 in maturing B cells, *Nat. Immunol.*, 3, 958, 2002.

[78] Coope, H.J., Atkinson, P.G., Huhse, B. et al., CD40 regulates the processing of NF-kappaB2 p100 to p52, *Embo. J.*, 21, 5375, 2002.

[79] Lin, L. and Ghosh, S., A glycine-rich region in NF-kappaB p105 functions as a processing signal for the generation of the p50 subunit, *Mol. Cell Biol.*, 16, 2248, 1996.

[80] Xiao, G., Harhaj, E.W., and Sun, S.C., NF-kappaB-inducing kinase regulates the processing of NF-kappaB2 p100, *Mol. Cell*, 7, 401, 2001.

[81] Ling, L., Cao, Z., and Goeddel, D.V., NF-kappaB-inducing kinase activates IKK-alpha by phosphorylation of Ser-176, *Proc. Natl. Acad. Sci. USA*, 95, 3792, 1998.

[82] Bonizzi, G., Bebien, M., Otero, D.C. et al., Activation of IKKalpha target genes depends on recognition of specific kappaB binding sites by RelB:p52 dimers, *Embo. J.*, 23, 4202, 2004.

[83] Xiao, G., Fong, A., and Sun, S.C., Induction of p100 processing by NF-kappaB-inducing kinase involves docking IkappaB kinase alpha (IKKalpha) to p100 and IKKalpha-mediated phosphorylation, *J. Biol. Chem.*, 279, 30099, 2004.

[84] Kallunki, T., Deng, T., Hibi, M. et al., c-Jun can recruit JNK to phosphorylate dimerization partners via specific docking interactions, *Cell*, 87, 929, 1996.

[85] Ducut Sigala, J.L., Bottero, V., Young, D.B. et al., Activation of transcription factor NF-kappaB requires ELKS, an IkappaB kinase regulatory subunit, *Science*, 304, 1963, 2004.

[86] Liao, G., Zhang, M., Harhaj, E.W. et al., Regulation of the NF-kappaB-inducing kinase by tumor necrosis factor receptor-associated factor 3-induced degradation, *J. Biol. Chem.*, 279, 26243, 2004.

[87] Grech, A.P., Amesbury, M., Chan, T. et al., TRAF2 differentially regulates the canonical and noncanonical pathways of NF-kappaB activation in mature B cells, *Immunity*, 21, 629, 2004.

[88] Martin, M.U. and Wesche, H., Summary and comparison of the signaling mechanisms of the Toll/interleukin-1 receptor family, *Biochim. Biophys. Acta*, 1592, 265, 2002.

[89] Yamamoto, M. and Akira, S., TIR domain-containing adaptors regulate TLR signaling pathways, *Adv. Exp. Med. Biol.*, 560, 1, 2005.

[90] Cao, Z., Xiong, J., Takeuchi, M. et al., TRAF6 is a signal transducer for interleukin-1, *Nature*, 383, 443, 1996.

[91] Ishida, T., Mizushima, S., Azuma, S. et al., Identification of TRAF6, a novel tumor necrosis factor receptor-associated factor protein that mediates signaling from an amino-terminal domain of the CD40 cytoplasmic region, *J. Biol. Chem.*, 271, 28745, 1996.

[92] Cao, Z., Henzel, W.J., and Gao, X., IRAK: a kinase associated with the interleukin-1 receptor, *Science*, 271, 1128, 1996.

[93] Thomas, J.A., Allen, J.L., Tsen, M. et al., Impaired cytokine signaling in mice lacking the IL-1 receptor-associated kinase, *J. Immunol.*, 163, 978, 1999.

[94] Li, S., Strelow, A., Fontana, E.J. et al., IRAK-4: a novel member of the IRAK family with the properties of an IRAK-kinase, *Proc. Natl. Acad. Sci. USA*, 99, 5567, 2002.

[95] Suzuki, N., Suzuki, S., Duncan, G.S. et al., Severe impairment of interleukin-1 and Toll-like receptor signalling in mice lacking IRAK-4, *Nature*, 416, 750, 2002.

[96] Li, X., Commane, M., Burns, C. et al., Mutant cells that do not respond to interleukin-1 (IL-1) reveal a novel role for IL-1 receptor-associated kinase, *Mol. Cell Biol.*, 19, 4643, 1999.

[97] Knop, J. and Martin, M.U., Effects of IL-1 receptor-associated kinase (IRAK) expression on IL-1 signaling are independent of its kinase activity, *FEBS Lett.*, 448, 81, 1999.

[98]   Takatsuna, H., Kato, H., Gohda, J. et al., Identification of TIFA as an adapter protein that links tumor necrosis factor receptor-associated factor 6 (TRAF6) to interleukin-1 (IL-1) receptor-associated kinase-1 (IRAK-1) in IL-1 receptor signaling, *J. Biol. Chem.*, 278, 12144, 2003.

[99]   Häcker, H., Redecke, V., Blagoev, B et al., Specificity in TLR signaling through distinct effector functions of TRAF3 and TRAF6, *Nature*, 439, 204, 2006.

[100] Devin, A., Cook, A., Lin, Y. et al., The distinct roles of TRAF2 and RIP in IKK activation by TNF-R1: TRAF2 recruits IKK to TNF-R1 while RIP mediates IKK activation, *Immunity*, 12, 419, 2000.

[101] Yamin, T.T. and Miller, D.K., The interleukin-1 receptor-associated kinase is degraded by proteasomes following its phosphorylation, *J. Biol. Chem.*, 272, 21540, 1997.

[102] Häcker, H., Vabulas, R.M., Takeuchi, O et al., Immune cell activation by bacterial CpG-DNA through myeloid differentiation marker 88 and tumor necrosis factor receptor-associated factor (TRAF)6, *J. Exp. Med.*, 192, 595, 2000.

[103] Gohda, J., Matsumura, T., and Inoue, J., Cutting edge: TNFR-associated factor (TRAF) 6 is essential for MyD88-dependent pathway but not toll/IL-1 receptor domain-containing adaptor-inducing IFN-beta (TRIF)-dependent pathway in TLR signaling, *J. Immunol.*, 173, 2913, 2004.

[104] Häcker, H., Redecke, V., Blagoev, B. et al., Specificity in TLR signaling through Distinct Effector Functions of TRAF3 and TRAF6, *Nature*, accepted for publication.

[105] Meylan, E., Burns, K., Hofmann, K. et al., RIP1 is an essential mediator of Toll-like receptor 3-induced NF-kappaB activation, *Nat. Immunol.*, 5, 503, 2004.

[106] Cusson-Hermance, N., Lee, T.H., Fitzgerald, K.A. et al., Rip1 mediates the Trif-dependent toll-like receptor 3 and 4-induced NF-kappaB activation but does not contribute to IRF-3 activation, *J. Biol. Chem.*, 2005.

[107] Covert, M.W., Leung, T.H., Gaston, J.E. et al., Achieving stability of lipopolysaccharide-induced NF-kappaB activation, *Science*, 309, 1854, 2005.

[108] Thome, M., CARMA1, BCL-10 and MALT1 in lymphocyte development and activation, *Nat. Rev. Immunol.*, 4, 348, 2004.

[109] Gaide, O., Favier, B., Legler, D.F. et al., CARMA1 is a critical lipid raft-associated regulator of TCR-induced NF-kappaB activation, *Nat. Immunol.*, 3, 836, 2002.

[110] Wang, D., Matsumoto, R., You, Y. et al., CD3/CD28 costimulation-induced NF-kappaB activation is mediated by recruitment of protein kinase C-theta, BCL10, and IkappaB kinase beta to the immunological synapse through CARMA1, *Mol. Cell Biol.*, 24, 164, 2004.

[111] Lee, K.Y., DíAcquisto, F., Hayden, M.S et al., PDK1 nucleates T cell receptor-induced signaling complex for NF-kappaB activation, *Science*, 308, 114, 2005.

[112] van Oers, N.S. and Chen, Z.J., Cell biology. Kinasing and clipping down the NF-kappaB trail, *Science*, 308, 65, 2005.

[113] Lucas, P.C., Yonezumi, M., Inohara, N. et al., Bcl10 and MALT1, independent targets of chromosomal translocation in malt lymphoma, cooperate in a novel NF-kappaB signaling pathway, *J. Biol. Chem.*, 276, 19012, 2001.

[114] Sun, L., Deng, L., Ea, C.K. et al., The TRAF6 ubiquitin ligase and TAK1 kinase mediate IKK activation by BCL10 and MALT1 in T lymphocytes, *Mol. Cell*, 14, 289, 2004.

[115] Chun, H.J., Zheng, L., Ahmad, M. et al., Pleiotropic defects in lymphocyte activation caused by caspase-8 mutations lead to human immunodeficiency, *Nature*, 419, 395, 2002.

[116] Su, H., Bidere, N., Zheng, L. et al., Requirement for caspase-8 in NF-kappaB activation by antigen receptor, *Science*, 307, 1465, 2005.

[117] Muzio, M., Chinnaiyan, A.M., Kischkel, F.C. et al., FLICE, a novel FADD-homol-ogous ICE/CED-3-like protease, is recruited to the CD95 (Fas/APO-1) death-inducing signaling complex, *Cell,* 85, 817, 1996.

[118] Yeh, W.C., Itie, A., Elia, A.J. et al., Requirement for Casper (c-FLIP) in regulation of death receptor-induced apoptosis and embryonic development, *Immunity,* 12, 633, 2000.

[119] Khush, R.S., Leulier, F., and Lemaitre, B., *Drosophila immunity*: Two paths to NF-kappaB, *Trends Immunol.,* 22, 260, 2001.

[120] Stoven, S., Ando, I., Kadalayil, L. et al., Activation of the Drosophila NF-kappaB factor Relish by rapid endoproteolytic cleavage, *Embo. Rep.,* 1, 347, 2000.

[121] Li, N. and Karin, M., Ionizing radiation and short wavelength UV activate NF-kappaB through two distinct mechanisms, *Proc. Natl. Acad. Sci. USA,* 95, 13012, 1998.

[122] Lee, S.J., Dimtchev, A., Lavin, M.F. et al., A novel ionizing radiation-induced sig-naling pathway that activates the transcription factor NF-kappaB, *Oncogene,* 17, 1821, 1998.

[123] Piret, B., Schoonbroodt, S., and Piette, J., The ATM protein is required for sustained activation of NF-kappaB following DNA damage, *Oncogene,* 18, 2261, 1999.

[124] Hur, G.M., Lewis, J., Yang, Q. et al., The death domain kinase RIP has an essential role in DNA damage-induced NF-kappaB activation, *Genes Dev.,* 17, 873, 2003.

4 Control of NF-κB Activity by Ubiquitination

Steven C. Ley and Yinon Ben-Neriah

CONTENTS

4.1 INTRODUCTION

Since the discovery of NF-κB transcription factors, research into the pathways controlling their activation has been intertwined with studies on the function of protein ubiquitination in controlling cell physiology (Figure 4.1). Baltimore and colleagues identified NF-κB as a specific DNA binding activity in crude nuclear

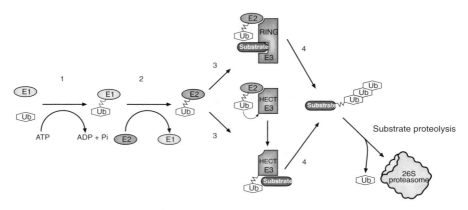

FIGURE 4.1 Schematic diagram of the ubiquitination pathway. The ubiquitin-proteasome is composed of four major enzyme types — E1, E2, E3, and the 26S proteasome — which function in a hierarchical fashion: (1) Ubiquitin is activated in an adenosine triphosphate (ATP)-dependent manner by a single ubiquitin-activating enzyme, E1. (2) The activated ubiquitin moiety is transferred to a number of different E2s. (3) The activated ubiquitin is attached to a Cys residue on homologous to the E6-AP carboxyl terminus (HECT) E3s and is then transferred to substrate or to a substrate-bound polyubiquitin chain. For RING E3s, ubiquitin is transferred directly from E2 to a substrate. Each E2 can interact with several E3s. Each E3 may target several different substrates, and certain substrates may be targeted by more than one E3. (4) The 26S proteasome recognizes polyubiquitin-modified substrates and proteolyzes them into short peptides, releasing ubiquitin for reuse.

extracts in 1986 and, shortly after its cytoplasmic inhibitor, IκB (inhibitor of NF-κB) [1,2,3]. This work suggested a model of NF-κB activation that required its liberation from the inhibitory effects of IκB. The connection between NF-κB and ubiquitin became apparent from research into the physiological mechanism of this release following agonist stimulation.

Using a simple detergent treatment of cell extracts, Baeuerle and Baltimore provided the first proof of concept for this model, showing that while inducing the dissociation from its inhibitor, NF-κB was activated [3]. Subsequent *in vivo* experiments using cell lines suggested that stimulus-induced IκB phosphorylation triggered release of associated NF-κB and could therefore be the critical event for the physiological activation of NF-κB [4]. However, it was later demonstrated that IκB phosphorylation was in fact insufficient for NF-κB activation [5,6]. In 1993, several groups (Siebenlist's, Greene's, Baldwin's, and Goodbourn's) [7,8,9,10] made the seminal observation that NF-κB activation was accompanied by IκB degradation. Furthermore, Baeuerle, Ben-Neriah, and colleagues showed that blockade of IκB degradation prevented NF-κB activation [11]. Elucidation of the mechanism of IκB proteolysis followed. Maniatis, Goldberg, and colleagues first showed that proteasome inhibitors inhibited NF-κB activation by blocking IκB degradation, implicating the proteasome in the activation process [12]. Soon after, the groups of Maniatis, Ben-Neriah, and Ciechanover demonstrated that signal-induced ubiquitination was required for the elimination of IκB from the NF-κB complex by the proteasome [13,14].

The motif regulating the degradation of IκBα was identified by sequence homology comparisons and site direction mutagenesis [13,15]. These studies characterized a conserved N-terminal sequence containing two serines, which were phosphorylated after phorbol ester stimulation, that are required for signal-induced IκBα ubiquitination and degradation by the proteasome. This information was critical for the subsequent molecular characterization of both the IκB kinase (IKK) complex (see Chapter 3 of this book) and the IκB ubiquitin ligase. Using a set of specific phosphopeptides, Yaron and colleagues showed that the phosphorylation-based motif (DpSGXXpS) of IκBα is sufficient for IκB recognition by components of the ubiquitin-system *in vitro* [16]. Mass spectroscopy identified βTrCP as a protein that could specifically interact with the IκB phosphopeptide motif [17]. Significantly, a *Drosophila* homolog of βTrCP, *slimb*, had previously been identified by Jiang and Struhl in a genetic screen as a likely E3 for β-catenin and Cubitus Interruptus (Ci) [18]. Several groups [19,20,21] then demonstrated that βTrCP is the substrate binding (receptor) subunit of an Skp1 cullin F-box (SCF)-type E3 ligase [22]. The structural basis for recognition of phospho-IκBα by the IκB E3 ligase was established by Pavletich and his colleagues, who solved the crystal structure of the substrate interacting pocket of βTrCP in complex with the cognate phosphopeptide from β-catenin [23]. The role of βTrCP in NF-κB activation is not limited to controlling IκBα degradation, and βTrCP was subsequently shown to control the proteasome-mediated degradation of IκBβ, IκBε [24,25], and NF-κB1 p105 [26,27,28], as well as the signal-induced processing of NF-κB2 p100 to p52 by the proteasome [29].

Recently, ubiquitination has been shown to have a novel role in NF-κB activation, which does not involve proteasome-mediated proteolysis of target proteins. Experiments by Chen and colleagues used *in vitro* reconstitution assays to investigate how IKK is activated [30] and discovered that the attachment of a unique type of polyubiquitin chain linked through K63 of ubiquitin to target proteins is an essential step in cytokine activation of IKK [31]. Research into NF-κB regulation therefore continues to reveal novel mechanisms by which ubiquitin regulates intracellular signaling, which are likely to have far-reaching implications in our understanding of the control of cell physiology.

4.2 SIGNAL-INDUCED IκB PROTEOLYSIS

The majority of stimuli that induce NF-κB activation target all three IκBs (IκBα, IκBβ, IκBε) for phosphorylation by the IKK complex (see Figure 4.2) [32]. IKK-mediated phosphorylation promotes IκB ubiquitination and subsequent degradation by the proteasome [33]. IκBα has been shown to be a better substrate for IKK when complexed with NF-κB dimers, providing a mechanism to ensure that NF-κB-free IκBα is protected from fortuitous, IKK-induced ubiquitination and degradation [33]. Mutation of the IKK phosphorylation sites (S^{32} or S^{36} of IκBα, or the equivalent sites of the other IκBs) renders IκBs resistant to signal-induced degradation and expression of such mutants of IκBα has been used to block NF-κB both in cell lines and *in vivo*.

Following its phosphorylation by IKK, IκBα is recognized by the F-box-WD repeat protein βTrCP (also called E3RS^{IκB} or Fbw1a) — the receptor subunit of the

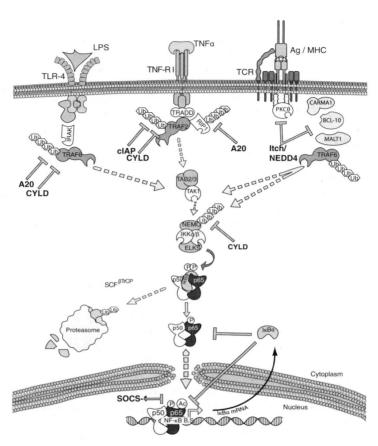

FIGURE 4.2 Receptor-induced NF-κB signaling pathways. This schematic diagram highlights the role of ubiquitination in activation of the canonical NF-κB signaling pathway by different receptor types. Ubiquitination is involved in two steps of the NF-κB activation process. transforming growth factor-beta-activated kinase 1 (TAK1) mitogen associated protein 3 (MAP 3)-kinase is activated by recruitment via TAB2/3 to K63-linked polyubiquitinated proteins, such as TRAF2/6 and NF-κB essential modifier (NEMO), and then subsequently activates the IKK complex. IKK phosphorylation of IκBα triggers its K48-linked polyubiquitination and proteasome-mediated degradation, releasing associated p50/p65 heterodimers to translocate into the nucleus and activate gene transcription. Cylindromatosis (CYLD) and A20 both negatively regulate NF-κB by removing K63-linked polyubiquitin from target proteins. A20 additionally functions as an E3 ligase, which catalyzes the addition of K48-linked polyubiquitin chains to RIP, triggering its proteasome-mediated degradation. Itch and NEDD4 E3 ligases polyubiquitinate BCL-10, inducing its proteolysis, possibly in lyzosomes. IKKa phosphorylates p65 triggering its polyubiquitination and proteasome-mediated degradation. Suppressor of cytokine signaling 1 (SOCS1) is thought to function as an E3 ligase for p65. IκBα is shown as a prototypical NF-κB regulated gene. IκBα forms a negative feedback loop to shut off NF-κB activation by removing p50/p65 dimers from DNA and transporting them back to the cytoplasm. Polyubiquitinated (Ub) and phosphorylated (P) proteins are indicated. Abbreviations are defined in the text.

RING E3 SCF$^{\beta TrCP}$ — through a short peptide stretch centered around the two inducibly phosphorylated serine (pS) residues [34,35,36,37]. This sequence, DpS-GXXpS, is conserved among all IκBs, from *Drosophila* to humans (although it is not restricted to IκBs) and represents one of the best-defined E3 recognition motifs [16]. Lysines 21 and 22, located 9–12 amino acids N-terminal to the βTrCP recognition site, are then selected by the concerted action of the SCF$^{\beta TrCP}$ complex and a specific E2, UbcH5, for the attachment of polyubiquitin trees. Ubiquitinated IκBα is selectively degraded by the proteasome, releasing associated NF-κB subunits [33]. IκBα is stabilized *in vivo* by mutating lysines 21 and 22, which blocks its ubiquitination [38]. This contrasts with other substrates of SCF-type E3s in which multiple lysines are targeted for ubiquitination [39]. It is possible that the spatial organization of the IκB/NF-κB complex hinders ubiquitin conjugation at other lysines.

The analysis of NF-κB activation in gut epithelial cells exposed to commensal gut bacteria has revealed further complexities in the regulation of IκBα degradation. Following stimulation, IκBα becomes phosphorylated and ubiquitinated but is not degraded, thereby blocking NF-κB activation and the production of proinflammatory cytokines [40]. Molecular cues that distinguish commensal bacteria from pathogenic bacteria with respect to NF-κB activation have not been identified. Rather, it appears that the unique properties of gut tissue are important. TLR9 stimulation at the apical surface (opposed to the gut) delivers a dominant inhibitory signal resulting in the stabilization of ubiquitinated IκBα. This signal overcomes other TLR signals, either apically or basolaterally delivered, which alone lead to ubiquitin-dependent IκB degradation (Lee, Ben-Neriah, and Raz, unpublished results). The molecular mechanism of this effect is not known, but it is independent of protein synthesis. It may entail posttranslational modification of proteasome-linked proteins, such as specific substrate-proteasome adaptors, or perhaps noncanonical ubiquitination of IκBα (i.e., other than Lys-48 conjugation) [41].

4.3 STRUCTURE-FUNCTION STUDIES OF βTRCP

IκB E3 is a four-subunit, SCF RING-type E3 [42]. SCFs comprise the largest E3 family currently known and have a central role in phosphorylation-mediated destruction of regulatory proteins involved in controlling the cell cycle, transcriptional pathways, and multiple aspects of development [22]. SCF complexes are composed of three common components, the scaffold protein Cul1, the ring protein Roc1/Rbx1/Hrt1, the adaptor protein Skp1, and a single-variable component, the substrate binding F-box protein, which for IκB E3 is βTrCP. More than 70 F-box proteins have been identified in humans and mice, while there are 326 putative F-box proteins in *C. elegans*.

F-box proteins harbor two protein–protein interaction modules: a ~40 amino acid F-box for Skp1 association and a ~280 aa, seven repeat WD40 domain, which constitutes the core substrate binding domain of the E3. Much of our understanding of the structural basis for phosphorylation-dependent substrate selection has emerged from studies of the F-box proteins βTrCP and Cdc4 [23,43]. The seven WD40 repeats of these proteins form a torus-like structure (see Figure 4.3), named a β-propeller, which is characteristic of this fold [44]. The cognate phosphopeptide binds to one

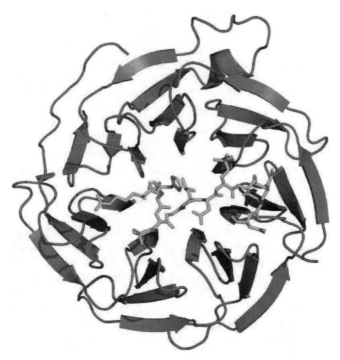

FIGURE 4.3 βTrCP bound to IκBα. Model of WD40 domain of βTrCP bound to phospho-rylated IκBa peptide (adapted from structure of βTrCP bound to phosphorylated β-catenin peptide by [23]).

face of the β-propeller of βTrCP, at a narrow extended channel, with the six residues of the destruction motif dipping into the channel. All seven WD40 repeats of βTrCP contribute contacts with the phosphopeptide, with maximal contacts maintained with the phosphate group of the N-terminal motif phosphoserine, aspartic acid, and the C-terminal phosphoserine (in that order). Of the six destruction motif residues, the aspartic acid and glycine are buried in the pocket, whereas the phosphoserines are relatively shallow. Four surface arginine residues lining the channel are involved in hydrogen bonding and electrostatic interaction with the phosphopeptide [23].

The IκBs (α, β, and ε), which are among the best characterized substrates of SCF^{βTrCP}, share their destruction motif (DpSGXXpS) with several other proteins, including β-catenin and the HIV protein Vpu1 [42]. Together these comprise the group of canonical (or typical) substrates that bind βTrCP with high affinity (kD of ~10 nM, D. Ceccarelli and F. Sicheri, personal communication), consistent with the structural characteristics of βTrCP binding pocket [23]. There are other atypical substrates of βTrCP, which contain variants of the destruction motif. For example, Cdc25a has a 4aa, instead of 2aa, spacer between the glycine and second phospho-serine [45], whereas p105 has a 3aa spacer [26] (see Section 4.4.1). The interaction of βTrCP with these substrates is harder to reconcile with the properties of the WD40-repeat binding pocket. hnRNP-U is particularly interesting since it interacts with βTrCP independently of phosphorylation and is not ubiquitinated or degraded, appearing to act as a pseudosubstrate for SCF^{βTrCP} [46]. hnRNP-U has no obvious βTrCP binding site and likely binds via a nonphosphorylated peptide rich in repetitive acidic residues [46,47]. This acidic repeat may substitute for the phosphodependent interaction of the canonical destruction motif or maintain electrostatic interactions with the surface arginine residues of the βTrCP binding pocket [23]. The low affinity interaction of βTrCP with hnRNP-U [46] may function as a safety mechanism, preventing weak association of βTrCP with irrelevant proteins. True substrates, such as phosphorylated IκB, which bind βTrCP with high affinity, may displace hnRNP-U from SCF^{βTrCP}, facilitating their specific ubiquitination and degradation.

βTrCP exists in two isoforms, 1 and 2, encoded by distinct genes [48]. The pocket structures of the two isoforms are very similar, suggesting that βTrCP 1 and 2 may have identical targets and functions. However, analyses of cells from two different *βTrCP1* knockout mouse strains raise the possibility that the two isoforms have some unique functions [49,50]. βTrCP1-deficient MEFs have markedly reduced growth rates compared to wild-type cells. This results from aberrant mitosis and increased apoptosis [50]. The mitotic defect appears to be caused by stabilization of Emi1, a mitosis inhibitor, which contains a canonical βTrCP destruction motif and may be specifically targeted by βTrCP1. Nakayama and colleagues noted a partial block in degradation of IκBα and β-catenin in splenocytes, thymocytes, and embryonic fibroblasts of the βTrCP1-deficient mice [49]. In contrast, Guardavaccaro and colleagues did not detect any reduction in IκB or β-catenin degradation in βTrCP1-deficient mouse embryo fibroblasts (MEFs) [50]. Taken together, these data may indicate that βTrCP 1 and 2 have redundant functions with respect to the regulation of IκBα and β-catenin ubiquitination. Consistent with this interpretation, there are no obvious physiological consequences of impaired NF-κB activation or aberrant Wnt activity in βTrCP1 knockout mice. There are also no clear defects suggesting aberrant mitosis, with the exception of reduced fertility reported in one of the knockout strains, that appears to result from defective spermatogenesis [50]. Assuming that βTrCP1 was completely eliminated in both studies, the paucity of developmental, physiologic, or clear pathologic sequella in the mutant mice suggests a redundant role for the two βTrCP isoforms, at least with respect to NF-κB and Wnt regulation. Furthermore, the elimination of one isoform may be compensated by upregulation of the other one (Davis and Ben-Neriah, unpublished data), as with many other genetic backup circuits [51]. It is also possible that Emi1 stability is coregulated by both βTrCP isoforms *in vivo*, perhaps by formation of a heterodimer [50]. Further conclusions about the possible functional divergence of βTrCP 1 and 2 await the generation of *βTrCP2$^{-/-}$* and *βTrCP1$^{-/-}$βTrCP2$^{-/-}$* mouse strains.

4.4 ACTIVATION OF NF-κB BY STRESS

Genotoxic stresses, such as ultraviolet (UV) light and DNA damage, activate NF-κB transcription factors to mediate both pro- and antiapoptotic functions [32,52]. Considerable research has been carried out into the mechanisms by which cellular stresses activate NF-κB. Chapter 3 briefly discusses some new data, suggesting a unique pathway from DNA damage to activation of the IKK complex, involving sumoylation and ubiquitinylation of NEMO. Here we will expand upon this discussion and include signaling pathways induced by other genotoxic stressors (Figure 4.4). However, it is difficult to draw general conclusions at this stage due to discrepancies between the stresses and cellular systems used by different laboratories.

As discussed in Chapter 3, DNA damage has recently been shown to induce a unique signaling pathway that relies on ubiquitin and ubiquitin-like modifications. Activation of NF-κB following exposure of cells to topoisomerase inhibitors, which cause DNA damage, is dependent on NEMO [53]. The zinc finger domain of NEMO,

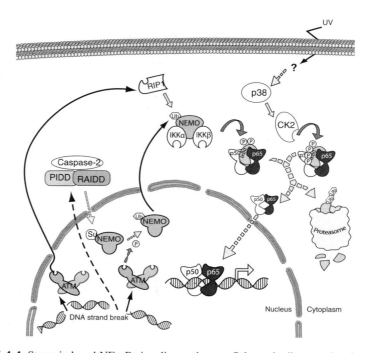

FIGURE 4.4 Stress-induced NF-κB signaling pathways. Schematic diagram showing proteins involved in NF-κB activation by UV irradiation and DNA damaging agents. UV has been reported to activate NF-κB via both IKK-dependent and -independent pathways. Only the IKK-independent CK2 pathway is shown, since the proteins involved in the IKK-dependent pathway have not been fully characterized. p38 acts as an allosteric activator of CK2, which phosphorylates (P) the C-terminal protein, glutamic acid, serine, and threonine (PEST) region of IκBa on multiple serine residues following UV irradiation. This triggers IκBa polyubiquitination by an unknown E3 ligase and degradation by the proteasome. DNA damage induces nuclear translocation and sumoylation of NEMO. Following desumoylation, NEMO is phosphorylated by the ATM kinase. This triggers NEMO polyubiquitination and subsequent translocation back to the cytoplasm, where it associates with and activates the IKK complex. RIP is essential for activation of NF-κB by DNA damaging agents, which induce it to form a complex with IKK. The complex of caspase-2/PIDD/RAIDD, termed the p53 induced protein with a death domain (PIDD)osome, is required for induction of NEMO sumoylation following genotoxic shock. Polyubiquitinated (Ub) and sumoylated (S) proteins are indicated. Abbreviations are defined in the text.

which is not necessary for tumor necrosis factor α (TNFα) activation of IKK [54], is required for DNA damage signals to activate the cytoplasmic IKK complex. The zinc fingers are needed to facilitate sumoylation of NEMO on lysine residues 277 and 309, which retains NEMO in the nucleus and is essential for activation of NF-κB by DNA damage [55]. Following desumoylation and then phosphorylation by ATM kinase, NEMO is monoubiquitinated on lysines 277 and 309, and translocates to the cytoplasm where it activates the IKK complex [56]. Blockade of Ubc13 function does not affect activation of NF-κB by genotoxic shock, suggesting that K63-linked ubiquitin is not involved in activation of IKK by this pathway (S. Miyamoto, personal communication). Consistent with this atypical mode of IKK activation, neither tumor necrosis factor receptor associated factor 2 (TRAF2) nor TRAF5 is required for NF-κB by DNA damaging agents [57]. Therefore, the mechanisms of both ubiquitination and sumoylation in this pathway remain to be defined.

Interestingly, the PIDDosome, a complex of caspase-2, PIDD, and RIP-associated TCH-1 homologous protein with a death domain (RAIDD) implicated in cellular responses to genotoxic stress [58], has recently been shown to activate NF-κB by inducing NEMO sumoylation (Figure 4.4; Tschopp, J., personal communication). Since the DNA damage response is often activated in early cancer [59,60], the common observation of NF-κB activation in cancer may be attributable to PIDDosome activation by the DNA damage checkpoint.

Activation of IKK in MEFs following treatment of cells with adriamycin or campthothecin requires receptor interacting protein (RIP) [57]. Coupling of these DNA damage signals to NF-κB activation involves the formation of a complex between IKK and RIP that is dependent on ATM kinase. The requirement for ATM in both nuclear ubiquitination of NEMO and formation of an IKK/RIP complex following DNA damage [55,57] suggests that these two events may be causally linked.

The activation of NF-κB by UV irradiation is not dependent on DNA damage [61], although the contribution of UV-induced DNA damage has complicated analyses of this cellular stress response. As one might expect, UV-induced NF-κB activation involves degradation of IκBα by the proteasome. However, it has been reported that this does not require phosphorylation of serines 32 and 36 of IκBα or activation of the IKK complex in HeLa cells, suggesting that UV controls IκBα degradation via an IKK-independent pathway [62,63]. In contrast, UV does not activate NF-κB in either IKKα/IKKβ-deficient MEFs or NEMO-deficient 70Z/3 B cells [54]. Furthermore, by expressing a dominant-negative mutant of βTrCP in HEK-293 cells, in order to trap the phosphorylated intermediate of IκBα, it has been shown that UV does induce the phosphorylation of IκBα on the same serines phosphorylated by IKK. Although these data are difficult to reconcile, it is possible that UV can induce IκBα proteolysis via both IKK-dependent and IKK-independent pathways, perhaps in a cell-type specific fashion.

The finding that zinc finger motifs of NEMO are required for UV activation of NF-κB [54] raises the interesting possibility that UV activates IKK by the same signaling pathway as DNA damaging agents — involving sumoylation and ubiquitination of NEMO [55]. However, the UV and DNA damage pathways are not identical, as RIP is only required for activation of NF-κB by DNA damage [57]. An IKK-independent pathway for UV activation of NF-κB has been characterized that involves casein kinase 2 (CK2) [64]. UV activates CK2 via p38 MAP kinase, which acts as an allosteric CK2 regulator [65]. Activated CK2 then directly phosphorylates IκBα on a cluster of serines in the C-terminal PEST region, triggering its degradation by the proteasome and activating NF-κB (Figure 4.4) [64]. IκBα can therefore be targeted for ubiquitin-dependent proteasomal degradation via two signaling pathways, which regulate distinct target phosphorylation sites on IκBα. The activation of NF-κB by the anticancer agent doxorubicin appears to be controlled by a different IKK-independent pathway. This induces IκBα degradation independently of phosphorylation of its N-terminus or PEST region and requires PI 3-kinase activity [66].

4.5 PROTEOLYSIS OF THE p105 AND p100 PRECURSOR PROTEINS

In contrast to the other members of the Rel family, p50 (NF-κB1) and p52 (NF-κB2) are synthesized as large inactive precursor proteins of 105 (p105) and 100 (p100) kDa, respectively (Chapter 2) [67]. Both precursors function as IκB proteins, which retain associated Rel subunits in the cytoplasm. Generation of the mature transcription factors, p50 and p52, from p105 and p100, respectively, involves ubiquitination and partial proteolysis by the proteasome (Figure 4.5). These proteolytic events, known as processing, remove their C-terminal halves, which include the IκB-like ankyrin repeat region.

4.5.1 NF-κB1 p105/p50

Processing of p105 to generate p50 occurs constitutively, and p50 generation is not usually modified following agonist stimulation [68]. However, cell stimulation with TNFα or lipopolysaccharide (LPS) increases the proteolysis of p105, leading predominantly to its complete degradation [10,69,70]. This releases associated Rel subunits to translocate from the cytoplasm into the nucleus and modulate expression of genes that regulate inflammatory responses [71].

4.5.1.1 Processing of NF-κB1 p105 to p50

The partial proteolysis of p105 by the proteasome to generate p50 is very unusual, as the proteasome normally completely degrades proteins [72]. Consequently, the mechanism by which p50 is produced from p105 has been intensely investigated. It has been suggested that p105 is processed cotranslationally by the proteasome and that synthesis of complete p105 molecules is not required for p50 generation [73]. However, several laboratories have demonstrated a clear precursor–product relationship between p105 and p50 in pulse-chase experiments [74,75,76,77] and it remains uncertain whether cotranslational p105 processing is physiologically important.

The central glycine-rich region (GRR), located between the RHD and ankyrin repeats, was the first motif in p105 shown to be required for processing to p50; a p105 deletion mutant lacking the GRR is unable to generate p50 [78], although p105 ubiquitination is not affected [79]. A related glycine-alanine repeat domain of EBNA1 (EBV nuclear antigen 1) blocks the proteolysis of EBNA1 by the proteasome [80]. Furthermore, insertion of the EBNA1 glycine-alanine repeat into different

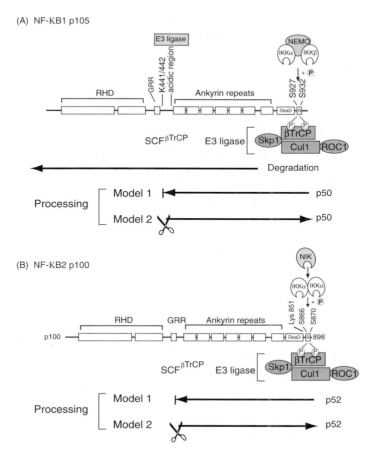

FIGURE 4.5 Proteolysis of NF-κB1 p105 and NF-κB2 p100. NF-κB1 p105 is constitutively proteolyzed by the proteasome to produce p50. (a) The GRR is essential for p50 production and has been proposed to function as a stop signal, which prevents entry of the Rel homology domain (RHD) into the proteasome (model 1). An alternative model proposes that the GRR facilitates internal cleavage by the proteasome, which then degrades the C-terminal fragment (model 2). In both models, the stability of the folded RHD is important in preventing its entry into the proteasome and proteolysis. The identity of the E3 ligase involved in p105 processing is not known but has been proposed to bind to an acidic region adjacent to the ankyrin repeats and ubiquitinate the neighboring lysine residues K441 and K442. Agonist stimulation induces the IKK complex to phosphorylate S927 and S932 in the p105 PEST region. This phosphorylated sequence then recruits the SCF$^{\beta TrCP}$ E3, which catalyzes the polyubiquitination of phospho-p105 on multiple lysine residues, triggering p105 proteolysis by the proteasome. Proteasome-mediated proteolysis of NF-κB2 p100 is triggered by receptor stimulation of the alternative NF-κB signaling pathway and generates p52. (b) Similar to p105, p100 processing requires the GRR and ubiquitination is mediated by the SCF$^{\beta TrCP}$ E3, which targets a single lysine residue (K851). S866 and S870 in the p100 PEST domain, which are required for NIK-induced recruitment of IKKα to p100, are thought to be directly phosphorylated by IKKα and form the binding site for SCF$^{\beta TrCP}$.

positions within IκBα prevents its signal-induced degradation, without affecting either IκBα phosphorylation or ubiquitination [81]. These data suggest that the glycine-alanine-rich region of EBNA1 may prevent the entry of the RHD into proteasome and it has been proposed that the p105 GRR may function in a similar fashion [79]. However, analysis of point mutants has indicated that production of p50 requires the correct folding of the RHD to protect it from proteasome-mediated unfolding and proteolysis [82]. Thus, although the GRR is necessary to prevent proteolysis of the RHD by the proteasome, it may not be sufficient. Blockade of entry into the proteasome is also thought to explain the inhibitory effect of p50 association with p105 on further processing to p50 [69,83]. This may be important for maintaining steady-state levels of p105 in unstimulated cells.

An alternative function for the GRR in p105 processing has recently been proposed by Rape and Jentsch [84], which is based on the observation that proteasomes can initiate proteolysis not only from N- or C-terminal ends of substrates but also from internal flexible polypeptide loops [85]. In p105, it is suggested that this function may be mediated by the GRR. Consistent with this hypothesis, it has been demonstrated that if the p105 GRR is placed between two stably folded domains, the proteasome cuts in the middle of the chimeric protein [78]. Rape and Jentsch suggest that the p105 GRR forms a hairpin loop, which inserts into the central cavity of the proteasome. Proteolysis is then initiated at the GRR and proceeds toward both ends of the polypeptide chain. The segment C-terminal to the GRR is completely degraded, whereas proteolysis comes to a halt when the proteasome reaches the tightly folded RHD, resulting in the generation of p50. Positive evidence to support this model is lacking, since it has yet to be shown that p105 actually undergoes endoproteolytic cleavage and C-terminal proteolytic fragments of p105 have not been detected in cells. Nevertheless this model is consistent with most reported experiments investigating the mechanism of p105 processing.

Analysis of chimeric proteins into which the p105 GRR has been transferred indicates that the GRR is not sufficient to promote proteolysis [79]. This appears to require an adjacent acidic region that has been proposed to function as a recognition site for an uncharacterized ubiquitin E3 ligase that ubiquitinates two nearby lysine residues (K441 and K442) and targets p105 to the proteasome [79]. Mutation of these p105 lysine residues to alanine blocks production of p50.

4.5.1.2   Signal-Induced Degradation of NF-κB1 p105

Following cellular stimulation with agonists, such as TNFα and LPS, p105 is rapidly phosphorylated by the classical IKK complex (IKKα/IKKβ/NEMO) on serines 927 and 932 in the p105 PEST region (Figure 4.6) [26,86]. Efficient phosphorylation of p105 by the IKK complex is ensured by the direct recruitment of IKKα and IKKβ to the p105 DD (death domain) [87,88]. The IKK-phosphorylated motif in p105 is directly recognized by βTrCP, the recognition component of the E3 SCF^{βTrCP}, which then catalyzes p105 ubiquitination, promoting subsequent p105 proteolysis by the proteasome [26,27,28,89]. Some studies have indicated that IKK-induced phosphorylation of p105 increases production of p50 [10,27,79]. However, the increase in p50 is usually small relative to the corresponding decrease in p105 and the major

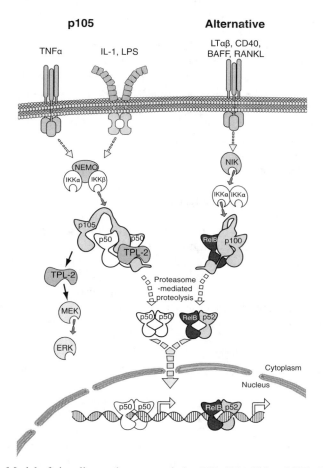

FIGURE 4.6 Model of signaling pathways regulating NF-κB1 p105 and NF-κB2 p100. The p105 pathway is involved in regulating immune and inflammatory responses. In this pathway, agonist stimulation induces the classical IKK complex to phosphorylate the p105 PEST region, triggering p105 polyubiquitination and subsequent proteolysis by the proteasome. This releases p50 homodimers to translocate into the nucleus and positively or negatively regulate gene expression. Signal-induced p105 proteolysis also releases TPL-2, resulting in its activation. p105-free tumor progression locus-2 (TPL-2) phosphorylates and activates MAP/ERK kinase (MEK), which in turn phosphorylates and activates ERK. The alternative NF-κB pathway that regulates the proteolysis of p100 is triggered by a subset of receptors that activate NF-κB. Receptor activation induces NF-κB inducing kinase (NIK) activity, in part by upregulating its expression, which phosphorylates and activates IKKα. IKKα directly phosphorylates p100, inducing its polyubiquitination and processing by the proteasome. This produces p52, which translocates into the nucleus in a complex with RelB to regulate the expression of genes important in controlling the organization of secondary lymphoid organs and humoral immune responses.

outcome of signal-induced p105 proteolysis is complete p105 degradation [28,90]. Consistently, knockdown of βTrCP blocks signal-induced p105 proteolysis but does not affect processing of p105 to p50 [26,27].

In vitro binding experiments with synthetic peptides indicate that βTrCP has a significantly lower affinity for phospho-p105 than phospho-IκBα [26]. This appears to be due to the introduction of an extra residue between the phosphorylated serine residues in the p105 βTrCP binding site (DpS^{927}GVGTpS^{932}) compared with the concensus βTrCP binding site (DpSGXXpS). This may reduce the efficiency of βTrCP-mediated ubiquitination of p105 compared to IκBα and explain why TNFα stimulation induces the proteolysis of only a fraction of cellular p105, whereas IκBα is normally completely degraded [86].

As noted above, p105 processing to p50 involves the ubiquitination by an unidentified E3 ligase of two lysine residues adjacent to the GRR region [79]. In contrast, SCF^{βTrCP} targets multiple lysine residues on IKK-phosphorylated p105 for ubiquitination, triggering its proteolysis [91]. Mutation of all 30 lysine residues in the C-terminal half of p105 is required to block βTrCP-mediated p105 ubiquitination. It is possible that the ubiquitination of multiple lysine residues by SCF^{βTrCP} overcomes the functions of the GRR and RHD in promoting processing of p105 to p50, thereby allowing complete p105 degradation.

Although p105 binds to p65, c-Rel, and p50 in unstimulated cells (Chapter 2) [74,92], analysis of p105-deficient mice indicates that p105 is particularly important for the cytoplasmic retention of p50 homodimers, which are bound weakly by IκBα [71]. Thus, signal-induced p105 proteolysis is thought to primarily trigger the nuclear translocation of p50 homodimers, which can function as transcriptional repressors but stimulate transcription when bound to the IκB-like nuclear proteins BCL-3 and IκBζ [93–96]. Recent data have demonstrated that p105 degradation in LPS-stimulated macrophages also releases the p105-associated kinase TPL-2, thereby activating its MEK kinase activity [89,97]. Thus signal-induced p105 proteolysis regulates the activation of the ERK MAP kinase cascade in addition to NF-κB. NF-κB agonists upregulate transcription of the NF-κB1 gene [98], which may be important for feedback regulation of NF-κB and ERK signaling pathways that are dependent on signal-induced p105 proteolysis.

TNFα-induced p105 proteolysis is dependent on expression of the serine/threonine kinase GSK-3β (glycogen synthase kinase 3β) [99]. GSK-3β forms a complex with p105 and phosphorylates p105 on serines 903 and 907 in the p105 PEST region *in vitro* [99,100]. Mutation of either of these residues to alanine impairs p105 proteolysis stimulated by TNFα. However, it is unclear how phosphorylation of serines 903 and 907 affects IKK-mediated phosphorylation of the p105 PEST region and subsequent recruitment of SCF^{βTrCP}.

4.5.2 Processing of NF-κB2 p100 to p52

In contrast to p105, processing of p100 to p52 is tightly regulated and in unstimulated cells the extent of p100 processing to p52 is limited [67]. However, following cell stimulation with a subset of NF-κB agonists, which includes lymphotoxin β (LT-β) [101,102], B cell activating factor (BAFF) [103,104] and CD40 ligand [105], p100 processing to p52 is induced (Figure 4.6). This results in the nuclear translocation of p52-RelB heterodimers, which bind to the promoters of genes containing a unique type of NF-κB binding site and trigger their transcription [106]. p52-RelB heterodimers are particularly important in regulating the expression of chemokine genes needed for organization of secondary lymphoid organs (Chapter 6).

The signaling pathway that regulates the inducible processing of p100 is known as the alternative NF-κB pathway, and requires IKKα, but not IKKβ or NEMO [101,102,107,108]. As discussed in Chapter 3 (Chapter 3, Figure 3.2b), stimulus-induced p100 processing requires the MAP 3-kinase NIK, which phosphorylates IKKα [109] and assists in its docking to p100, facilitating efficient p100 phosphorylation by IKKα [110,111]. IKKα binding to p100 depends on serines 866 and 870, which are known to be essential for signal-induced p100 processing [90,111]. While it is likely that IKKα directly phosphorylates these serines, this has yet to be shown. *In vitro* kinase assays have suggested that IKKα can phosphorylate p100 serine 872 and also four serines in the N-terminal half of p100 [111]; however, the physiological significance of these phosphorylation sites is unknown.

Nevertheless, phosphorylation of p100 by IKKα triggers p100 ubiquitination and subsequent partial proteolysis by the 26S proteasome to generate p52 [90,105,111,112]. Processing of p100 to p52 requires a central GRR [105,113,114], which probably has a similar function to the GRR in p105. RNA interference experiments have indicated that ubiquitination and processing of p100 induced by overexpressed NIK requires SCF^{βTrCP} [29,42], which also catalyses IκBα and p105 ubiquitination [42]. NIK expression induces βTrCP binding to p100, which requires p100 serines 866 and 870 that are part of a sequence related to the βTrCP-binding site on IκBα and are required for binding to IKKα (see above) [29]. Although it has yet to be shown using synthetic phosphopeptides, βTrCP is probably recruited to p100 via direct binding to phosphorylated serines 866 and 870. Lysine 855 serves as the major ubiquitin-anchoring residue on p100 for processing triggered by overexpressed NIK [115]. Mutation of lysine 855 stabilizes p100 and renders it resistant to βTrCP-mediated processing [115]. Finally, NIK may also promote the ubiquitination of p100 by triggering its direct recruitment to the proteasome by inducing interaction of the p100 DD with S9, a non-ATPase subunit in the 19S "lid" on the 26S proteasome [29].

4.5.2.1 p100 Processing and Cell Transformation

Although p100 processing only occurs at low levels in most normal cells, high levels of p52 are found in various transformed cells and deregulated processing of p100 may be important in the transformation mechanism [116]. Leukemic T cells infected with the human T cell leukaemia virus type I (HTLV-I) display elevated levels of p100 processing that are induced by the viral oncoprotein kinase Tax [112]. Similar to NIK, Tax specifically targets IKKα to p100, triggering phosphorylation-dependent ubiquitination and processing of p100. This recruitment, although independent of NIK, also requires serine 866 and 870 of p100 [117]. One of the transforming proteins of EBV, LMP1 (latent membrane protein 1), also induces the constitutive processing of p100 in both epithelial and B cells [118–121]. LMP1

promotes p100 processing via a NIK/IKKα pathway, similar to that used by endogenous cellular receptors.

Constitutive processing of p100 has been detected in various lymphomas that
contain *nfkb2* gene rearrangements [122]. Such genetic alterations always result in
the generation of C-terminally truncated p100 proteins in which a processing inhibitory domain that encompasses the p100 DD is removed [90]. The mechanisms for
signal-induced and constitutive processing of p100 are not identical and it has been
suggested that constitutive p100 processing may occur cotranslationally [29,113],
in contrast to signal-induced p100 proteolysis, which is clearly a posttranslational
process. Analysis of mutant cell lines has also indicated that while IKKα is required
for constitutive processing of p100 deletion mutants, NIK is not [123]. In addition,
the subcellular localization of signal-induced and constitutive p100 processing is
distinct. Constitutive p100 processing requires its nuclear/cytoplasmic shuttling,
while NIK-induced processing is only partially dependent on p100 nuclear localization [124]. Intriguingly, βTrCP is not necessary for constitutive processing of oncogenic p100 mutants [29]. Similarly, knockdown of βTrCP by RNA interference
indicates that the ubiquitination and processing of p100 induced by Tax is only
mediated in part by βTrCP [117]. These observations suggest the existence of a
second E3, perhaps located in the nucleus, which promotes the ubiquitination of
p100. The identity of this putative E3 and the specific lysine residues that are involved
in constitutive p100 processing are not known.

4.6 REGULATION OF IKK ACTIVATION BY UBIQUITIN CONJUGATION

The role of ubiquitination in triggering the proteolysis of IκBs by the proteasome
is well established. However, recent studies have identified a novel role for ubiquitination in the activation of NF-κB, which is independent of proteasome-mediated
proteolysis (Figure 4.2) [31]. A more detailed analysis of the mechanisms of IKK
activation is provided in Chapter 3.

4.6.1 Activation of IKK by IL-1R and TLR4

One of the initial events in IL-1R and TLR signaling involves homotypic interaction
between the Toll-IL-1R (TIR) domains of the MyD88 adaptor protein and the
activated receptor (see Chapter 3 for further discussion of TLR signaling) [125].
MyD88 then binds to interleukin-1 receptor-associated kinase 4 (IRAK4) and
IRAK1, which in turn recruit TRAF6, forming a signaling complex that is released
from the receptor into the cytoplasm to activate downstream kinases, including the
IKK complex and JNK [126]. Genetic studies have demonstrated that TRAF6 is
essential for activation of the IKK complex and jun N-terminal kinase (JNK) by
both the IL-1 receptor (IL-1R) and Toll-like receptors (TLRs) [127]. Using a cell-
free system and column purification to elucidate the mechanism by which TRAF6
activates IKK, Deng and colleagues identified two active protein complexes [30].
The first of these contained a ubiquitin-conjugating E2 enzyme complex consisting
of Ubc13 and a Ubc-like protein Uev1A (Mms2). Ubc13/Uev1A in conjunction with

TRAF6, which functions as an E3 by virtue of its RING finger domain, catalyze the formation of a unique type of polyubiquitin chain linked through K63 of ubiquitin. K63-linked ubiquitin chains adopt a more extended conformation than K48-linked ubiquitin chains [128] and do not target proteins for proteasome-mediated proteolysis. Blockade of K63-linked polyubiquitination impaired TRAF6 activation of IKK *in vitro* and expression of a catalytically inactive mutant of Ubc13 in cells reduced NF-κB activation by IL-1β, TNFα and over-expressed TRAF6, suggesting an important functional role for this modification in NF-κB activation *in vivo*. The targets of K63 polyubiquitination further support a functional role in this pathway, and include TRAF6 itself and NEMO [129,130]. Similarly, TRAF2, which also contains a RING finger domain, can catalyze K63-linked polyubiquitination of itself and RIP [131].

The second TRAF6-regulated IKK activating complex identified consisted of TAK1 and the adaptor proteins TAB1 and TAB2 [129,132]. As discussed in Chapter 3, multiple studies, variously using RNA interference, pharmacological inhibition, and gene targeting, have confirmed that TNFα and IL-1β activation of NF-κB and JNK require TAK1 [129,133–136]. TAB2 and TAB3 can bind to K63-linked polyubiquitin chains via a novel zinc finger (ZnF) domain at their C-termini, and mutation of this domain abolishes the ability of TAB2 and TAB3 to bind polyubiquitin chains and to activate TAK1 and IKKα *in vitro* or NF-κB in cell lines [130]. Conversely, replacing the TAB2/3 ZnF with a heterologous ubiquitin binding domain reconstitutes activation of TAK1, IKK, and NF-κB, strongly suggesting that binding to ubiquitin is integral to the signaling function of TAB2/3. Based on these data, it has been proposed that IL-1R or TLR stimulation induces the K63-linked polyubiquitination of TRAF6 and NEMO, which then recruit TAB2/3-TAK1 complexes leading to IKK activation.

4.6.2 ACTIVATION OF IKK BY THE T CELL ANTIGEN RECEPTOR

Costimulation of T cells via the T cell antigen receptor (TCR) and CD28 activates the IKK complex to promote NF-κB activation [137]. TCR/CD28 costimulation involves recruitment of two signaling cassettes to lipid rafts- the IKK complex [138,139] and the ternary CARMA1 (CARD11)/BCL-10/MALT1 complex [140]. The CARMA1/BCL-10/MALT1 complex may stimulate IKK via K63-linked polyubuquitination [141,142]. Activation of IKK by overexpressed BCL-10 in cell-free experiments involves induction of K63-linked polyubiquitination of NEMO and requires both mucosal-associated lymploid tissue (MALT1) and the Ubc13/Uev1A E2 [141,142]. Furthermore, mutant NEMO in which the target lysine for BCL-10-induced ubiquitination (K399) is changed to alanine does not support optimal NF-κB activation when expressed in a NEMO-deficient Jurkat T cell line [141]. These data suggest that activation of NF-κB in T cells is, at least partially, dependent on K63-linked ubiquitination of BCL-10.

The identity of the E3 that ubiquitinates NEMO in T cells is controversial. In one study, it is suggested that MALT1 can directly act as an E3 [141] although this protein does not have any motifs known to encode E3 domains. The second study shows that BCL-10 induces the oligomerization of MALT1, which then binds to and activates the E3 ligase activity of TRAF6 [142]. Although RNA interference experiments are consistent with a role for TRAF6, together with TRAF2, in TCR activation of IKK in Jurkat T cells, there are currently no genetic data to confirm a role for TRAFs in TCR signaling in primary T cells.

4.7 UBIQUITIN-MEDIATED MECHANISMS FOR TERMINATING NF-κB ACTIVITY

To limit the potentially damaging effects of excessive production of proinflammatory mediators, cells have evolved mechanisms to ensure that the duration of NF-κB activation is tightly controlled. The best characteristic of these is the NF-κB-dependent transcriptional induction of IκBα [143,144,145], which enters the nucleus free of associated NF-κB [146]. IκBα then dissociates NF-κB from DNA and exports it back to the cytoplasm, thereby turning off NF-κB-dependent transcription [147,148]. Recent research has demonstrated that several other important ubiquitin-dependent cellular control mechanisms to terminate NF-κB activation exist, which are described in Section 4.7 (see Figure 4.2).

4.7.1 DE-UBIQUITINATION OF CYTOPLASMIC REGULATORY PROTEINS

As discussed above, *in vitro* reconstitution experiments have identified a nondegradative role for ubiquitination in the activation of the IKK complex [30]. This regulatory ubiquitination involves the attachment of K63-linked ubiquitin to target proteins, including TRAFs (TRAF2 and TRAF6) [129] and NEMO [130], which then trigger IKK activation. The physiological significance of this model of IKK activation was initially unclear due to its reliance on *in vitro* assays and the inhibitory effects on NF-κB activation of overexpressed, catalytically inactive Ubc13. However, strong genetic support for the hypothesis that K63-linked ubiquitination is important in regulating NF-κB activation was provided by the demonstration that cylindromatosis (CYLD), a tumor suppressor protein linked to predisposition to cylindroma benign skin tumors [149], downregulates NF-κB activation [150,151,152].

A role for CYLD in NF-κB activation was suggested when CYLD was identified as NEMO-interacting protein and also in an RNA interference screen for deubiquitination enzymes that could block NF-κB activation [150,151,152]. CYLD contains two sequence motifs similar to the cysteine and histidine boxes found in the UBP ubiquitin-specific protease subfamily of deubiquitination enzymes [149]. Cell transfection experiments indicate that CYLD facilitates the disassembly of the K63-linked polyubiquitin chains on TRAF2, TRAF6, and NEMO but does affect K48-linked ubiquitination of IκBα or β-catenin [150,151,152]. As a consequence of this activity, overexpressed CYLD impairs NF-κB activation by multiple agonists, including TNFα, IL-1β, and LPS [150,151,152]. Furthermore, CYLD knockdown by RNA interference enhances NF-κB activation by IL-1β and LPS [153]. CYLD depletion has been reported to have no effect on TNFα activation of NF-κB. However, JNK is hyperactivated in CYLD-deficient cells following TNFα stimulation [153]. Consistently, TNFα-induced polyubiquitination of TRAF2 has been shown to couple to JNK, rather than NF-κB activation [154].

Truncations of CYLD found in patients with cylindromatosis (discussed in detail in Chapter 9), result in a phenotype that is restricted to epidermal body regions. Therefore, although overexpression experiments in cell lines suggest a general role for CYLD in controlling NF-κB, NF-κB activation may be regulated by CYLD in a cell- and stimulus-restricted fashion under physiological conditions. CYLD knock-out mice will be important to determine when and where CYLD controls NF-κB and/or JNK activity.

CYLD constitutively suppresses the ubiquitination of TRAF2 in unstimulated cells [155], providing an explanation for the basal activation of NF-κB and JNK associated with CYLD knockdown [153]. After cellular stimulation, CYLD is rapidly phosphorylated in a NEMO-dependent fashion. This prevents CYLD from inhibiting TRAF2 ubiquitination and activation of downstream signaling events. *In vitro* kinase assays suggest that CYLD may be a direct target of the IKK complex [155]. CYLD expression is also upregulated by NF-κB, which may serve to limit NF-κB activation in a negative feedback loop [156]. The regulation of NF-κB and CYLD is therefore closely interlinked.

Further evidence of a role for K63-linked ubiquitination in the regulation of NF-κB was provided by recent studies on the zinc-finger protein A20 [157], which was shown to turn off NF-κB activation by removing K63-linked ubiquitin from the adaptor protein RIP [131]. RIP, which is rapidly recruited to the activated TNFR1 complex and is a direct substrate for K63-linked polyubiquitination mediated by TRAF2 [131], is essential for activation of the IKK complex by TNFR1 (Chapter 3) [158,159]. Upon prolonged stimulation RIP disappears from the TNFR1 complex. This has been proposed to be regulated by A20 [131], which first removes K63-linked ubiquitin on RIP via its N-terminal OTU deubiquitination enzyme domain [160]. The C-terminal zinc finger domain of A20 then targets RIP for proteasomal degradation by catalyzing the K48-linked polyubiquitination of RIP, thereby terminating NF-κB activation. *In vitro* and transfection experiments have indicated that A20 can also remove K63 polyubiquitin chains from TRAF6 [161].

A20 expression is very low or undetectable in unstimulated cells but is rapidly transcriptionally induced by NF-κB [157]. A20 therefore functions as a deubiquit-inase in an autoregulatory circuit shutting down NF-κB activation. An important role for A20 in the correct regulation of NF-κB *in vivo* is consistent with the phenotype of A20[-/-] mice. IKK activity in A20-deficient cells is enhanced and pro-longed following stimulation with either TNFα or LPS and consequently A20[-/-] mice develop severe inflammation in multiple organs and are hypersensitive to LPS-induced toxic shock [161,162].

Deubiquitination of NF-κB regulatory proteins is also used by pathogens to evade the host immune response. *Yersinia pestis*, which is the causative agent for bubonic plague, injects a protein called YopJ into host cells during infection. YopJ encodes a cysteine protease originally shown to remove SUMO-1 from target proteins in mammalian cells. More recently, it has been shown that YopJ is a promiscuous deubiquitinating protein that removes ubiquitin moieties from critical proteins, including TRAF2, TRAF6, and IκBα [163]. In contrast to CYLD, YopJ removes both K63-linked polyubiquitin chains that activate IKK and K-48-linked chains that promote the degradation of IκBα. Consequently, expression of YopJ, but not a point mutant lacking a conserved cysteine within the protease domain, blocks the activation of IKK and MAP kinases [164].

4.7.2 Ubiquitin-Mediated Proteolysis of Upstream Regulatory Proteins

In several examples the duration of NF-κB activation has been shown to be controlled by proteolysis of specific upstream signaling proteins, in addition to RIP. This involves the addition of ubiquitin to target proteins by specific E3s.

One mechanism to shut off receptor activation of NF-κB is the proteolysis of activating receptors. The C-terminus of TNFRII interacts with the ankyrin repeat and SOCS box (ASB) family member, ASB3 [165]. ASB3 recruits the E3 adaptors Elongins-B/C to TNFRII, leading to ubiquitination of the cytoplasmic tail on multiple lysines, which triggers TNFRII proteolysis by the proteasome. Downregulation of ASB3 by RNA interference blocks TNFRII degradation induced by TNFα stimulation and potentiates TNFRII-mediated cytotoxicity. In an analogous fashion, Triad3A, a RING finger E3, interacts with the cytoplasmic tail of TLR9 and induces TLR9 ubiquitination and proteolytic degradation [166]. Knockdown of Triad3A by RNA interference blocks CpG-induced degradation of TLR9 and enhances activation of NF-κB by CpG.

Following stimulation of TNFRII, TRAF2 is ubiquitinated (presumably with K48-linked ubiquitin) and its levels decrease due to proteasome-mediated proteolysis [167]. Degradation of TRAF2 has been suggested to be triggered by cIAP-1 (cellular inhibitor of apoptosis 1) based on the inhibitory effects of cIAP-1 RING mutants on TRAF2 ubiquitination and proteolysis [167]. cIAP-1 can recruit E2 ubiquitin-conjugating enzymes via its RING domain and function as an E3 [168]. Recent experiments have indicated that ubiquitination of TRAF2 by c-IAP-1 involves their colocation with the ER-associated E2 Ubc6 in a perinuclear compartment [169]. cIAP-1 and cIAP-2 have also been reported to interact with RIP, triggering its ubiquitination [170]. Genetic support for these functions of IAPs is currently lacking, but it is possible that these proteins, whose expression is regulated by NF-κB [171], are part of a negative feedback loop to terminate receptor activation of NF-κB by triggering proteolysis of specific intracellular signaling proteins.

Downregulation of NF-κB activation by the TCR involves the TCR-induced ubiquitination and degradation of BCL-10 [172]. Overexpression experiments suggest that the HECT domain ubiquitin ligases NEDD4 and Itch promote BCL-10 ubiquitination. The possible role of Itch in BCL10 degradation is particularly interesting since Itch-deficient mice display an activated phenotype that is consistent with stabilization of BCL-10 [173]. In anergic T cells, NEDD4 and Itch have also been implicated in triggering the ubiquitination and proteolysis of PLCγ1 (phospholipase Cγ1) and PKCθ, which are both essential for activation of NF-κB by the TCR [174]. The type of ubiquitin linkage attached to BCL-10 has not been determined; however, degradation of PLCγ1, PKCθ, and BCL-10 is not mediated by the proteasome and, at least for BCL-10, involves transient localization to lysosomal vesicles [172,174]. It is, therefore, possible that BCL-10 is not ubiquitinated by K48-linked ubiquitin.

4.7.3 UBIQUITIN-MEDIATED PROTEOLYSIS OF REL PROTEINS

Recent experiments have uncovered a new role for ubiquitin-mediated proteolysis of NF-κB subunits in turning off NF-κB activity [175,176]. After TNFα stimulation, promoter-bound p65 becomes polyubiquitinated (presumably with K48-linked ubiquitin) and is then degraded in the nucleus by the proteasome. *In vitro* experiments have suggested that the cytokine signal inhibitor SOCS-1 acts as a p65 ubiquitin ligase [175], but SOCS-1 deficient mice have no apparent signs of NF-κB hyperactivation [177]. Proteasome-mediated proteolysis of p65 is a major mechanism of NF-κB response termination in IκBα$^{-/-}$ cells. However, even in wild-type cells, in which resynthesized IκBα removes NF-κB from target genes, degradation of Rel subunits contributes to the downregulation of NF-κB activity [176].

LPS stimulation also induces proteolysis of p65 by the proteasome and this has been shown to be triggered by phosphorylation of serine 536 in the C-terminal activation domain of p65 by IKKα (Chapter 5) [178]. The turnover of c-Rel is also controlled by its IKKα-mediated phosphorylation in LPS-stimulated cells. An important role for IKKα-induced Rel subunit degradation in TLR signaling is consistent with analyses of knock-in mice expressing a mutant IKKα, IKKα (AA), which cannot be activated. These mice display an exacerbated inflammatory response to infection with group B *Streptococcus* or following LPS injection. The IKK complex therefore appears to have evolved to carry out two opposing roles to ensure the rapid but transient nuclear localization of NF-κB transcription factors in response to TLR stimulation. IKKβ induces the degradation of IκBs, switching on NF-κB activation, whereas IKKα switches off NF-κB activation by triggering the degradation of p65 and c-Rel. Both of these steps involve ubiquitin-dependent proteolysis of the target proteins by the proteasome. It is presently unknown whether TNFα, and other NF-κB agonists, also regulate the proteolysis of p65 and c-Rel via IKKα-mediated phosphorylation.

4.8 THERAPEUTIC IMPLICATIONS FOR UBIQUITIN-CONTROL OF NF-κB SIGNALING

Aberrant NF-κB regulation has been implicated in autoimmune diseases and in certain types of cancer (Chapter 8). Consequently, many pharmaceutical companies have programs to develop drugs that can inhibit NF-κB or the signaling pathways leading to its activation [179,180]. Given their central role of regulating NF-κB activation, the proteins involved in ubiquitination of IκBs and their subsequent proteolysis are attractive targets for drug development (see Chapter 10).

The successful treatment of a hematological malignancy using a proteasome inhibitor has validated the ubiquitin-proteasome system as therapeutic target [181]. Bortezomib (formerly known as PS-341) binds directly with and inhibits the proteasome enzymatic complex and has recently been shown to have significant therapeutic activity in the treatment of advanced multiple myeloma (MM) [182]. In a large clinical trial, approximately one third of patients with relapsed and refractory MM showed significant clinical benefit with bortezomib, which was subsequently approved by the U.S. Food and Drug Administration (FDA).

Bortezomib's effects on cells are partly mediated through NF-κB inhibition [182,183], but also through additional mechanisms that include endoplasmic reticulum stress [184], c-Jun N-terminal kinase activation, altered growth factor expression [181], and blockade in cell cycle progression via the stabilizing effects on cell cycle inhibitory proteins, such as cyclin-dependent kinase inhibitors and p53 [185]. As a result, continuous proteasomal inhibition is toxic, and bortezomib must be administered periodically [185]. Therefore, a multifront approach, synergizing current treatment modalities with targeting the ubiquitin-proteasome pathway, is likely to be more successful. Ongoing multicenter studies are now investigating the efficacy of bortezomib in conjunction with standard chemotherapy and radiotherapy in a number of malignancies [181].

An alternative approach to inhibiting NF-κB is to block IκB ubiquitination. This would be expected to have fewer unwanted side effects than global proteasome inhibition and may be particularly attractive for cancer therapy [186]. The validity of this approach has been established using IκB "superrepressor" constructs in cells and recently in a murine model of hepatitis-associated cancer [187,188,189]. The potential use of various repressor constructs is discussed in Chapter 10. Another approach to preventing IκB ubiquitination is to impair βTrCP recruitment. Cell-penetrating IκB phosphopeptides have been used for this purpose in cell lines [16], but their efficacy has yet to be demonstrated *in vivo*. It might also be possible to develop small-molecule inhibitors that structurally mimic the ligase recognition motif or specifically inactivate $SCF^{\beta TrCP}$ (through allosteric inhibition or disassembly).

One obvious drawback with targeting βTrCP is the potential to affect $SCF^{\beta TrCP}$ targets other than IκBs [190]. A critical issue is likely to be whether $SCF^{\beta TrCP}$ is the only E3 that contributes to the degradation of each substrate. For example, depletion of both βTrCP isoforms by RNA interference stabilizes the entire IκB cytoplasmic pool, but only a small fraction of β-catenin [50] (M. Davis and Y. Ben-Neriah, unpublished data). Therefore, an alternative pathway may play a redundant role in controlling β-catenin proteolysis. Consistent with this hypothesis, a distinct E3 complex containing the Siah RING finger protein has been shown to recognize and promote the degradation of β-catenin, independently of $SCF^{\beta TrCP}$ [191,192]. The redundancy of βTrCP with respect to regulation of β-catenin proteolysis might be expected to be advantageous in βTrCP-targeted drug therapy of cancer, since β-catenin functions as a tumor promoter. However, the generation of mice lacking both βTrCP 1 and 2 will be crucial to determine whether βTrCP blockade is likely to have unwanted side effects via other target proteins.

Expression of a catalytically inactive mutant of Ubc13 in cell lines has been shown to block NF-κB by a number of different stimuli [30]. This raises the possibility that pharmacological blockade of K63-linked ubiquitination might be a valid approach to inhibiting NF-κB activation, perhaps by targeting the catalytic activity of the Ubc13/Uev1A E2 ligase with small-molecule inhibitors. It might also be possible to target TRAFs, which function as E3s in this pathway, by preventing their interaction with activating receptors using membrane-permeable peptides.

Currently, selective inhibition of IκB ubiquitination appears to be a technically feasible goal. Yet, systemic inhibition of NF-κB is not without risk due to its crucial role in regulating immune responses [193]. This is a particularly important consideration when NF-κB blockade is considered in combination with standard chemotherapy, which itself compromises the immune system and exacerbates tissue damage [188]. Furthermore, NF-κB inhibition may in fact be a "double-edged sword," as in the presence of a carcinogen, and perhaps even under chemotherapeutic poisoning, NF-κB inhibition may facilitate, rather than prevent, tumor development [194].

4.9 SYNOPSIS

An intense research effort during the past decade has generated a detailed scheme of the signaling pathways regulating the activation of NF-κB (Figures 4.1, 4.2, and 4.3). A striking feature of this scheme is the key regulatory role of ubiquitination at multiple steps and the coupling of this modification to signal-induced phosphorylation. The canonical pathway involves ubiquitin-mediated activation of the classical IKK complex, which then targets the three NF-κB inhibitory proteins (IκBα, β, and ε) and p105 for βTrCP-mediated ubiquitination and proteasomal destruction. Stimulation of the alternative pathway by a subset of NF-κB agonists targets p100 for βTrCP-mediated processing, generating p52-RelB heterodimers. Recent genetic data have highlighted the important role for removal of K63-linked ubiquitin in shutting off NF-κB activity and impairment of this control in cylindromatosis results in increased NF-κB activation, which may be important in tumor progression. K48-linked ubiquitination has also been shown to be important in downregulating NF-κB activity by promoting proteasome-mediated proteolysis of p65 and cRel.

Several key questions remain regarding the role of ubiquitination in NF-κB activation:

- How many NF-κB agonists utililize TRAF-mediated ubiquitination to activate IKK and also what is the structural basis for ubiquitination-dependent activation of IKK?
- Where are IκB ubiquitination and proteolysis taking place? Some experimental evidence suggests that ubiquitination and proteolysis of both IκB and p100 are controlled by a cytoplasmic/nuclear shuttling mechanism.
- Are there any specific proteasome adaptor molecules that ensure degradation of IκBs, while protecting associated NF-κB from proteolysis by the proteasome?
- What are the E3s involved in controlling the ubiquitination and proteolysis of Rel subunits in the nucleus?
- How is the expression and activity of the key ubiquitin system enzymes involved in NF-κB activation (i.e., βTrCP, Ubc13/Uev1a, CYLD, and A20) controlled and is the proteasome itself regulated?
- What is the mechanism by which phosphorylated and ubiquitinated IκB is protected from proteasomal degradation in gut epithelial cells?

As NF-κB research is entering its third decade, the regulation and functions of ubiquitination are still likely to remain a major focus, and the answers to many of these questions should not be long in coming.

ACKNOWLEDGMENTS

The authors would like to thank Sankar Ghosh, Shigeki Miyamoto, and Jurg Tschopp for communication of unpublished results and Shao-Cong Sun for advice on the p100 processing section. The help of the NIMR Photographics department in preparing the drawings and of Phil Walker in preparing Figure 4.3 are also gratefully acknowledged.

Relevant research in the authors' laboratories was supported by the U.K. Medical Research Council (to S.C.L.) and the Israel Science Foundation-Program for Centres of Excellence, Prostate Cancer Foundation-Israel, and the Human Frontiers Science Program (to Y B-N).

REFERENCES

[1]   Sen, R. and Baltimore, D., Inducibility of kappa immunoglobulin enhancer-binding protein NF-kappaB by a posttranslational mechanism, *Cell,* 47, 921, 1986.

[2]   Sen, R. and Baltimore, D., Multiple nuclear factors interact with the immunoglobulin enhancer sequences, *Cell,* 46, 705, 1986.

[3]   Baeuerle, P.A. and Baltimore, D., Activation of DNA-binding activity in an apparently cytoplasmic precursor of the NF-kappaB transcription factor, *Cell,* 53, 211, 1988.

[4]   Ghosh, S. and Baltimore, D., Activation *in vitro* of NF-κB by phosphorylation of its inhibitor IκB, *Nature,* 344, 678, 1990.

[5]   Alkalay, I., Yaron, A., Hatzubai, A. et al., *In vivo* stimulation of IκB phosphorylation is not sufficient to activate NF-κB, *Mol. Cell. Biol.,* 15, 1294, 1995.

[6]   DiDonato, J.A., Mercurio, F., and Karin, M., Phosphorylation of IkappaBalpha precedes but is not sufficient for its dissociation from NF-kappaB, *Mol. Cell. Biol.,* 15, 1302, 1995.

[7]   Brown, K., Park, S., Kanno, T. et al., Mutual regulation of the transcriptional activator NF-kappaB and its inhibitor, IkappaBalpha, *Proc. Natl. Acad. Sci. USA,* 90, 2532, 1993.

[8]   Sun, S.-C., Ganchi, P.A., Ballard, D.W. et al., NF-κB controls expression of inhibitor IκBα: Evidence for an inducible autoregulatory pathway, *Science,* 259, 1912, 1993.

[9]   Beg, A.A., Finco, T.S., Nantermet, P.V. et al., Tumor necrosis factor and interleukin-1 lead to phosphorylation and loss of I kappa B alpha: A mechanism for NF-kappa B activation, *Mol. Cell Biol.,* 13, 3301, 1993.

[10] Mellits, K.H., Hay, R.T., and Goodbourn, S., Proteolytic degradation of MAD3 (IκBα) and enhanced processing of the NF-κB precursor p105 are obligatory steps in the activation of of NF-κB, *Nuc. Acid. Res.,* 21, 5059, 1993.

[11] Henkel, T., Machleidt, T., Alkalay, I. et al., Rapid proteolysis of IkappaBalpha is necessary for activation of transcription factor NF-kappaB, *Nature,* 365, 182, 1993.

[12] Palombella, V.J., Rando, O.J., Goldberg, A.L. et al., The ubiquitin-proteasome pathway is required for processing the NF-κB1 precursor protein and activation of NF-κB, *Cell,* 78, 773, 1994.

[13] Chen, Z., Hagler, J., Palombella, V.J. et al., Signal-induced site-specific phosphorylation targets IκBα to the ubiquitin-proteasome pathway, *Genes Dev.,* 9, 1586, 1995.

[14] Alkalay, I., Yaron, A., Hatzubai, A. et al., Stimulation-dependent IκBα phosphorylation marks the NF-κB inhibitor for degradation via the ubiquitin-proteasome pathway, *Proc. Natl. Acad. Sci. USA,* 92, 10599, 1995.

[15] Brown, K., Gerstberger, S., Carlson, L. et al., Control of IκBα proteolysis by site-specific signal-induced phosphorylation, *Science, 267*, 1485, 1995.

[16] Yaron, A., Gonen, H., Alkalay, I. et al., Inhibition of NF-κB cellular function via specific targeting of the IkappaB-ubiquitin ligase, *Embo J., 16*, 6486, 1997.

[17] Yaron, A., Hatzubai, A., Davis, M. et al., Identification of the receptor component of the IkappaBalpha-ubiquitin ligase, *Nature, 396*, 590, 1998.

[18] Jiang, J. and Struhl, G., Regulation of the Hedgehog and Wingless signalling pathways by the F-box/WD40-repeat protein Slimb, *Nature, 391*, 493, 1998.

[19] Spencer, E., Jiang, J., and Chen, Z.J., Signal-induced ubiquitination of IκBα by the F-box protein Slimb/βTrCP, *Genes Dev., 13*, 284, 1999.

[20] Tan, P., Fuchs, S.Y., Chen, A. et al., Recruitment of a ROC1-CUL1 ubiquitin ligase by Skp1 and HOS to catalyze the ubiquitination of IkappaBalpha, *Mol. Cell., 3*, 527, 1999.

[21] Winston, J.T., Strack, P., Beer-Romero, P. et al., The SCFbeta-TrCP-ubiquitin ligase complex associates specifically with phosphorylated destruction motifs in IkappaB-alpha and beta-catenin and stimulates IkappaBalpha ubiquitination *in vitro*, *Genes Dev., 13*, 270, 1999.

[22] Deshaies, R.J., SCF and Cullin/Ring H2-based ubiquitin ligases, *Ann. Rev. Cell. Dev. Biol., 15*, 435, 1999.

[23] Wu, G., Xu, G., Schulman, B.A. et al., Structure of a βTrCP1-Skp1-β-catenin complex: Destruction motif binding and lysine specificity of the SCF^{βTrCP1} ubiquitin ligase, *Genes Dev., 11*, 1445, 2003.

[24] Shirane, M., Hatakeyama, S., Hattori, K. et al., Common pathway for the ubiquitination of IκBα, IκBβ, and IκBε mediated by the F-box protein FWD1, *J. Biol. Chem., 274*, 28169, 1999.

[25] Wu, C. and Ghosh, G., βTrCP mediates the signal-induced ubiquitination of IκBβ, *J. Biol. Chem., 274*, 29591, 1999.

[26] Lang, V., Janzen, J., Fischer, G.Z. et al., βTrCP-mediated proteolysis of NF-κB1 p105 requires phosphorylation of p105 serines 927 and 932, *Mol. Cell. Biol., 23*, 402, 2003.

[27] Orian, A., Gonen, H., Bercovich, B. et al., SCF^{βTrCP} ubiquitin ligase-mediated processing of NF-κB p105 requires phosphorylation of its C-terminus by IκB kinase, *Embo J., 19*, 2580, 2000.

[28] Heissmeyer, V., Krappmann, D., Hatada, E.N. et al., Shared pathways of IκB kinase-induced SCF(βTrCP)-mediated ubiquitination and degradation for the NF-κB precursor p105 and IκBa, *Mol. Cell. Biol., 21*, 1024, 2001.

[29] Fong, A. and Sun, S.-C., Genetic evidence for the essential role of β-transducin repeat-containing protein in the inducible processing of NF-κB2/p100, *J. Biol. Chem., 277*, 22111, 2002.

[30] Deng, L., Wang, C., Spencer, E. et al., Activation of the IκB kinase complex by TRAF6 requires a dimeric ubiquitin-conjugating enzyme complex and a unique polyubiquitin chain, *Cell, 103*, 351, 2000.

[31] Sun, L. and Chen, Z.J., The novel functions of ubiquitination in signaling, *Curr. Opin. Cell Biol., 16*, 1, 2004.

[32] Hayden, M.S. and Ghosh, S., Signaling to NF-κB, *Genes Dev., 18*, 2195, 2004.

[33] Karin, M. and Ben-Neriah, Y., Phosphorylation meets ubiquitination: The control of NF-κB activity, *Ann. Rev. Immunol., 18*, 621, 2000.

[34] Winston, J.T., Strack, P., Beer-Romero, P. et al., The SCF$^{β-TrCP}$-ubiquitin ligase complex associates specifically with phosphorylated destruction motifs in IκBα and β-catenin and stimulates IκBα ubiquitination *in vitro*, *Genes Dev., 13*, 270, 1999.

[35] Spencer, E., Jiang, J., and Chen, Z.J., Signal-induced ubiquitination of IκBα by the F-box protein Slimb/βTrCP, *Genes Dev.,* 13, 284, 1999.

[36] Hattori, K., Hatakeyama, S., Shirane, M. et al., Molecular dissection of the interactions among IkappaBalpha, FWD1, and Skp1 required for ubiquitin-mediated proteolysis of IkappaBalpha, *J. Biol. Chem.,* 274, 29641, 1999.

[37] Kroll, M., Margottin, F., Kohl, A. et al., Inducible degradation of IκBα by the proteasome requires interaction with the F-box protein h-βTrCP, *J. Biol. Chem.,* 274, 7941, 1999.

[38] Baldi, L., Brown, K., Franzoso, G. et al., Critical role for lysines 21 and 22 in signal-induced, ubiquitin-mediated proteolysis of IkappaBalpha, *J. Biol. Chem.,* 271, 376, 1996.

[39] Petroski, M.D. and Deshaies, R.J., Context of multiubiquitin chain attachment influences the rate of Sic1 degradation, *Mol. Cell,* 11, 1435, 2003.

[40] Neish, A.S., Gewirtz, A.T., Zeng, H. et al., Prokaryotic regulation of epithelial responses by inhibition of IκBα ubiquitination, *Science,* 289, 1560, 2000.

[41] Peng, J., Schwartz, D., Elias, J.E. et al., A proteomics approach to understanding protein ubiquitination, *Nat Biotechnol,* 21, 921, 2003.

[42] Ben-Neriah, Y., Regulatory functions of ubiquitination in the immune system, *Nature Immunol.,* 3, 20, 2002.

[43] Orlicky, S., Tang, X., Willems, A. et al., Structural basis for phosphodependent substrate selection and orientation by the SCFCdc4 ubiquitin ligase, *Cell,* 112, 243, 2003.

[44] Neer, E.J., Schmidt, C.J., Nambudripad, R. et al., The ancient regulatory-protein family of WD-repeat proteins, *Nature,* 371, 297, 1994.

[45] Busino, L., Donzelli, M., Chiesa, M. et al., Degradation of Cdc25A by βTrCP during S phase and in response to DNA damage, *Nature,* 426, 87, 2003.

[46] Davis, M., Hatzubai, A., Anderson, J.S. et al., Pseudosubstrate regulation of the SCF(βTrCP) ubiquitin ligase by hnRNP-U, *Genes Dev.,* 16, 439, 2002.

[47] Kanemori, Y., Uto, K., and Sagata, N., βTrCP recognizes a previously undescribed nonphosphorylated destruction motif in Cdc25A and Cdc25B phosphatases, *Proc. Natl. Acad. Sci. USA,* 102, 6279, 2005.

[48] Chiaur, D.S., Murthy, S., Cenciarelli, C. et al., Five human genes encoding F-box proteins: Chromosome mapping and analysis in human tumors, *Cytogenet Cell Genet,* 88, 255, 2000.

[49] Nakayama, K., Hatakeyama, S., Maruyama, S. et al., Impaired degradation of inhibitory subunit of NF-κB (IκB) and β-catenin as a result of targeted disruption of the βTrCP1 gene, *Proc. Natl. Acad. Sci. USA,* 100, 8752, 2003.

[50] Guardavaccaro, D., Kudo, Y., Boulaire, J. et al., Control of meiotic and mitotic progression by the F box protein βTrCP1 *in vivo, Dev. Cell,* 4, 799, 2003.

[51] Kafri, R., Bar-Even, A., and Pilpel, Y., Transcription control reprogramming in genetic backup circuits, *Nat. Genet.,* 37, 295, 2005.

[52] Campbell, K.J., Rocha, S., and Perkins, N.D., Active repression of anti-apoptotic gene expression by RelA (p65) NF-κB, *Mol. Cell,* 13, 853, 2004.

[53] Huang, T.T., Wuerzberger-Davis, S.M., Seufzer, B.J. et al., NF-κB activation by camptothecin: A linkage between nuclear DNA damage and cytoplasmic signaling events, *J. Biol. Chem.,* 275, 9501, 2000.

[54] Huang, T.T., Feinberg, S.L., Suryanarayanan, S. et al., The zinc finger domain of NEMO is selectively required for NF-κB activation by UV radiation and topoisomerase inhibitors, *Mol. Cell Biol.,* 22, 5813, 2002.

[55] Huang, D.B., Wuerzberger-Davis, S.M., Wu, Z.-H. et al., Sequential modification of NEMO/IKKγ by SUMO-1 and ubiquitin mediates NF-κB activation by genotoxic shock, *Cell,* 115, 565, 2003.

[56] Abraham, R.T., Cell cycle checkpoint signaling through the ATM and ATR kinases, *Genes Dev.,* 15, 2177, 2001.

[57] Hur, G.M., Lewis, J., Yang, Q. et al., The death domain kinase RIP has an essential role in DNA damage-induced NF-κB activation, *Genes Dev.,* 17, 873, 2003.

[58] Tinel, A. and Tschopp, J., The PIDDosome, a protein complex implicated in activation of caspase-2 in response to genotoxic stress, *Science,* 304, 843, 2004.

[59] Bartkova, J., Horejsi, Z., Koed, K. et al., DNA damage response as a candidate anticancer barrier in early human tumorigenesis, *Nature,* 434, 864, 2005.

[60] Gorgoulis, V.G., Vassiliou, L.V., Karakaidos, P. et al., Activation of the DNA damage checkpoint and genomic instability in human precancerous lesions, *Nature,* 434, 907, 2005.

[61] Devary, Y., Rosette, C., DiDonato, J.A. et al., NF-κB activation by ultraviolet light is not dependent on a nuclear signal, *Science,* 261, 1442, 1993.

[62] Bender, K., Gottlicher, M., Whiteside, S. et al., Sequential DNA damage-independent and -dependent activation of NF-κB by UV, *Embo J.,* 17, 5170, 1998.

[63] Li, N. and Karin, M., Ionizing radiation and short wavelength UV activate NF-κB through two distinct mechanisms, *Proc. Natl. Acad. Sci. USA,* 95, 13012, 1998.

[64] Kato, T., Delhase, M., Hoffman, A. et al., CK2 is a C-terminal IkappaB kinase responsible for NF-κB activation during the UV response, *Mol. Cell,* 12, 829, 2003.

[65] Sayed, M., Kim, S.O., Salh, B.S. et al., Stress-induced activation of protein kinase CK2 by direct interaction with p38 mitogen-activated protein kinase, *J. Biol. Chem.,* 275, 16569, 200.

[66] Tergaonkar, V., Bottero, V., Ikawa, M. et al., IκB kinase-independent IκBα degradation pathway: Functional NF-κB activity and implications in cancer therapy, *Mol. Cell. Biol.,* 23, 8070, 2003.

[67] Beinke, S. and Ley, S.C., Functions of NF-κB1 and NF-κB2 in immune cell biology, *Biochem. J.,* 382, 393, 2004.

[68] Karin, M. and Ben-Neriah, Y., Phosphorylation meets ubiquitination: The control of NF-κB activity, *Ann. Rev. Immunol.,* 18, 621, 2000.

[69] Harhaj, E.W., Maggirwar, S.B., and Sun, S.C., Inhibition of p105 processing by NF-κB proteins in transiently transfected cells, *Oncogene,* 12, 2385, 1996.

[70] Syrovets, T., Jendrach, M., Rohwedder, A. et al., Plasmin-induced expression of cytokines and tissue factor in human monocytes involves AP-1 and IKK-β-mediated NF-κB activation, *Blood,* 97, 3941, 2001.

[71] Ishikawa, H., Claudio, E., Dambach, D. et al., Chronic inflammation and susceptibility to bacterial infections in mice lacking the polypeptide (p) 105 precursor (NF-κB1) but expressing p50, *J. Exp. Med.,* 187, 985, 1998.

[72] Ciechanover, A., Orian, A., and Schwartz, A.L., Ubiquitin-mediated proteolysis: Biological regulation via destruction, *Bioessays,* 22, 442, 2000.

[73] Lin, L., DeMartino, G.N., and Greene, W.C., Cotranslational biogenesis of NF-κB p50 by the 26S proteasome, *Cell,* 92, 819, 1998.

[74] Mercurio, F., DiDonato, J.A., Rosette, C. et al., p105 and p98 precursor proteins play an active role in NF-κB-mediated signal transduction, *Genes Dev.,* 7, 705, 1993.

[75] Fan, C.-M. and Maniatis, T., Generation of p50 subunit of NF-κB by processing of p105 through an ATP-dependent pathway, *Nature,* 354, 395, 1991.

[76] Donald, R., Ballard, D.W., and Hawiger, J., Proteolytic processing of NF-κB/IκB in human monocytes, *J. Biol. Chem.,* 270, 9, 1995.

[77] Belich, M.P., Salmeron, A., Johnston, L.H. et al., TPL-2 kinase regulates the proteolysis of the NF-κB inhibitory protein NF-κB1 p105, *Nature,* 397, 363, 1999.

[78] Lin, L. and Ghosh, S., A glycine-rich region in NF-κB p105 functions as a processing signal for the generation of the p50 subunit, *Mol. Cell. Biol.*, 16, 2248, 1996.

[79] Orian, A., Schwartz, A.L., Israel, A. et al., Structural motifs involves in ubiquitin-mediated processing of the NF-κB precursor p105: Roles of the glycine-rich region and a downstream ubiquitination domain, *Mol. Cell. Biol.*, 19, 3664, 1999.

[80] Levitskaya, J., Shapiro, A., Leonchiks, A. et al., Inhibition of ubiquitination/protea-some-dependent protein degradation by the Gly-Ala repeat domain of the Epstein-Barr virus nuclear antigen 1, *Proc. Natl. Acad. Sci. USA*, 94, 12616, 1997.

[81] Sharipo, A., Imreh, M., Leonchiks, A. et al., A minimal glycine-alanine repeat prevents the interaction of ubiquitinated IκBα with the proteasome: A new mechanism for selective inhibition of proteolysis, *Nature Med.*, 4, 939, 1998.

[82] Lee, C., Schwartz, M.P., Prakash, S. et al., ATP-dependent proteases degrade their substrates by processively unraveling them for the degradation signal, *Mol. Cell*, 7, 627, 2001.

[83] Cohen, S., Orian, A., and Ciechanover, A., Processing of p105 is inhibited by docking of p50 active subunits to the ankyrin repeat domain, and inhibition is alleviated by signaling via the carboxyl-terminal phosphorylation/ubiquitin-ligase binding domain, *J. Biol. Chem.*, 276, 26769, 2001.

[84] Rape, M. and Jentsch, S., Productive RUPture: Activation of transcription factors by proteasomal processing, *Biochim. Biophys. Acta*, 1695, 209, 2004.

[85] Liu, C.W., Corboy, M.J., DeMartino, G.N. et al., Endoproteolytic activity of the proteasome, *Science*, 299, 408, 2003.

[86] Salmeron, A., Janzen, J., Soneji, Y. et al., Direct phosphorylation of NF-κB p105 by the IκB kinase complex on serine 927 is essential for signal-induced p105 proteolysis, *J. Biol. Chem.*, 276, 22215, 2001.

[87] Heissmeyer, V., Krappmann, D., Wulczyn, F.G. et al., NF-κB p105 is a target on IkB kinases and controls signal induction of BCL-3-p50 complexes, *EMBO J.*, 18, 4766, 1999.

[88] Beinke, S., Belich, M.P., and Ley, S.C., The death domain of NF-κB1 p105 is essential for signal-induced p105 proteolysis, *J. Biol. Chem.*, 277, 24162, 2002.

[89] Beinke, S., Robinson, M.J., Salmeron, A. et al., Lipopolysaccharide activation of the TPL-2/MEK/Extracellular signal-regulated kinase mitogen-activated protein kinase cascade is regulated by IκB kinase-induced proteolysis of NF-κB1 p105, *Mol. Cell. Biol.*, 24, 9658, 2004.

[90] Xiao, G., Harhaj, E.W., and Sun, S.-C., NF-kappaB-inducing kinase regulates the processing of NF-κB2 p100, *Mol. Cell*, 7, 401, 2001.

[91] Cohen, S., Achbert-Weiner, H., and Ciechanover, A., Dual effects of IκB kinase β-mediated phosphorylation on p105 fate: SCF(βTrCP)-dependent degradation and SCF(βTrCP)-independent processing, *Mol. Cell. Biol.*, 24, 475, 2004.

[92] Rice, N.R., MacKichan, M.L., and Israel, A., The precursor of NF-κB p50 has IκB-like functions, *Cell*, 71, 243, 1992.

[93] Yamamoto, M., Yamazaki, S., Uematsu, S. et al., Regulation of Toll/IL-1-receptor-mediated gene expression by the inducible nuclear protein IκBζ, *Nature*, 430, 218, 2004.

[94] Bours, V., Franzoso, G., Azarenko, V. et al., The oncoprotein Bcl-3 directly transactivates through κB motifs via association with DNA-binding p50B homodimers, *Cell*, 72, 729, 1993.

[95] Franzoso, G., Bours, V., Park, S. et al., The candidate oncoprotein Bcl-3 is an antagonist of p50/NF-κB mediated inhibition, *Nature*, 359, 339, 1992.

[96] Fujita, T., Nolan, G.P., Liou, H.-C. et al., The candidate proto-oncogene *bcl-3* encodes a transcriptional coactivator that activates through p50 homodimers, *Genes Dev.*, 7, 1354, 1993.

[97] Waterfield, M., Jin, W., Reiley, W. et al., IKKβ is an essential component of the Tpl-2 signaling pathway, *Mol. Cell Biol.*, 24, 6040, 2004.

[98] Ten, R.M., Paya, C.V., Israel, N. et al., The characterization of the promoter of the gene encoding the p50 subunit of NF-kappaB indicates that it participates in its own regulation, *Embo J.*, 11, 195, 1992.

[99] Demarchi, F., Bertoli, C., Sandy, P. et al., Glycogen synthase-3beta regulates NF-κB1/p105 stability, *J. Biol. Chem.*, 278, 39583, 2003.

[100] Demarchi, F., Verardo, R., Varnum, B. et al., Gas6 anti-apoptotic signaling requires NF-κB activation, *J. Biol. Chem.*, 276, 31738, 2001.

[101] Yilmaz, Z.B., Weih, D., Sivakumar, V. et al., RelB is required for Peyer's patch development: Differential regulation of p52-RelB by lymphotoxin and TNF, *Embo J.*, 22, 121, 2003.

[102] Dejardin, E., Droin, N.M., Delhase, M. et al., The lymphotoxin-β receptor induces different pathways of gene expression via two NF-κB pathways, *Immunity*, 17, 525, 2002.

[103] Claudio, E., Brown, K., Park, S. et al., BAFF-induced NEMO-independent processing of NF-κB2 in maturing B cells, *Nat. Immunol.*, 3, 958, 2002.

[104] Kayagaki, N., Yan, M., Seshasayee, D. et al., BAFF/BLys receptor 3 binds the B cell survival factor BAFF ligand through a discrete surface loop and promotes processing of NF-κB2, *Immunity*, 10, 515, 2002.

[105] Coope, H.J., Atkinson, P.G.P., Huhse, B. et al., CD40 regulates the processing of NF-κB2 p100 to p52, *Embo J.*, 21, 5375, 2002.

[106] Bonizzi, G., Bebien, M., Otero, D.C. et al., Activation of IKKα target genes depends on recognition of specific κB binding sites by RelB:p52 dimers, *Embo J.*, 23, 4202, 2004.

[107] Pomerantz, J.L. and Baltimore, D., Two pathways to NF-κB, *Mol. Cell*, 10, 693, 2002.

[108] Muller, J.R. and Siebenlist, U., Lymphotoxin-β receptor induces sequential activation of distinct NF-κB factors via separate signaling pathways, *J. Biol. Chem.*, 278, 12006, 2003.

[109] Ling, L., Cao, Z., and Goeddel, D.V., NF-κB-inducing kinase activates IKK-α by phosphorylation of Ser-176, *Proc. Natl. Acad. Sci. USA*, 95, 3792, 1998.

[110] Senftleben, U., Cao, Y., Xiao, G. et al., Activation by IKKα of a second evolutionary conserved, NF-κB signaling pathway, *Science*, 293, 1495, 2001.

[111] Xiao, G., Fong, A., and Sun, S.-C., Induction of p100 processing by NF-κB-inducing kinase involves docking of IκB kinase α (IKKα) to p100 and IKKα-mediated phosphorylation, *J. Biol. Chem.*, 279, 30099, 2004.

[112] Xiao, G., Cvijic, M.E., Fong, A. et al., Retroviral oncoprotein Tax induces processing of NF-κB2/p100 in T cells: Evidence for the involvement of IKKα, *Embo J.*, 20, 6805, 2001.

[113] Heusch, M., Lin, L., Geleiunas, R. et al., The generation of *nfkb2* p52: Mechanism and efficiency, *Oncogene*, 18, 6201, 1999.

[114] Betts, J.C. and Nabel, G.J., Differential regulation of NF-κB2 (p100) processing and control by amino-terminal sequences, *Mol. Cell. Biol.*, 16, 6363, 1996.

[115] Amir, R.E., Haecker, H., Karin, M. et al., Mechanism of processing of the NF-κB2 p100 precursor: Identification of the specific polyubiquitin chain-anchoring lysine residue and analysis of the role of NEDD8-modification on the SCF(βTrCP) ubiquitin ligase, *Oncogene*, 23, 2540, 2004.

[116]   Sun, S.-C. and Xiao, G., Deregulation of NF-κB and its upstream kinases in cancer, *Cancer Metastasis Rev.*, 22, 405, 2003.

[117]   Qu, Z., Qing, G., Rabson, A. et al., Tax deregulation of NF-κB2 p100 processing involves both βTrCP-dependent and -independent mechanisms, *J. Biol. Chem.*, 279, 44563, 2004.

[118]   Atkinson, P.G.P., Coope, H.J., Rowe, M. et al., Latent membrane protein 1 of Epstein-Barr virus stimulates processing of NF-κB p100 to p52, *J. Biol. Chem.*, 279, 51134, 2003.

[119]   Luftig, M.A., Yasui, T., Soni, V. et al., Epstein-Barr virus latent infection membrane protein 1 TRAF-binding site induces NIK/IKKα-dependent non-canonical NF-κB activation, *Proc. Natl. Acad. Sci. USA*, 101, 141, 2004.

[120]   Saito, N., Courtois, G., Chiba, A. et al., Two carboxy-terminal activation regions of Epstein-Barr virus latent membrane protein 1 activate NF-κB through distinct signaling pathways in fibroblast cell lines, *J. Biol. Chem.*, 278, 46565, 2003.

[121]   Eliopoulos, A.G., Caamano, J., Flavell, J. et al., Epstein-Barr virus-encoded latent infection membrane protein 1 regulates the processing of p100 NF-κB2 to p52 via an IKKγ/NEMO-independent signalling pathway, *Oncogene*, 22, 7557, 2003.

[122]   Rayet, B. and Gelinas, C., Aberrant *rel/nfkb* genes and activity in human cancer, *Oncogene*, 18, 6938, 1999.

[123]   Qing, G. and Xiao, G., Essential role of IkappaB kinase in the constitutive processing of NF-κB2 p100, *J. Biol. Chem.*, 280, 9765, 2005.

[124]   Liao, G. and Sun, S.-C., Regulation of NF-κB2/p100 processing by its nuclear shuttling, *Oncogene*, 22, 4868, 2003.

[125]   Takeda, K. and Akira, S., TLR signaling pathways, *Sem. Immunol.*, 16, 3, 2004.

[126]   Janssens, S. and Beyaert, R., Functional diversity and regulation of different interleukin-1 receptor-associated kinase (IRAK) family members, *Mol. Cell.*, 11, 293, 2003.

[127]   Lomaga, M.A., Yeh, W.C., Sarosi, I. et al., TRAF6 deficiency results in osteopetrosis and defective interleukin-1, CD40 and LPS signaling, *Genes Dev.*, 13, 1015, 1999.

[128]   Varadan, R., Assfalg, M., Haririnia, A. et al., Solution conformation of Lys63-linked di-ubiquitin chain provides clues to functional diversity of polyubiquitin signaling, *J. Biol. Chem.*, 279, 7055, 2004.

[129]   Wang, C., Deng, L., Hong, M. et al., TAK1 is a ubiquitin-dependent kinase of MKK and IKK, *Nature*, 412, 346, 2001.

[130]   Kanayama, A., Seth, R.B., Sun, L. et al., TAB2 and TAB3 activate the NF-kappaB pathway through binding to polyubiquitin chains, *Mol. Cell*, 15, 535, 2004.

[131]   Wertz, I.E., O'Rourke, K.M., Zhou, H. et al., De-ubiquitination and ubiquitin ligase domains of A20 downregulate NF-κB signalling, *Nature*, 430, 694, 2004.

[132]   Takaesu, G., Kishida, S., Hiyama, A. et al., TAB2, a novel adaptor protein, mediates activation of TAK1 MAP KKK by linking TAK1 to TRAF6 in the IL-1 signal transduction pathway, *Mol. Cell*, 5, 649, 2000.

[133]   Takaesu, G., Surabhi, R.M., Park, K.-J. et al., TAK1 is critical for IκB kinase-mediated activation of the NF-κB pathway, *J. Mol. Biol.*, 326, 105, 2003.

[134]   Ninomiya-Tsuji, J., Kajino, T., Ono, K. et al., A resorcylic acid lactone, 5Z-7-oxozeaenol, prevents inflammation by inhibiting the catalytic activity of TAK1 MAPK kinase kinase, *J. Biol. Chem.*, 278, 18485, 2003.

[135]   Sato, S., Sanjo, H., Takeda, K. et al., Essential function for the kinase TAK1 in innate and adaptive immune responses, *Nat. Immunol.*, 6, 1087, 2005.

[136]   Shim, J.H., Xiao, C., Paschal, A.E. et al., TAK1, but not TAB1 or TAB2, plays an essential role in multiple signaling pathways *in vivo*, *Genes Dev.*, 19, 2668, 2005.

[137]   Weil, R. and Israel, A., T cell receptor and B cell receptor mediated activation of NF-κB in lymphocytes, *Curr. Opin. Immunol.,* 16, 1, 2004.

[138]   Wang, D., Matsumoto, R., You, Y. et al., CD3/CD28 costimulation-induced NF-kappaB activation is mediated by recruitment of protein kinase C-theta, Bcl10, and IkappaB kinase beta to the immunological synapse through CARMA1, *Mol. Cell Biol.,* 24, 164, 2004.

[139]   Gaide, O., Favier, B., Legler, D.F. et al., CARMA1 is a critical lipid raft-associated regulator of TCR-induced NF-kappaB activation, *Nat. Immunol.,* 3, 836, 2002.

[140]   Thome, M., CARMA1, BCL-10 and MALT1 in lymphocyte development and activation, *Nat. Rev. Immunol.,* 4, 348, 2004.

[141]   Zhou, H., Wertz, I., O'Rourke, K et al., Bcl10 activates the NF-κB pathway through ubiquitination of NEMO, *Nature,* 427, 167, 2003.

[142]   Sun, L., Deng, L., Ea, C.-K. et al., The TRAF6 ubiquitin ligase and TAK1 kinase mediate IKK activation by BCL10 and MALT1 in T lymphocytes, *Mol. Cell,* 14, 289, 2004.

[143]   Chiao, P.J., Miyamoto, S., and Verma, I.M., Autoregulation of I kappa B alpha activity, *Proc. Natl. Acad. Sci. USA,* 91, 28, 1994.

[144]   Ito, C.Y., Kazantsev, A.G., and Baldwin, A.S., Three NF-kappaB sites in the I kappa B-alpha promoter are required for induction of gene expression by TNF alpha, *Nuc. Acid Res.,* 22, 3787, 1994.

[145]   Scott, M.L., Fujita, T., Liou, H.C. et al., The p65 subunit of NF-kappaB regulates I kappa B by two distinct mechanisms, *Genes Dev.,* 7, 1266, 1993.

[146]   Arenzana-Seisdedos, F., Thompson, J., Rodriguez, M. et al., Inducible nuclear expression of newly synthesized I kappa B alpha negatively regulates DNA-binding and transcriptional activities of NF-kappaB, *Mol. Cell. Biol.,* 15, 2689, 1995.

[147]   Arenzana-Seisdedos, F., Turpin, P., Rodriguez, M. et al., Nuclear localization of IkappaBalpha promotes active transport of NF-kappaB from the nucleus to the cytoplasm, *J. Cell Sci.,* 110, 369, 1997.

[148]   Tam, W.F., Lee, L.H., Davis, L. et al., Cytoplasmic sequestration of rel proteins by IκBα requires CRM1-dependent nuclear export, *Mol. Cell. Biol.,* 20, 2269, 2000.

[149]   Bignell, G.R., Warren, W., Seal, S. et al., Identification of the familial cylindromatosis tumour-suppressor gene, *Nat. Genet.,* 25, 160, 2000.

[150]   Kovalenko, A., Chable-Bessia, C., Cantarella, G. et al., The tumour suppressor CYLD negatively regulates NF-κB signalling by deubiquitination, *Nature,* 424, 801, 2003.

[151]   Trompouki, E., Hatzivassiliou, E., Tsichritzis, T. et al., CYLD is a deubiquitinating enzyme that negatively regulates NF-κB activation by TNFR family members, *Nature,* 424, 793, 2003.

[152]   Brummelkamp, T.R., Nijman, S.M.B., Dirac, A.M.G. et al., Loss of the cylindromatosis tumour suppressor inhibits apoptosis by activating NF-κB, *Nature,* 424, 797, 2003.

[153]   Reiley, W., Zhang, M., and Sun, S.-C., Negative regulation of JNK signaling by the tumor suppressor CYLD, *J. Biol. Chem.,* 279, 55161, 2004.

[154]   Habelhah, H., Takahashi, S., Cho, S.G. et al., Ubiquitination and translocation of TRAF2 is required for activation of JNK but not p38 or NF-κB, *Embo J.,* 23, 322, 2004.

[155]   Reiley, W., Zhang, M., Xuefeng, W. et al., Regulation of the deubiquitinating enzyme CYLD by IκB kinase gamma-dependent phosphorylation, *Mol. Cell. Biol.,* 25, 3886, 2005.

[156]   Jono, H., Lim, J.H., Chen, L.-F. et al., NF-κB is essential for induction of CYLD, the negative regulator of NF-κB, *J. Biol. Chem.,* 279, 36171, 2004.

[157]   Beyaert, R., Heyninck, K., and van Huffel, S., A20 and A20-binding proteins as cellular inhibitors of nuclear factor-κB-dependent gene expression and apoptosis, *Biochem. Pharm.,* 60, 1143, 2000.

[158] Zhang, S.Q., Kovalenko, A., Cantarella, G. et al., Recruitment of the IKK signalosome to the p55 TNF receptor: RIP and A20 bind to NEMO (IKKγ) upon receptor stimulation, *Immunity,* 12, 301, 2000.

[159] Devin, A., Cook, A., Lin, Y. et al., The distinct roles of TRAF2 and RIP in IKK activation by TNF-R1: TRAF2 recruits IKK to TNF-R1 while RIP mediates IKK activation, *Immunity,* 12, 419, 2000.

[160] Evans, P.C., Ovaa, H., Hamon, M. et al., Zinc-finger protein A20, a regulator of inflammation and cell survival, has de-ubiquitinating activity, *Biochem. J.,* 378, 727, 2004.

[161] Boone, D.L., Turer, E.E., Lee, E.G. et al., The ubiquitin-modifying enzyme A20 is required for termination of Toll-like receptor responses, *Nat. Immunol.,* 5, 1052, 2004.

[162] Lee, E.G., Boone, D.L., Chai, S. et al., Failure to regulate TNF-induced NF-κB and cell death responses in A20-deficient mice, *Science,* 289, 2350, 2000.

[163] Zhou, H., Monack, D.M., Kayagaki, N. et al., Yersinia virulence factor YopJ acts as a deubiquitinase to inhibit NF-kappaB activation, *J. Exp. Med.,* 202, 1327, 2005.

[164] Orth, K., Xu, Z., Mudgett, M.B. et al., Disruption of signaling by *Yersinia* effector YopJ, a ubiquitin-like protein protease, *Science,* 290, 1594, 2000.

[165] Chung, A.S., Guan, Y.-J., Yuan, Z.-L. et al., Ankyrin repeat and SOCS box 3 (ASB3) mediates ubiquitination and degradation of tumor necrosis factor receptor II, *Mol. Cell. Biol.,* 25, 4716, 2005.

[166] Chuang, T.-H. and Ulevitch, R.J., Triad3A, an E3 ubiquitin-protein ligase regulating Toll-like receptors, *Nat. Immunol.,* 5, 495, 2004.

[167] Li, X., Yang, Y., and Ashwell, J.D., TNF-RII and c-IAP1 mediate ubiquitination and degradation of TRAF2, *Nature,* 416, 345, 2002.

[168] Vaux, D.L. and Silke, J., IAPs, RINGs and ubiquitylation, *Nat. Rev. Mol. Cell Biol.,* 6, 287, 2005.

[169] Wu, C.-J., Conze, D.B., Li, X. et al., TNFα induced c-IAP1/TRAF2 complex translocation to a Ubc6-containing compartment and TRAF2 ubiquitination, *EMBO J.,* 24, 1886, 2005.

[170] Park, S.M., Yoon, J., and Lee, T.H., Receptor interacting protein is ubiquitinated by cellular inhibitor of apoptosis proteins (c-IAP1and c-IAP2) *in vitro, FEBS Lett.,* 566, 151, 2004.

[171] Wang, C.Y., Mayo, M.W., Kornuluk, R.G. et al., NF-κB antiapoptosis: Induction of TRAF1 and TRAF2 and c-IAP1 and c-IAP2 to suppress caspase-8 activation, *Science,* 281, 1680, 1998.

[172] Scharschmidt, E., Wegener, E., Heissmeyer, V. et al., Degradation of BCL10 induced by T cell activation negatively regulates NF-κB signaling, *Mol. Cell. Biol.,* 24, 3860, 2004.

[173] Fang, D., Elly, C., Gao, B. *et al.,* Dysregulation of T lymphocyte function in itchy mice: A role for Itch in TH2 differentiation, *Nat. Immunol.,* 3, 281, 2002.

[174] Heissmeyer, V., Macian, F., Im, S.-H. et al., Calcineurin imposes T cell unresponsiveness through targeted proteolysis of signaling proteins, *Nat. Immunol.,* 5, 255, 2004.

[175] Ryo, A., Suizu, F., Yoshida, Y. et al., Regulation of NF-kappaB signaling by Pin1-dependent prolyl isomerization and ubiquitin-mediated proteolysis of p65/RelA, *Mol. Cell,* 12, 1413, 2003.

[176] Saccani, S., Marazzi, I., Beg, A.A. et al., Degradation of promoter-bound p65/RelA is essential for the prompt termination of the nuclear factor κB response, *J. Exp. Med.,* 200, 107, 2004.

[177]   Marine, J.C., Topham, D.J., McKay, C. et al., SOCS1 deficiency causes a lymphocyte-dependent perinatal lethality, *Cell,* 98, 609, 1999.

[178]   Lawrence, T., Bebien, M., Liu, G.Y. et al., IKKα limits macrophage NF-κB activation and contributes to the resolution of inflammation, *Nature,* 434, 1138, 2005.

[179]   Nakanishi, C. and Toi, M., Nuclear factor-κB inhibitors as sensitizers to anticancer drugs, *Nat. Rev. Cancer,* 5, 297, 2005.

[180]   Verma, I.M., Nuclear factor (NF)-κB proteins: Therapeutic targets, *Annu. Rheum. Dis.,* 63 Suppl 2, ii57, 2004.

[181]   Rajkumar, S.V., Richardson, P.G., Hideshima, T. et al., Proteasome inhibition as a novel therapeutic target in human cancer, *J. Clin. Oncol.,* 23, 630, 2005.

[182]   Hideshima, T., Chauhan, D., Richardson, P. et al., NF-κB as a therapeutic target in multiple myeloma, *J. Biol. Chem.,* 277, 16639, 2002.

[183]   Russo, S.M., Tepper, J.E., Baldwin, A.S., Jr. et al., Enhancement of radiosensitivity by proteasome inhibition: Implications for a role of NF-kappaB, *Int. J. Radiat. Oncol. Biol. Phys.,* 50, 183, 2001.

[184]   Lee, A.H., Iwakoshi, N.N., Anderson, K.C. et al., Proteasome inhibitors disrupt the unfolded protein response in myeloma cells, *Proc. Natl. Acad. Sci. USA,* 100, 9946, 2003.

[185]   Adams, J., Proteasome inhibition: A novel approach to cancer therapy, *Trends Mol. Med.,* 8, S49, 2002.

[186]   Fuchs, S.Y., Spiegelman, V.S., and Kumar, K.G., The many faces of βTrCP E3 ubiquitin ligases: Reflections in the magic mirror of cancer, *Oncogene,* 23, 2028, 2004.

[187]   Cusack, J.C., Jr., Liu, R., and Baldwin, A.S., Jr., Inducible chemoresistance to 7-ethyl-10-[4-(1-piperidino)-1-piperidino]-carbonyloxycamptothe cin (CPT-11) in colorectal cancer cells and a xenograft model is overcome by inhibition of nuclear factor-κB activation, *Cancer Res.,* 60, 2323, 2000.

[188]   Lavon, I., Goldberg, I., Amit, S. et al., High susceptibility to bacterial infection, but no liver dysfunction, in mice compromised for hepatocyte NF-κB activation, *Nat. Med.,* 6, 573, 2000.

[189]   Pikarsky, E., Porat, R.M., Stein, I. et al., NF-κB functions as a tumour promoter in inflammation-associated cancer, *Nature,* 431, 461, 2004.

[190]   Spiegelman, V.S., Tang, W., Katoh, M. et al., Inhibition of HOS expression and activities by Wnt pathway, *Oncogene,* 21, 856, 2002.

[191]   Liu, J., Stevens, J., Rote, C.A. et al., Siah-1 mediates a novel beta-catenin degradation pathway linking p53 to the adenomatous polyposis coli protein, *Mol. Cell,* 7, 927, 2001.

[192]   Matsuzawa, S.I. and Reed, J.C., Siah-1, SIP, and Ebi collaborate in a novel pathway for β-catenin degradation linked to p53 responses, *Mol. Cell,* 7, 915, 2001.

[193]   Caamano, J. and Hunter, C.A., NF-κB family of transcription factors: Central regulators of innate and adaptive immune functions, *Clin. Microbiol. Rev.,* 15, 414, 2002.

[194]   Kamata, H., Honda, S.-I., Maeda, S. et al., Reactive oxygen species promote TNFα-induced death and sustained JNK activation by inhibiting MAP kinase phosphatases, *Cell,* 120, 649, 2005.

5 Regulation of Nuclear NF-κB Action: A Key Role for Posttranslational Modification

Lin-Feng Chen and Warner C. Greene

CONTENTS

5.1   INTRODUCTION

The NF-κB/Rel family of transcription factors plays a central role in governing diverse biological processes, ranging from the inflammatory and immune responses to cellular proliferation, differentiation, and survival. Since its discovery in 1986, NF-κB has been one of the most intensely studied transcription factors in eukaryotic biology. Each of the seven mammalian NF-κB/Rel-related proteins contains an N-terminal Rel homology domain (RHD), as discussed in Chapter 2, which mediates DNA binding, dimerization, and interaction with the IκB family proteins. P65, c-Rel, and RelB contain C-terminal transactivation domains (TADs), while p50 and p52 proteins lack these regions. As such, the binding of p50 and p52 homodimers can repress κB-specific transcription involving the recruitment of different corepressors [1]. The prototypical NF-κB complex is a heterodimer of p50 and p65. However, NF-κB/Rel family members can form different homo- or heterodimers that likely confer a degree of target gene specificity [2].

NF-κB is activated by a dizzying array of stimuli, including proinflammatory cytokines, growth factors, DNA damaging agents, and bacterial or viral proteins [3]. Likely reflecting the transcriptional strength of p65 and the need to tightly control its action, NF-κB is subject to regulation at multiple levels within cells. NF-κB activation occurs in two phases that take place in different cellular compartments. The first phase corresponds to the proximal cytoplasmic events that lead to the activation of the IKKs, as discussed in Chapter 3, and in turn the phosphorylation, polyubiquitylation, and proteasome-mediated degradation of IκBs (classical pathway) or the posttranslational processing of p100 to p52 (nonclassical pathway) discussed in Chapter 4. These first-phase events set the stage for the second phase, which is initiated by the rapid translocation of liberated NF-κB/Rel complexes into the nucleus. This second phase is less well-studied but involves a series of interesting posttranslational modifications, like phosphorylation and acetylation, targeting either the NF-κB subunits themselves or the histones associated with the chromatin of NF-κB target genes. These second-phase events are important as they regulate both the strength and duration of NF-κB action. In this chapter, we will describe recent progress involving this second phase of NF-κB activation.

5.2 PHOSPHORYLATION AS A KEY REGULATOR OF NF-κB ACTION

While the phosphorylation of IκBα by IκB kinase (IKK) is a pivotal early event in the activation of NF-κB, phosphorylation of the p65 subunit of NF-κB is also required for NF-κB to achieve its full transcriptional potential [4,5,6,7,8]. The first indication that phosphorylation might play a role in the biology of NF-κB emerged from studies of tumor necrosis factor α (TNFα)-stimulated cells in which phosphorylation of p65 was detected. Importantly, this posttranslational modification was associated with enhanced binding of NF-κB to its cognate enhancer [9]. Constitutive phosphorylation of p65 has been detected in endothelial cells and B lymphocytes [10], cells that are notable for their constitutive expression of nuclear NF-κB [9]. However, p65 phosphorylation was more frequently found to be an inducible modification. Multiple sites of phosphorylation in p65 have been identified and associated with the action of multiple stimulus-coupled kinases that act both in the cytoplasm and in the nucleus (Figure 5.1).

Especially important is the action of the catalytic subunit of protein kinase A (PKA$_c$), which phosphorylates serine 276 located in the RHD of p65. This event modifies several properties of NF-κB, including its DNA binding, dimerization, and transcriptional activity [11,12,13]. Phosphorylation of serine 276, however, is not the exclusive domain of PKA$_c$. Rather, different agonists appear to mobilize different kinases to modify this key phospho-acceptor. In cells stimulated with lipopolysaccharide (LPS), PKAc functions as the lead kinase [13,14]; in cells stimulated with TNF-α, however, serine 276 is modified by the mitogen and stress-activated kinase-1 (MSK-1) [15]. Interestingly, while PKA$_c$ acts on p65 in the cytoplasm, MSK-1 acts after p65 has translocated into the nucleus. Through the action of yet another kinase, protein kinase C (PKC)-ζ, TNF-α induces phosphorylation of serine 311 in

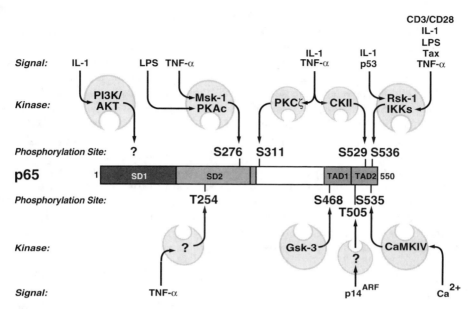

FIGURE 5.1 Targeting of p65 by multiple protein kinases induced by different stimuli. Six serine and two threonine phosphorylation sites have been identified in p65. These include serines 276 and 311 in the RHD (composed of SD1 and SD2) and serines 468, 529, 535, and 536 in the transactivation (TA) domain. Unidentified phosphorylation sites may also exist. Phosphorylation of serine 276 is mediated by PKA$_c$ and MSK-1, which are activated by LPS and TNF-α, respectively. Serine 311 is phosphorylated by PKC, which is activated by TNF-α or IL-1. TNF-α and IL-1 also stimulate casein kinase II (CKII), promoting the phosphorylation of serine 529. Serine 536 is targeted for phosphorylation either by the IKKs, which are activated by TNF-α, LPS, IL-1, CD3/CD28, and human T cell leukaemia virus (HTLV-1) Tax protein, or by RSK-1, which is activated by IL-1 or p53. Serine 468 is phosphorylated by GSK-3 and serine 535 by CaMKIV. The kinases mediating the phosphorylation of threonine 254 and 505 have not been identified. These findings emphasize the plethora of stimulus-coupled kinases that can successfully target p65 for phosphorylation at different acceptor sites.

the RHD of p65, a modification that also enhances the overall transcriptional activity of p65 [16,17].

Another key phosphorylation event involves serine 536 of p65. Phosphorylation at this site is, likewise, catalyzed by multiple kinases that are activated by different stimuli [18,19,20,21,22]. Serine 536 was first shown to be a target of the IKKs *in vitro* [19,23]. The production of antiphospho serine 536 antibodies revealed that this site is phosphorylated after stimulation with TNF-α, LPS [18,19], T cell costimulation [20], lymphotoxin β [21], phorbol myristate acetate (PMA)/ionomycin, or interleukin (IL)-1β [22]. Serine 536 is also targeted for phosphorylation during p53 activation of NF-κB, mediated by the kinase RSK-1. Phosphorylation at this site diminishes the ability of p65 to interact with IκBα [24].

Other sites of phosphorylation involving the TADs of p65 have been identified. Serine 529 is phosphorylated by casein kinase II (CKII) after IL-1 or TNF-α treatment [25,26,27], and serine 535 is phosphorylated by calmodulin-dependent kinase IV (CaMKIV) [28,29]. Phosphorylation at these two sites enhances the transcriptional activity of NF-κB.

P65 is also phosphorylated by a number of other kinases, including glycogen synthase kinase-3β (GSK-3β), NF-κB-activating kinase, TANK-binding kinase-1 (TBK1)/TRAF2-associated kinase, and AKT/phosphatidylinositol 3-kinase (PI3K) [4,5,6,7,8]. GSK-3β phosphorylates p65 *in vitro* and *in vivo* at serine 468 in TAD2 of p65. Of note, this phosphorylation event reduces the basal activity of NF-κB in unstimulated cells [7,30]. TBK1 phosphorylates serine 536 *in vitro* and *in vivo* [22], potentiating p65 transcriptional activity [31]. The involvement of PI3K and AKT in the phosphorylation of p65 remains rather controversial. Certain studies suggest that PI3K/AKT directly phosphorylates p65 at serine 536 in response to IL-1; others suggest that PI3K/AKT serves as an intermediate kinase that activates IKK2, which in turn phosphorylates serine 536 [6,8,32]. Still other studies argue against a direct or indirect role for phosphorylation of serine 536 by PI3K; addition of wortmanin, a PI3K inhibitor, does not block phosphorylation of this site following LPS or TNF-α stimulation [18,22].

Selective threonine residues in p65 are also phospho-acceptor sites. For example, TNF-α stimulation of cells leads to phosphorylation of threonine 254 in the RHD. This promotes the interaction of p65 with Pin1, which in turn inhibits p65 binding to IκBα and thus enhances the nuclear accumulation and action of p65 [33]. Conversely, phosphorylation of threonine 505 in response to p14[ARF] leads to inhibition of p65-dependent transcriptional activity [34,35]. In both of these cases, the kinases involved remain unidentified. These findings underscore the multiplicity of phosphorylation events involving different domains of p65 that are mediated, often in a rather redundant manner, by different kinases, which in turn are activated by distinct stimuli (Figure 5.1). While the functional consequences of phosphorylation at these different sites have been deciphered, often the underlying mechanism remains unclear.

Phosphorylation at these sites likely promotes the assembly of p65 with different cellular cofactors that, in turn, influence the function of p65. Precedent for such a scenario is found in the phosphorylation of serine 133 of the cyclic AMP response element-binding protein cAMP response element binding (CREB) protein and serine 15 in p53 that commonly facilitates the recruitment of the transcriptional coactivator p300/CBP [36,37]. Similarly, phosphorylation of p65 on serine 276 induced by PKA$_c$ [13,14] or MSK-1 [15] enhances the interaction of p65 with p300 and CREB binding protein (CBP). This phosphorylated NF-κB complex also more effectively displaces inhibitory complexes of p50 homodimers and histone deacetylase-1 (HDAC1) that are often bound to κB enhancers under basal conditions [13,14]. Phosphorylation of serine 311 by PKCζ and serine 536 by the IKK complex likewise enhance the assembly of p65 with p300 [17,38]. Phosphorylation of serine 536 additionally regulates the interaction of the TATA-binding protein-associated factor II 31 (TAF$_{II}$31) and the corepressor amino-terminal enhancer of split (AES) to the TAD of p65 [22]. The phosphorylation of p65 by CaMKIV at serine 535 results in the recruitment of CBP and the concomitant release of the corepressor, silencing mediator of retinoic acid and thyroid hormone receptor (SMRT) [28,29]. Although unproven, it seems likely that phosphorylation of serines 529 and 468 similarly regulate the interaction of p65 with selected coactivators or corepressors, leading to changes in its transcriptional activity [39–41].

Kinase redundancy in the phosphorylation of specific sites in p65 is frequently observed; however, the kinases involved are often induced by distinct stimuli or in a cell-type-specific manner. For example, serine 276 of p65 is phosphorylated by PKA_c in response to LPS or by MSK1 after TNF-α stimulation. Similarly, serine 536 is phosphorylated by either IKKα or IKKβ. IKKβ mediates this phosphorylation when monocytes or macrophages are stimulated with LPS [18], when human hepatic stellate cells are triggered by CD40 [42], or when T cells are stimulated with CD3/CD28 [43]. Conversely, IKKα, but not IKKβ, phosphorylates serine 536 in response to ligand activation of the lymphotoxin β receptor signaling pathway [21] or after stimulation with the Tax oncoprotein of human T-leukemia virus (HTLV)-1 [44]. However, both IKKα and IKKβ are involved in the phosphorylation of serine 536 after stimulation with TNF-α or IL-1 [19,22].

How phosphorylation specificity is achieved when these different signaling pathways are activated is not well understood. Different patterns of p65 phosphorylation could certainly influence the recruitment and divestment of different transcriptional cofactors, which in turn could mediate distinct profiles of gene expression. Precedence for such a scenario is found with steroid receptor coactivator 3 (SRC-3). Phosphorylation of SRC-3 at different sites differentially regulates the interaction with downstream transcriptional activators and coactivators and in this way integrates diverse signaling pathways [45].

5.3 NF-κB AND ACETYLATION

The assembly of NF-κB with different coactivator and corepressor proteins is crucial in orchestrating its ultimate biological effects. For example, NF-κB can associate with a series of different coactivators such as p300/CBP, p300/CBP-associated factor (PCAF), and members of the p160 nuclear receptor coactivator family, including SRC-1 and -3, each of which exhibits intrinsic histone acetyltransferase (HAT) activity [46,47]. The interaction of the p65 subunit of NF-κB with either CBP or p300 sharply increases the transcriptional activity of the NF-κB complex through both modification of chromatin structure [48,49,50,51] and, as discussed in greater detail below, the direct acetylation of p65 [52]. Similarly, SRC-1 can also function as a coactivator with NF-κB, but in this case, binding is mediated by the p50 subunit [48,53]. Another p160 family member, SRC-3, functions as an NF-κB coactivator and intriguingly is phosphorylated by the IKKs [45,54]. The arginine methyltransferase CARM1/PRMT4 is a novel transcriptional coactivator of NF-κB that acts synergistically with the p300/CBP and SRC-2 acetyltransferases [55].

In addition to its selective interaction with various coactivators, NF-κB also can physically associate with a family of corepressors, including specific HDACs [14,52,56,57]. Recruitment of these HDACs to the enhancers of various NF-κB target genes can lead to modification of the surrounding histone tails, promoting transcriptionally repressive changes in chromatin structure. In addition, HDAC3 promotes the deacetylation of p65, which diminishes DNA binding and transcriptional activity and increases binding to IκBα [52]. Similarly, sirtuin (silent mating type information regulation 2 homolog 1 [SIRT1]), the mammalian homologue of the yeast silencing information regulator (SIR)-2, physically interacts with p65 and inhibits transcription by deacetylating p65 [57]. The association of NF-κB with HDAC1 and HDAC2 inhibits the expression of many NF-κB-regulated genes, including the IL-8 gene [56]. In contrast to HDAC3 and SIRT1, HDAC1 and HDAC2 appear to operate solely by promoting changes in chromatin structure rather than through deacetylation of NF-κB. Silencing mediator for retinoid and thyroid hormone receptor (SMRT) and the nuclear corepressors (NCoRs) also bind to NF-κB and may in fact form a bridge for the binding of HDAC3 and SIRT1 [58,59,60,61].

As noted, coactivators and corepressors regulate gene expression by modifying the N-terminal tails of histones through their HAT or HDAC activities [62]. Acetylated histones are associated with transcriptionally active regions of chromatin, while deacetylated histones are associated with repressive chromatin [63,64,65]. In addition to targeting histones for modification, the HATs and HDACs also directly regulate the acetylation status of a variety of transcription factors, including p53, GATA-1, MyoD, Stat3, and E2F. Acetylation has been implicated in the regulation of many functions of these transcription factors, including dimerization, subcellular localization, assembly with DNA, interaction with other cofactors, and overall transcriptional activity [66,67,68,69,70].

5.3.1 ACETYLATION OF p65

Acetylation plays a key role in modulating the nuclear action of NF-κB. Acetylation of p65 has now been demonstrated both *in vivo* and *in vitro* [38,52]. Endogenous p65 is acetylated in a stimulus-coupled manner — after activation of cells with TNF-α or PMA, for example. The p300/CBP and PCAF acetyltransferases are of particular importance in the acetylation of p65 both *in vivo* and *in vitro* [38,52,71]. Like the histone proteins, acetylated p65 is subject to deacetylation by HDACs. In the case of p65, HDAC3 and SIRT1 mediate this function in association with the NCoR/SMRT complex [52,57,61,71].

Three major acetylation sites have been identified within p65: lysines 218, 221, and 310 (Figure 5.2) [72]. Modification of these three lysines induces specific changes in the action of NF-κB. For example, acetylation of lysine 221 enhances the DNA binding properties of NF-κB and impairs the assembly of p65 with IκBα in conjuction with the acetylation of lysine 218. Conversely, acetylation of lysine 310 is required for full transcriptional activity of p65 but exerts no effect on its DNA binding or IκBα assembly properties. More importantly, acetylation of lysine 310 is required to activate endogenous NF-κB genes. For example, TNF-α-induced expression of the E-selectin gene is markedly lower in p65-deficient cells reconstituted with p65-K310R than in cells reconstituted with wild-type p65 [38]. Furthermore, deacetylation of lysine 310 by SIRT1 correlates with a loss of NF-κB-regulated gene expression and sensitization of cells to TNF-α-induced apoptosis [57]. Lysines 122 and 123 are also subject to acetylation by p300/CBP and PCAF, although these modifications reduce both NF-κB-mediated transcription and binding of p65 to κB-enhancer DNA (Figure 5.2) [71]. It is interesting to consider the possibility that

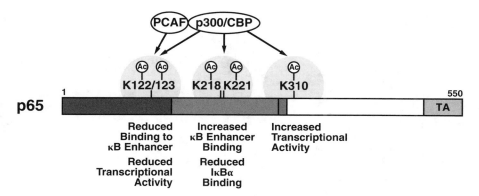

FIGURE 5.2 Acetylation of lysines by the p300/CBP and PCAF acetyltranferases regulates distinct functions of p65. Five primary sites of p65 acetylation have been identified: lysines 122, 123, 218, 221, and 310. Acetylation of lysine 221 by p300/CBP enhances the DNA binding affinity of p65 for the κB-enhancer and together with acetylation of lysine 218 sharply impairs the binding of IκBα to p65. In contrast, acetylation of lysine 310 does not modulate either DNA binding or IκBα assembly, but instead leads to enhanced transcriptional activity. It is not clear whether acetylated lysine 310 serves as a platform for binding of a transcriptional coactivator. Conversely, acetylation of lysines 122 and 123 by both p300 and p300/CBP-PCAF reduces the binding of p65 to the κB enhancer and impairs the overall transcriptional activity of p65. These findings emphasize how the acetylation of different lysine residues in p65 can shape its action in very specific and different ways.

p300/CBP- and PCAF-mediated acetylation of p65 may occur in a cell-type-specific manner, leading to very different transcriptional outcomes [71,72].

The biological importance of the reversible acetylation of p65 is highlighted by its role in regulating the assembly of p65 with its specific inhibitor, IκBα [52]. Stimulus-coupled activation of NF-κB is associated with the near-complete depletion of IκBα. However, among many target genes, NF-κB upregulates expression of the IκBα gene, leading to resynthesis of this inhibitor [73]. Newly synthesized IκBα proteins can shuttle in and out the nucleus and participate in the retrieval of nuclear NF-κB, a response that terminates the NF-κB-mediated transcriptional response (Figure 5.3) [74,75]. This retrieval function of IκBα is regulated by the acetylation status of p65. Specifically, when p65 is acetylated at lysines 218 and 221, it cannot bind effectively to IκBα. However, deacetylation of p65 at these sites by HDAC3 greatly enhances its binding to IκBα and leads, in turn, to a rapid chromosomal region maintenance (CRM) 1-dependent nuclear export of the NF-κB complex [52]. This export response marks the end of the transcriptional response and helps to replenish the depleted cytoplasmic pool of latent NF-κB/IκBα complexes, thereby readying the cell for the next NF-κB-inducing stimulus. As such, the reversible acetylation of p65 serves as an intranuclear molecular switch that shapes both the strength and duration of the NF-κB response [52]. In support of this model, treatment of human leukemia cells with HDAC inhibitors like superoylanilide hydroxamic acid (SAHA) or MS-275 markedly increases the levels of acetylated p65, enhances the nuclear accumulation of p65, and diminishes the association of p65 with IκBα [76].

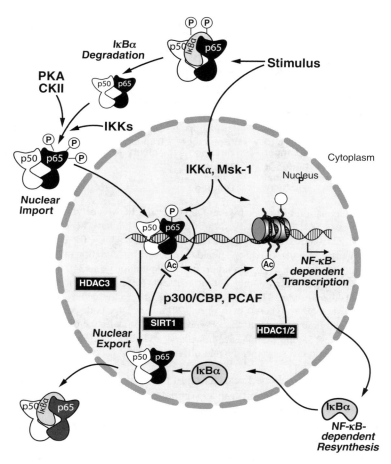

FIGURE 5.3 Phosphorylation and acetylation of p65 and histone tails regulate NF-κB target gene expression. After the degradation of IκBα, p65 is subject to phosphorylation by various kinases, either in the cytoplasm (e.g., PKA$_c$, CKII, and IKKs) or in the nucleus (e.g., MSK1, IKK1) in response to different stimuli. The nuclear kinases also can modify the tails of histones and promote changes in chromatin structure. Phosphorylation of p65 enhances its ability to bind to transcriptional coactivators p300/CBP and PCAF, which in turn acetylate both p65 and histones surrounding the promoter of NF-κB-responsive genes. These two types of post-translational changes importantly participate in the regulation of NF-κB target gene expression. Acetylated p65 is subsequently deacetylated by HDAC3 and SIRT1. Deacetylation of lysine 310 in p65 by HDAC3 or SIRT1 inhibits the transcriptional response of NF-κB. Deacetylation of lysines 218 and 221 in p65 by HDAC3 allows more efficient binding of RelA to newly synthesized IκBα proteins, whose expression is induced by NF-κB. IκBα binding to deacety-lated p65 leads to the rapid CRM-1-dependent nuclear export of the NF-κB complex, thereby terminating the NF-κB transcriptional response. Return of NF-κB to the cytoplasm bound to IκBα helps replenish the depleted cytoplasmic pool of latent NF-κB complexes, thereby readying the cell for the next NF-κB-inducing signal. Additionally, HDAC1/2-mediated deacetylation of histone tails contributes to termination of the NF-κB transcriptional response by altering the chromatin structure surrounding NF-κB-responsive genes.

While site-specific phosphorylation and acetylation of p65 regulate many different functions of NF-κB, acetylation of p65 is, in fact, regulated by prior phosphorylation of NF-κB. How is this linkage established? Phosphorylation of p65 on either serines 276 or 536 enhances the interaction of p65 with p300/CBP [38,51], providing a potential link between phosphorylation and acetylation. Indeed, blockade of phosphorylation of serine 276 or 536 of p65 reduces acetylation of lysine 310 [38]. Although p65 is phosphorylated both in the cytoplasm and in the nucleus, it appears to be acetylated only in the nucleus, likely reflecting the principally nuclear localization of the relevant acetyltransferases. Consistent with a contingent relationship between phosphorylation and acetylation, phosphorylation of p65 is detected within 5 minutes after TNF-α stimulation, while acetylation is not detected until 10 minutes after stimulation. Whether phosphorylation of p65 on other serines similarly regulates acetylation is uncertain. However, phosphorylation of serine 311 is associated with improved binding of p300/CBP to p65 [17], and thus a link to subsequent acetylation of p65 appears likely.

5.3.2 Acetylation of p50

The p50 subunit of NF-κB undergoes acetylation at three sites: lysines 431, 440, and 441. As with p65, these acetylations are stimulus-coupled. For example, addition of TNF-α or LPS consistently induces the acetylation of p50 [77,78]. These posttranslational modifications lead to improved DNA binding by p50 and enhanced transcriptional activity. For example, acetylation of p50 increases NF-κB-dependent transcription from the HIV LTR and enhances the expression of several NF-κB target genes, including cyclooxygenase and nitric oxide synthase in TNF-α or LPS-stimulated cells [77,78,79]. It is currently unknown if other members of the NF-κB/Rel family are regulated by acetylation. Interestingly, except for lysine 310 in p65, the other lysine residues targeted for acetylation in p50 and p65 are highly conserved in all NF-κB/Rel family members, including the Dorsal protein of *Drosophila*. Thus, acetylation may play a more general role in regulating the activity of the entire family of Rel proteins.

5.4 HISTONE ACETYLATION REGULATES NF-κB GENE ACTIVATION

In addition to direct posttranslational modification of p65, stimulus-coupled acetylation of histones present in chromatin surrounding NF-κB target genes is likely important in governing the overall transcriptional response. In eukaryotic cells, DNA is packaged into chromatin with varying degrees of compaction. The fundamental structural units of chromatin, the nucleosomes, are composed of histone octamers, with a central heterotetramer of histones H3 and H4 and two heterodimers of histones H2A and H2B. Genes residing in densely compacted heterochromatin are maintained in a transcriptionally inactive state, likely because the promoters are inaccessible to RNA polymerase II (polII) and other cofactors [80]. To promote increased accessibility of various promoters to components of the cell's transcription and replication machinery, chromatin remodeling must occur. Two different kinds of enzymatic complexes mediate remodeling: ATP-dependent chromatin-remodeling complexes, such as switching/sucrose nonfermenting (SWI/SNF), use ATP to change the position of specific nucleosomes or to alter the three-dimensional structure of the nucleosome. The other enzymes, including HATs, HDACs, kinases, and methyltransferases, modify the N-terminal tails of the histone proteins, thereby influencing transcriptional activity in local chromatin regions. In fact, the N-terminal histone tails are subject to multiple posttranslational modifications, including acetylation, phosphorylation, methylation, ubiquitylation, and ADP ribosylation. In general, acetylated histone tails are found in transcriptionally active regions of chromatin, while deacetylated histones accumulate in transcriptionally repressed regions [64,81,82].

The acetylation of local histone tails can be an important step governing the activation of NF-κB target genes. For example, IL-1β stimulation is associated with the acetylation of histone H4 in the granulocyte macrophage-colony stimulating factor (GM-CSF) promoter [83,84]. Similarly, LPS stimulation leads to hyperacetylation of histones H3 and H4 in the promoters of the IL-8, to macrophage inflammatory protein (MIP)-1α, and IL-12p40 genes, and to acetylation of H4 in the NF-κB-dependent IL-6 promoter [14,85,86]. TNF-α stimulation results in the hyperacetylation of H3 in the LTR of HIV [87] and promoter of the E-selectin gene in endothelial cells [88]. These increased levels of H3 and H4 acetylation correlate with increased recruitment of RNA polymerase II to these various transcription units and an overall increase in gene expression [85,86,87]. Viral infection can also promote the hyperacetylation of histones surrounding the NF-κB-dependent interferon-β-composite enhancer, thereby helping to mobilize the cellular antiviral defenses [89].

Importantly, high basal levels of histone H4 acetylation and fully accessible NF-κB binding sites are characteristic of the promoter regions of NF-κB regulated genes that undergo rapid upregulation, among them IκBα, MIP-2, and manganese superoxide dismutase (MnSOD). These genes appear to be preprogrammed for rapid activation through appropriate epigenetic modifications of the surrounding chromatin structure [86]. Thus, the different patterns of histone acetylation may determine the speed with which transcription of different NF-κB target genes is initiated. It is not known whether these patterns occur in a cell-type-specific or stimulus-specific manner or contribute to target gene specificity.

Conversely, impaired NF-κB action is associated with HDAC-mediated histone deacetylation. For example, glucocorticoids inhibit NF-κB-dependent gene activation through deacetylation of key histone residues, either by facilitating the recruitment of HDACs or by directly repressing HAT activity [83,84]. Similarly, *Drosophila* jun N-terminal kinase (JNK) inhibits the NF-κB pathway by recruiting dHDAC1 to the promoters of various NF-κB target genes [90]. In addition, p50 homodimers can function as κB-specific repressors by recruiting HDAC-1-containing complexes or SMRT-HDAC3 complexes, which likely promote histone deacetylation in regions where these homodimers bind [14,61]. The phosphorylation state of p65 appears to function as a switch that controls the association of NF-κB with the cofactors responsible for modifying histone acetylation in these conditions [14]. These findings underscore how dynamic changes in the pattern of histone acetylation mediated by HATs and HDACs help to regulate the expression of NF-κB target genes.

5.5 HISTONE PHOSPHORYLATION ACTIVATES NF-κB TARGET GENES

Phosphorylation is an important posttranslational modification of histones [91]. Like histone acetylation, histone phosphorylation is associated with transcriptional activation. Indeed, phosphorylation of serine 10 of histone H3 is a strong predictor of inducible gene expression [92,93,94,95,96]. For example, in mammalian cells, histone H3 surrounding the c-fos promoter undergoes rapid phosphorylation during the immediate-early response elicited by epidermal growth factor [95,97].

A strong link also exists between the phosphorylation of H3 and the activation of NF-κB target genes. Many proinflammatory stimuli that activate NF-κB also induce the phosphorylation of histone H3 on serine 10 [98]. Often, specific sites are phosphorylated and acetylated in tandem. For example, LPS and TNF-α induce both the phosphorylation of serine 10 in the N-terminal tail of histone H3 and the acetylation of the adjacent lysine 14 in the chromatin surrounding various cytokine and chemokine genes. Histone H3 phosphorylation similarly correlates with the effective recruitment of NF-κB to the κB enhancers residing within the upstream regulatory regions of the IL-6, IL-8, monocyte chemotactic protein (MCP), and IL-12p40 genes [98]. Thus, NF-κB activation is associated with the effective recruitment of both histone kinases and HATs to the promoter regions of target genes. The resulting phosphorylation and acetylation of the surrounding histone tails creates an environment that is favorable for the recruitment of RNA polII to these promoters [98]. However, phosphorylation of histone H3 is not required for activation of the promoter of every NF-κB target gene; activation of the NF-κB-responsive TNF-α and MIP-1α genes, for example, is not associated with H3 phosphorylation [98].

Phosphorylation of histone H3 serine 10 may be correlated with the activation of the p38 mitogen-activated protein kinase; however, the effect of p38 appears indirect. Specifically, p38 is not recruited to the promoters of NF-κB target genes and requires the subsequent activity of downstream kinases [98]; IKKα may be one of these kinases. IKKα is readily detected in the nucleus and promotes the phosphorylation of serine 10 of histone H3 both *in vivo* and *in vitro* [99,100]. Further, the kinetics of serine 10 phosphorylation closely correlate with the recruitment of IKKα to the promoter, as revealed in chromatin immunoprecipitation assays. IKKα likely functions as part of a larger nuclear complex containing p65 and CBP. All three of these proteins are recruited to the IL-6 promoter as a unit after TNF-α stimulation. P65 may play a key nucleating role in the assembly of this complex and in recruiting IKKα and CBP to specific promoters [99,100]. These results suggest a novel nuclear function for IKKα in modifying histone function through the phosphorylation of serine 10 in histone H3, which is critical for the activation of many NF-κB-dependent target genes.

However, some redundancy in this phosphorylation event appears likely. Loss of IKKα expression is not associated with defective phosphorylation of serine 10 in histone H3 in the cIAP-2 and IL-8 promoter after treatment with laminin [61]. In addition, some NF-κB target genes, such as IκBα, are activated normally after LPS stimulation in the absence of IKKα recruitment to the promoter [86]. Furthermore, cells from "knock-in" mice expressing a catalytically inactive form of IKKα respond normally to TNF-α, IL-1, and LPS. These findings bolster the notion that other kinases can phosphorylate serine 10 [101,102]. MSK1 and PKA, which phosphorylate key serine residues of p65, may also function as histone H3 serine 10 kinases in the presence of different stimuli [103,104]. In addition, IKKβ could represent an additional histone H3 serine 10 kinase. IKKβ effectively phosphorylates histone H3 *in vitro* and is recruited to the promoter of some NF-κB target genes after activation [61]. Indeed, it is interesting to consider the possibility that different kinases are preferentially recruited to the promoters of different NF-κB target genes in a stimulus-specific or cell-context-specific manner. Such recruitment of these enzymes could provide an additional level of control contributing to selective activation of different genes in response to different extracellular stimuli.

5.6 CONCLUSIONS

Our understanding of the multilayered regulation of NF-κB is expanding quite rapidly. Control of NF-κB is not limited to the early cytoplasmic events that lead to IκBα degradation and translocation of NF-κB into the nucleus. Rather, the action of this transcription factor is also regulated by later events involving the posttranslational phosphorylation and acetylation of NF-κB itself and the modification of histones surrounding cellular genes whose expression is induced by NF-κB. The recognition that these nuclear posttranslational events shape the strength and duration of the NF-κB transcriptional response has opened an exciting new chapter in NF-κB biology (Figure 5.3).

It is now clear that both the phosphorylation and acetylation of the p65 subunit of NF-κB are required to generate a fully active NF-κB complex. In the absence of these modifications, NF-κB displays significantly impaired transcriptional activity. Emerging evidence now suggests that these modifications can occur in an ordered manner with phosphorylation triggering subsequent acetylation of p65 through recruitment of acetyltransferases such as p300 and CBP. The p65 subunit might also be regulated by less well-studied posttranslational modifications such as ubiquitylation and methylation [105]. The composite effects of these modifications are a biologically rich area for future NF-κB research.

However, nuclear posttranslational changes are not limited to NF-κB. Indeed, posttranslational modifications of histones are associated with changes in chromatin structure that influence the expression of NF-κB responsive genes. As we have discussed, the phosphorylation and acetylation of histone tails in chromatin surrounding NF-κB target genes can markedly enhance their transcription. Similarly, methylation of certain arginine or lysine residues in the histones can promote activation of NF-κB target genes [55,85,88], while methylation at other sites is associated with transcriptional repression. Each of these different posttranslational modifications of the histone tails creates a specific "mark" in the chromatin that may regulate the binding or divestment of specific cofactors, which combine to shape the ultimate transcriptional response. In aggregate, these epigenetic modifications have been termed the "histone code" [106,107]. Of note, like the modification of p65, this code is often constructed sequentially. For example, phosphorylation of serine

10 in H3 enhances the acetylation at lysine 14 [95,108,109]. a finding that mirrors the phosphorylation triggering acetylation of p65.

By analogy to the histone code, "transcription factor codes" may also exist [110,111,112,113]. The modification of p65 by phosphorylation or acetylation at specific sites could, in fact, dictate specific biological responses, reflecting the gain or loss of select cofactors whose association with p65 is regulated by its state of modification. These modifications and the specific factor binding they induce could also confer a degree of target gene selectivity. We predict this will be an exciting area of future NF-κB research.

ACKNOWLEDGMENTS

L.F.C. is the recipient of an Arthritis Foundation Investigator Award. This work was supported in part by a National Institutes of Health grant (RO1 CA89001-02) to W.C.G., a National Institutes of Health training grant (AI07305) to L.F.C., and by funds from the J. David Gladstone Institutes and benefited from core facilities provided through the UCSF-GIVI Center for AIDS Research (National Institutes of Health Grant P30 MH59037). We thank J. Carroll for assistance in the preparation of the figures and S. Ordway and G. Howard for editorial assistance. Portions of this chapter were based on prior publications including L.F. Chen and W.C. Greene "Shaping the nuclear action of NF-κB" *Nat. Rev. Mol. Cell. Biol.* 5(5), 392–401 2004, and L.F. Chen and W.C. Greene Regulation of distinct biological activities of the NF-κB transcription factor complex by acetylation. *J. Mol. Med.* 81(9), 549–557, 2003.

REFERENCES

[1]   Ghosh, S., May, M.J., and Kopp, E.B., NF-κB and Rel proteins: Evolutionarily conserved mediators of immune responses, *Annu. Rev. Immunol.* 16 (2), 225–260, 1998.

[2]   Baldwin, A.S., Jr., The NF-kappaB and IkappaB proteins: New discoveries and insights, *Annu. Rev. Immunol.* 14, 649–683, 1996.

[3]   Ghosh, S. and Karin, M., Missing pieces in the NF-kappaB puzzle, *Cell* 109 Suppl, S81-96, 2002.

[4]   Bonnard, M., Mirtsos, C., Suzuki, S. et al., Deficiency of T2K leads to apoptotic liver degeneration and impaired NF-kappaB-dependent gene transcription, *Embo J.* 19 (18), 4976–4985, 2000.

[5]   Hoeflich, K.P., Luo, J., Rubie, E.A. et al., Requirement for glycogen synthase kinase-3beta in cell survival and NF-kappaB activation, *Nature* 406 (6791), 86–90, 2000.

[6]   Sizemore, N., Leung, S., and Stark, G.R., Activation of phosphatidylinositol 3-kinase in response to interleukin-1 leads to phosphorylation and activation of the NF-kappaB p65/RelA subunit, *Mol. Cell. Biol.* 19 (7), 4798–4805, 1999.

[7]   Schwabe, R.F. and Brenner, D.A., Role of glycogen synthase kinase-3 in TNF-alpha-induced NF-kappaB activation and apoptosis in hepatocytes, *Am. J. Physiol. Gastrointest. Liver Physiol.* 283 (1), G204–211, 2002.

[8]   Madrid, L.V., Mayo, M.W., Reuther, J.Y. et al., Akt stimulates the transactivation
      potential of the RelA/p65 Subunit of NF-kappaB through utilization of the Ikappa B
      kinase and activation of the mitogen-activated protein kinase p38, *J. Biol. Chem.* 276
      (22), 18934–18940, 2001.

[9]   Naumann, M. and Scheidereit, C., Activation of NF-kappaB *in vivo* is regulated by
      multiple phosphorylations, *Embo J.* 13 (19), 4597–4607, 1994.

[10] Anrather, J., Csizmadia, V., Soares, M.P. et al., Regulation of NF-kappaB RelA
      phosphorylation and transcriptional activity by p21(ras) and protein kinase Czeta in
      primary endothelial cells, *J. Biol. Chem.* 274 (19), 13594–13603, 1999.

[11] Mosialos, G. and Gilmore, T.D., v-Rel and c-Rel are differentially affected by muta-
      tions at a consensus protein kinase recognition sequence, *Oncogene* 8 (3), 721–730,
      1993.

[12] Ganchi, P.A., Sun, S.C., Greene, W.C. et al., A novel NF-kappaB complex containing
      p65 homodimers: Implications for transcriptional control at the level of subunit
      dimerization, *Mol. Cell. Biol.* 13 (12), 7826–7835, 1993.

[13] Zhong, H., SuYang, H., Erdjument-Bromage, H. et al., The transcriptional activity
      of NF-kappaB is regulated by the IkappaB-associated PKA_c subunit through a cyclic
      AMP-independent mechanism, *Cell* 89 (3), 413–424, 1997.

[14] Zhong, H., May, M.J., Jimi, E. et al., The phosphorylation status of nuclear
      NF-kappaB determines its association with CBP/p300 or HDAC-1, *Mol. Cell* 9 (3),
      625–636, 2002.

[15] Vermeulen, L., De Wilde, G., Damme, P.V. et al., Transcriptional activation of the
      NF-kappaB p65 subunit by mitogen- and stress-activated protein kinase-1 (MSK1),
      Embo J. 22 (6), 1313–1324, 2003.

[16] Leitges, M., Sanz, L., Martin, P. et al., Targeted disruption of the ζPKC gene results
      in the impairment of the NF-kappaB pathway, *Mol. Cell* 8 (4), 771–780, 2001.

[17] Duran, A., Diaz-Meco, M.T., and Moscat, J., Essential role of RelA Ser311 phos-
      phorylation by ζPKC in NF-kappaB transcriptional activation, *Embo J.* 22 (15),
      3910–3918, 2003.

[18] Yang, F., Tang, E., Guan, K. et al., IKK beta plays an essential role in the phospho-
      rylation of RelA/p65 on serine 536 induced by lipopolysaccharide, *J. Immunol.* 170
      (11), 5630–5635, 2003.

[19] Sakurai, H., Suzuki, S., Kawasaki, N. et al., TNF-alpha-induced IKK phosphorylation
      of NF-kappaB p65 on serine 536 is mediated through TRAF2, TRAF5 and TAK1
      signaling pathway, *J. Biol. Chem.*, 2003.

[20] Mattioli, I., Dittrich-Breiholz, O., Livingstone, M. et al., Comparative analysis of
      T-cell costimulation and CD43 activation reveals novel signaling pathways and target
      genes, *Blood* 104 (10), 3302–3304, 2004.

[21] Jiang, X., Takahashi, N., Matsui, N. et al., The NF-kappaB activation in lymphotoxin
      beta receptor signaling depends on the phosphorylation of p65 at serine 536, *J. Biol.
      Chem.* 278 (2), 919–926, 2003.

[22] Buss, H., Dorrie, A., Schmitz, M.L. et al., Constitutive and interleukin-1-inducible
      phosphorylation of p65 NF-{kappa}B at serine 536 is mediated by multiple protein
      kinases including I{kappa}B kinase (IKK)-{alpha}, IKK{beta}, IKK{epsilon},
      TRAF family member-associated (TANK)-binding kinase 1 (TBK1), and an unknown
      kinase and couples p65 to TATA-binding protein-associated factor II31-mediated
      interleukin-8 transcription, *J. Biol. Chem.* 279 (53), 55633–55643, 2004.

[23] Sakurai, H., Chiba, H., Miyoshi, H. et al., IκB kinases phosphorylate NF-κB p65
      subunit on serine 536 in the transactivation domain, *J. Biol. Chem.* 274 (43),
      30353–30356, 1999.

[24] Bohuslav, J., Chen, L.F., Kwon, H. et al., p53 induces NF-κB activation by an IκB kinase-independent mechanism involving phosphorylation of p65 by ribosomal S6 kinase 1, *J. Biol. Chem.* 279 (25), 26115–26125, 2004.

[25] Bird, T.A., Schooley, K., Dower, S.K. et al., Activation of nuclear transcription factor NF-kappaB by interleukin-1 is accompanied by casein kinase II-mediated phosphorylation of the p65 subunit, *J. Biol. Chem.* 272 (51), 32606–32612, 1997.

[26] Wang, D. and Baldwin, A.S., Jr., Activation of nuclear factor-kappaB-dependent transcription by tumor necrosis factor-alpha is mediated through phosphorylation of RelA/p65 on serine 529, *J. Biol. Chem.* 273 (45), 29411–29416, 1998.

[27] Wang, D., Westerheide, S.D., Hanson, J.L. et al., TNF-α-induced phosphorylation of RelA/p65 on Ser529 is controlled by casein kinase II, *J. Biol. Chem.* 275 (42), 32592–32597, 2000.

[28] Jang, M.K., Goo, Y.H., Sohn, Y.C. et al., Ca2+/calmodulin-dependent protein kinase IV stimulates nuclear factor-kappaB transactivation via phosphorylation of the p65 subunit, *J. Biol. Chem.* 276 (23), 20005–20010, 2001.

[29] Bae, J.S., Jang, M.K., Hong, S. et al., Phosphorylation of NF-kappaB by calmodulin-dependent kinase IV activates anti-apoptotic gene expression, *Biochem. Biophys. Res. Commun.* 305 (4), 1094–1098, 2003.

[30] Buss, H., Dorrie, A., Schmitz, M.L. et al., Phosphorylation of serine 468 by GSK-3beta negatively regulates basal p65 NF-kappaB activity, *J. Biol. Chem.* 279 (48), 49571–49574, 2004.

[31] Fujita, F., Taniguchi, Y., Kato, T. et al., Identification of NAP1, a regulatory subunit of IkappaB kinase-related kinases that potentiates NF-kappaB signaling, *Mol. Cell. Biol.* 23 (21), 7780–7793, 2003.

[32] Madrid, L.V., Wang, C.Y., Guttridge, D.C. et al., AKT suppresses apoptosis by stimulating the transactivation potential of the RelA/p65 subunit of NF-kappaB, *Mol. Cell. Biol.* 20 (5), 1626–1638, 2000.

[33] Ryo, A., Suizu, F., Yoshida, Y. et al., Regulation of NF-kappaB signaling by Pin1-dependent prolyl isomerization and ubiquitin-mediated proteolysis of p65/RelA, *Mol. Cell* 12 (6), 1413–1426, 2003.

[34] Rocha, S., Garrett, M.D., Campbell, K.J. et al., Regulation of NF-kappaB and p53 through activation of ATR and Chk1 by the ARF tumour suppressor, *Embo J.*, 2005.

[35] Rocha, S., Campbell, K.J., and Perkins, N.D., p53- and Mdm2-independent repression of NF-kappa B transactivation by the ARF tumor suppressor, *Mol. Cell* 12 (1), 15–25, 2003.

[36] Parker, D., Ferreri, K., Nakajima, T. et al., Phosphorylation of CREB at Ser-133 induces complex formation with CREB-binding protein via a direct mechanism, *Mol. Cell. Biol.* 16 (2), 694–703, 1996.

[37] Lambert, P.F., Kashanchi, F., Radonovich, M.F. et al., Phosphorylation of p53 serine 15 increases interaction with CBP, *J. Biol. Chem.* 273 (49), 33048–33053, 1998.

[38] Chen, L.L., Williams, S.A., Mu, Y. et al., NF-kappaB RelA phosphorylation regulates RelA acetylation, 2005.

[39] Schmitz, M.L., Stelzer, G., Altmann, H. et al., Interaction of the COOH-terminal transactivation domain of p65 NF-kappaB with TATA-binding protein, transcription factor IIB, and coactivators, *J. Biol. Chem.* 270 (13), 7219–7226, 1995.

[40] Xu, X., Prorock, C., Ishikawa, H. et al., Functional interaction of the v-Rel and c-Rel oncoproteins with the TATA-binding protein and association with transcription factor IIB, *Mol. Cell. Biol.* 13 (11), 6733–6741, 1993.

[41] Blair, W.S., Bogerd, H.P., Madore, S.J. et al., Mutational analysis of the transcription activation domain of RelA: Identification of a highly synergistic minimal acidic activation module, *Mol. Cell Biol.* 14 (11), 7226–7234, 1994.

[42]   Schwabe, R.F., Schnabl, B., Kweon, Y.O. et al., CD40 activates NF-kappaB and c-Jun
        N-terminal kinase and enhances chemokine secretion on activated human hepatic
        stellate cells, *J. Immunol.* 166 (11), 6812–6819, 2001.
[43]   Mattioli, I., Sebald, A., Bucher, C. et al., Transient and selective NF-kappaB p65
        serine 536 phosphorylation induced by T cell costimulation is mediated by IkappaB
        kinase beta and controls the kinetics of p65 nuclear import, *J. Immunol.* 172 (10),
        6336–6344, 2004.
[44]   O'Mahony, A.M., Montano, M., Van Beneden, K. et al., Human T-cell lymphotropic
        virus type 1 tax induction of biologically active NF-{kappa}B requires I{kappa}B
        kinase-1-mediated phosphorylation of RelA/p65, *J. Biol. Chem.* 279 (18),
        18137–18145, 2004.
[45]   Wu, R.C., Qin, J., Yi, P. et al., Selective phosphorylations of the SRC-3/AIB1 coac-
        tivator integrate genomic reponses to multiple cellular signaling pathways, *Mol. Cell*
        15 (6), 937–949, 2004.
[46]   Chrivia, J.C., Kwok, R.P., Lamb, N. et al., Phosphorylated CREB binds specifically
        to the nuclear protein CBP, *Nature* 365 (6449), 855–859, 1993.
[47]   Eckner, R., Ewen, M.E., Newsome, D. et al., Molecular cloning and functional
        analysis of the adenovirus E1A-associated 300-kD protein (p300) reveals a protein
        with properties of a transcriptional adaptor, *Genes Dev.* 8 (8), 869–884, 1994.
[48]   Sheppard, K.A., Rose, D.W., Haque, Z.K. et al., Transcriptional activation by
        NF-kappaB requires multiple coactivators, *Mol. Cell. Biol.* 19 (9), 6367–6378, 1999.
[49]   Vanden Berghe, W., De Bosscher, K., Boone, E. et al., The NF-κB engages CBP/p300
        and histone acetyltransferase activity for transcriptional activation of the interleukin-
        6 gene promoter, *J. Biol. Chem.* 274 (45), 32091–32098, 1999.
[50]   Gerritsen, M.E., Williams, A.J., Neish, A.S. et al., CREB-binding protein/p300 are
        transcriptional coactivators of p65, *Proc. Natl. Acad. Sci. USA* 94 (7), 2927–2932, 1997.
[51]   Zhong, H., Voll, R.E., and Ghosh, S., Phosphorylation of NF-kappa B p65 by PKA
        stimulates transcriptional activity by promoting a novel bivalent interaction with the
        coactivator CBP/p300, *Mol. Cell* 1 (5), 661–671, 1998.
[52]   Chen, L.F., Fischle, W., Verdin, E. et al., Duration of nuclear NF-κB action regulated
        by reversible acetylation, *Science* 293 (5535), 1653–1657, 2001.
[53]   Na, S.Y., Lee, S.K., Han, S.J. et al., Steroid receptor coactivator-1 interacts with the
        p50 subunit and coactivates NF-κB-mediated transactivations, *J. Biol. Chem.* 273
        (18), 10831–10834, 1998.
[54]   Wu, R.C., Qin, J., Hashimoto, Y. et al., Regulation of SRC-3 (pCIP/ACTR/AIB-
        1/RAC-3/TRAM-1) coactivator activity by IκB kinase, *Mol. Cell. Biol.* 22 (10),
        3549–3561, 2002.
[55]   Covic, M., Hassa, P.O., Saccani, S. et al., Arginine methyltransferase CARM1 is a
        promoter-specific regulator of NF-kappaB-dependent gene expression, *Embo J.* 24
        (1), 85–96, 2005.
[56]   Ashburner, B.P., Westerheide, S.D., and Baldwin, A.S., Jr., The p65 (RelA) subunit
        of NF-kappaB interacts with the histone deacetylase (HDAC) corepressors HDAC1
        and HDAC2 to negatively regulate gene expression, *Mol. Cell. Biol.* 21 (20),
        7065–7077, 2001.
[57]   Yeung, F., Hoberg, J.E., Ramsey, C.S. et al., Modulation of NF-kappaB-dependent
        transcription and cell survival by the SIRT1 deacetylase, *Embo J.*, 2004.
[58]   Lee, S.K., Kim, J.H., Lee, Y.C. et al., Silencing mediator of retinoic acid and thyroid
        hormone receptors, as a novel transcriptional corepressor molecule of activating
        protein-1, NF-κB, and serum response factor, *J. Biol. Chem.* 275 (17), 12470–12474,
        2000.

[59]   Baek, S.H., Ohgi, K.A., Rose, D.W. et al., Exchange of N-CoR corepressor and Tip60 coactivator complexes links gene expression by NF-kappaB and beta-amyloid precursor protein, *Cell* 110 (1), 55–67, 2002.

[60]   Gao, Z., Chiao, P., Zhang, X. et al., Coactivators and corepressors of NF-kappaB in IkappaBalpha gene promoter, *J. Biol. Chem.*, 2005.

[61]   Hoberg, J.E., Yeung, F., and Mayo, M.W., SMRT derepression by the IkappaB kinase alpha: a prerequisite to NF-kappaB transcription and survival, *Mol. Cell.* 16 (2), 245–255, 2004.

[62]   Berger, S.L., Gene activation by histone and factor acetyltransferases, *Curr. Opin. Cell. Biol.* 11 (3), 336–341, 1999.

[63]   Imhof, A., Yang, X.J., Ogryzko, V.V. et al., Acetylation of general transcription factors by histone acetyltransferases, *Curr. Biol.* 7 (9), 689–692, 1997.

[64]   Kuo, M.H. and Allis, C.D., Roles of histone acetyltransferases and deacetylases in gene regulation, *Bioessays* 20 (8), 615–626, 1998.

[65]   Agalioti, T., Lomvardas, S., Parekh, B. et al., Ordered recruitment of chromatin modifying and general transcription factors to the IFN-beta promoter, *Cell* 103 (4), 667–678, 2000.

[66]   Yuan, Z.L., Guan, Y.J., Chatterjee, D. et al., Stat3 dimerization regulated by reversible acetylation of a single lysine residue, *Science* 307 (5707), 269–273, 2005.

[67]   Gu, W. and Roeder, R.G., Activation of p53 sequence-specific DNA binding by acetylation of the p53 C-terminal domain, *Cell* 90 (4), 595–606, 1997.

[68]   Boyes, J., Byfield, P., Nakatani, Y. et al., Regulation of activity of the transcription factor GATA-1 by acetylation, *Nature* 396 (6711), 594–598, 1998.

[69]   Martínez-Balbás, M.A., Bauer, U.M., Nielsen, S.J. et al., Regulation of E2F1 activity by acetylation, *Embo J.* 19 (4), 662–671, 2000.

[70]   Sartorelli, V., Puri, P.L., Hamamori, Y. et al., Acetylation of MyoD directed by PCAF is necessary for the execution of the muscle program, *Mol. Cell* 4 (5), 725–734, 1999.

[71]   Kiernan, R., Bres, V., Ng, R.W. et al., Post-activation turn-off of NF-κB-dependent transcription is regulated by acetylation of p65, *J. Biol. Chem.* 278 (4), 2758–2766, 2003.

[72]   Chen, L.F., Mu, Y., and Greene, W.C., Acetylation of RelA at discrete sites regulates distinct nuclear functions of NF-κB, *Embo J.* 21 (23), 6539–6548, 2002.

[73]   Sun, S.-C., Ganchi, P.A., Ballard, D.W. et al., NF-κB controls expression of inhibitor IκBα: Evidence for an inducible autoregulatory pathway, *Science* 259 (5103), 1912–1915, 1993.

[74]   Arenzana-Seisdedos, F., Thompson, J., Rodriguez, M.S. et al., Inducible nuclear expression of newly synthesized IκBα negatively regulates DNA-binding and transcriptional activities of NF-κB, *Mol. Cell. Biol.* 15 (5), 2689–2696, 1995.

[75]   Arenzana-Seisdedos, F., Turpin, P., Rodriguez, M. et al., Nuclear localization of IκBα promotes active transport of NF-κB from the nucleus to the cytoplasm, *J. Cell. Sci.* 110 (Pt 3), 369–378, 1997.

[76]   Dai, Y., Rahmani, M., Dent, P. et al., Blockade of histone deacetylase inhibitor-induced RelA/p65 acetylation and NF-kappaB activation potentiates apoptosis in human leukemia cells through a process mediated by oxidative damage, XIAP downregulation and JNK1 activation, *Mol. Cell. Biol.*, 25(13), 5429–5444, 2005.

[77]   Deng, W.G. and Wu, K.K., Regulation of inducible nitric oxide synthase expression by p300 and p50 acetylation, *J. Immunol.* 171 (12), 6581–6588, 2003.

[78]   Deng, W.G., Zhu, Y., and Wu, K.K., Up-regulation of p300 binding and p50 acetylation in TNF-α -induced cyclooxygenase-2 promoter activation, *J. Biol. Chem.* 278 (7), 4770–4777, 2003.

[79] Furia, B., Deng, L., Wu, K. et al., Enhancement of NF-κB acetylation by coactivator p300 and HIV-1 Tat proteins, *J. Biol. Chem.* 277 (7), 4973–4980, 2002.

[80] Horn, P.J. and Peterson, C.L., Molecular biology. Chromatin higher order folding — wrapping up transcription, *Science* 297 (5588), 1824–1827, 2002.

[81] Imhof, A. and Wolffe, A.P., Transcription: Gene control by targeted histone acetylation, *Curr. Biol.* 8 (12), R422–424, 1998.

[82] Sterner, D.E. and Berger, S.L., Acetylation of histones and transcription-related factors, *Microbiol. Mol. Biol. Rev.* 64 (2), 435–459, 2000.

[83] Ito, K., Jazrawi, E., Cosio, B. et al., p65-activated histone acetyltransferase activity is repressed by glucocorticoids: Mifepristone fails to recruit HDAC2 to the p65-HAT complex, *J. Biol. Chem.* 276 (32), 30208–30215, 2001.

[84] Ito, K., Barnes, P.J., and Adcock, I.M., Glucocorticoid receptor recruitment of histone deacetylase 2 inhibits interleukin-1β-induced histone H4 acetylation on lysines 8 and 12, *Mol. Cell. Biol.* 20 (18), 6891–6903, 2000.

[85] Saccani, S. and Natoli, G., Dynamic changes in histone H3 Lys 9 methylation occurring at tightly regulated inducible inflammatory genes, *Genes Dev.* 16 (17), 2219–2224, 2002.

[86] Saccani, S., Pantano, S., and Natoli, G., Two waves of nuclear factor kappaB recruitment to target promoters, *J. Exp. Med.* 193 (12), 1351–1359, 2001.

[87] Williams, S.A., Chen, L.F., Kwon, H. et al., Prostratin antagonizes HIV latency by activating NF-kappaB, *J. Biol. Chem.* 279 (40), 42008–42017, 2004.

[88] Edelstein, L.C., Pan, A., and Collins, T., Chromatin modification and the endothelial-specific activation of the E-selectin gene, *J. Biol. Chem.* 280 (12), 11192–11202, 2005.

[89] Parekh, B.S. and Maniatis, T., Virus infection leads to localized hyperacetylation of histones H3 and H4 at the IFN-beta promoter, *Mol. Cell* 3 (1), 125–129, 1999.

[90] Kim, T., Yoon, J., Cho, H. et al., Downregulation of lipopolysaccharide response in Drosophila by negative crosstalk between the AP1 and NF-kappaB signaling modules, *Nat. Immunol.* 6 (2), 211–218, 2005.

[91] Berger, S.L., Histone modifications in transcriptional regulation, *Curr. Opin. Genet. Dev.* 12 (2), 142–148, 2002.

[92] Wei, Y., Yu, L., Bowen, J. et al., Phosphorylation of histone H3 is required for proper chromosome condensation and segregation, *Cell* 97 (1), 99–109, 1999.

[93] Wei, Y., Mizzen, C.A., Cook, R.G. et al., Phosphorylation of histone H3 at serine 10 is correlated with chromosome condensation during mitosis and meiosis in Tetrahymena, *Proc. Natl. Acad. Sci. USA* 95 (13), 7480–7484, 1998.

[94] Mahadevan, L.C., Willis, A.C., and Barratt, M.J., Rapid histone H3 phosphorylation in response to growth factors, phorbol esters, okadaic acid, and protein synthesis inhibitors, *Cell* 65 (5), 775–783, 1991.

[95] Cheung, P., Tanner, K.G., Cheung, W.L. et al., Synergistic coupling of histone H3 phosphorylation and acetylation in response to epidermal growth factor stimulation, *Mol. Cell* 5 (6), 905–15, 2000.

[96] Clayton, A.L., Rose, S., Barratt, M.J. et al., Phosphoacetylation of histone H3 on c-fos- and c-jun-associated nucleosomes upon gene activation, *Embo J.* 19 (14), 3714–3726, 2000.

[97] Sassone-Corsi, P., Mizzen, C.A., Cheung, P. et al., Requirement of Rsk-2 for epidermal growth factor-activated phosphorylation of histone H3, *Science* 285 (5429), 886–891, 1999.

[98] Saccani, S., Pantano, S., and Natoli, G., p38-Dependent marking of inflammatory genes for increased NF-kappaB recruitment, *Nat. Immunol.* 3 (1), 69–75, 2002.

[99]   Yamamoto, Y., Verma, U.N., Prajapati, S. et al., Histone H3 phosphorylation by IKK-α is critical for cytokine-induced gene expression, *Nature* 423 (6940), 655–659, 2003.

[100]   Anest, V., Hanson, J.L., Cogswell, P.C. et al., A nucleosomal function for IkappaB kinase-alpha in NF-kappaB-dependent gene expression, *Nature* 423 (6940), 659–663, 2003.

[101]   Cao, Y., Bonizzi, G., Seagroves, T.N. et al., IKKα provides an essential link between RANK signaling and cyclin D1 expression during mammary gland development, *Cell* 107 (6), 763–775, 2001.

[102]   Sil, A.K., Maeda, S., Sano, Y. et al., IkappaB kinase-alpha acts in the epidermis to control skeletal and craniofacial morphogenesis, *Nature* 428 (6983), 660–664, 2004.

[103]   Soloaga, A., Thomson, S., Wiggin, G.R. et al., MSK2 and MSK1 mediate the mitogen- and stress-induced phosphorylation of histone H3 and HMG-14, *Embo J.* 22 (11), 2788–2797, 2003.

[104]   Salvador, L.M., Park, Y., Cottom, J. et al., Follicle-stimulating hormone stimulates protein kinase A-mediated histone H3 phosphorylation and acetylation leading to select gene activation in ovarian granulosa cells, *J. Biol. Chem.* 276 (43), 40146–40155, 2001.

[105]   Saccani, S., Marazzi, I., Beg, A.A. et al., Degradation of promoter-bound p65/RelA is essential for the prompt termination of the nuclear factor kappaB response, *J. Exp. Med.* 200 (1), 107–113, 2004.

[106]   Jenuwein, T. and Allis, C.D., Translating the histone code, *Science* 293 (5532), 1074–1080, 2001.

[107]   Strahl, B.D. and Allis, C.D., The language of covalent histone modifications, *Nature* 403 (6765), 41–45, 2000.

[108]   Lo, W.S., Duggan, L., Tolga, N.C. et al., Snf1 — a histone kinase that works in concert with the histone acetyltransferase Gcn5 to regulate transcription, *Science* 293 (5532), 1142–1146, 2001.

[109]   Lo, W.S., Trievel, R.C., Rojas, J.R. et al., Phosphorylation of serine 10 in histone H3 is functionally linked *in vitro* and *in vivo* to Gcn5-mediated acetylation at lysine 14, *Mol. Cell* 5 (6), 917–926, 2000.

[110]   Rosenfeld, M.G. and Glass, C.K., Coregulator codes of transcriptional regulation by nuclear receptors, *J. Biol. Chem.* 276 (40), 36865–36868, 2001.

[111]   Tansey, W.P., Transcriptional activation: Risky business, *Genes Dev.* 15 (9), 1045–1050, 2001.

[112]   Gamble, M.J. and Freedman, L.P., A coactivator code for transcription, *Trends Biochem. Sci.* 27 (4), 165–167, 2002.

[113]   Chen, L.F. and Greene, W.C., Shaping the nuclear action of NF-κB, *Nat. Rev. Mol. Cell. Biol.* 5 (5), 392–401, 2004.

6 NF-κB in the Innate Immune System

Matthew S. Hayden and Sankar Ghosh

CONTENTS

6.1   INTRODUCTION

As discussed in Chapter 1, NF-κB was discovered from studies on the immune system. These studies led to the appreciation of the multiple roles played by NF-κB during the initiation, maintenance, and resolution of the immune response. The mechanisms of NF-κB signaling have been discussed in detail in Chapters 3, 4, and 5; therefore, in this chapter and the next, we will focus on the biological role of NF-κB in the immune system. All of the mammalian NF-κB, IκB, and IKK family members have been knocked-out in mice and analysis of perturbations in the immune responses of these mice have underscored the important role of NF-κB in this crucial aspect of physiology.

The immune system can be broadly divided into two arms — innate and adaptive. The innate immune system includes barriers that prevent pathogen entry; germline encoded pattern recognition receptors (PRRs) that identify pathogens; soluble antimicrobial effectors; and various cellular components with antimicrobial

and proinflammatory activities. In addition, through the activation of antigen presenting cells and the liberation of cytokines and chemokines, the innate immune
system is responsible for initiating and shaping the adaptive immune response.
Regulation of NF-κB activity occurs during each of these steps in the innate immune
response. In particular, NF-κB is a key effector in the response generated by the
activation of PRRs, of which Toll-like receptors (TLRs) are best studied. The focus
of the current chapter is on the role of NF-κB in innate responses downstream of
PRR signaling, and the role of NF-κB in the development and maintenance of cells
that mediate these responses.

6.2   NF-κB IN BARRIER FUNCTION

Perhaps the most ancient and conserved aspect of host defense is the existence of
physical barriers to pathogen invasion. As with other components of host defense, the
barrier systems of mammals have evolved into complex and highly specialized tissues.
These epithelial barriers comprise the first and, arguably, the most important component of the innate immune system. In the absence of these barriers, or upon their
failure, the host can be rapidly overwhelmed by a variety of microbes to which it is
constantly exposed. Like the components of the hematopoietic system, cells of the
skin and mucosal epithelium undergo considerable turnover throughout the life of the
organism. NF-κB plays a crucial role in these replenishing cell populations; however,
as we shall see, the contribution of NF-κB to these processes is not always as expected.

The generation of IκBα deficient mice initially revealed the importance of NF-κB
in skin physiology [1]. These mice exhibit gross abnormalities in epidermal morphology including acanthosis (thickening of the skin) with hyperkeratosis, as well
as widespread inflammatory infiltration. While this inflammatory infiltrate might have
been predicted from what was previously known about NF-κB function, the epidermal phenotype was unexpected. Following these initial findings, subsequent studies
with additional genetic targeting of the NF-κB pathway have provided further support
for a specialized role for NF-κB in skin. Overexpression of p65 results in decreased
proliferation of the basal layer of the epidermis while inactivation of NF-κB using
DN-IκBα results in extensive hyperproliferation [2]. Notably, the hyperproliferation
observed upon suppression of NF-κB was most pronounced in layers of the epidermis
that are normally populated by differentiated keratinocytes. Thus NF-κB may facilitate terminal differentiation of keratinocytes during epidermal growth by regulating
the expression of some epidermal differentiation genes [3]. However, keratinocyte-
specific deletion of IKKβ or NFκ-B essential modifier (NEMO) demonstrates that
the observed defects are secondary to a failure to maintain inflammatory homeostasis
in the skin and sensitization to tumor necrosis factor (TNF)-induced apoptosis, rather
than an intrinsic requirement for NF-κB in keratinocyte development [4,5].

Interestingly, a quite distinct role has been proposed for IKKα in skin development. The $IKK\alpha^{-/-}$ mice die perinatally due to multiple morphological defects, particularly in epidermal and skeletal development [6,7,8]. These initial studies
indicated that IKKα had a marginal role in classical NF-κB signaling pathways, but
instead had some role in development, as knockout mice exhibit altered limb bud
morphology [6,7]. The epidermal layer of the skin in IKKα deficient mice fails to

differentiate properly, although this defect is independent of NF-κB activity. The *IKKα*$^{-/-}$ mouse can be rescued by knocking-in or reconstituting *in vitro* with IKKα in which both of the activation loop serines are mutated to alanines (IKKαAA) [9,10]. As an aside, it appears that IKKα in keratinocytes is also responsible for limb bud development, as selective expression of IKKα or kinase dead IKKα in basal keratinocytes rescues the morphogenetic defects of the *IKKα*$^{-/-}$ mice [11]. The mechanism of NF-κB-independent control of keratinocyte differentiation immediately downstream of IKKα is unknown. However, it is dependent on the presence of a previously unrecognized nuclear localization signal (NLS) in IKKα (aa232-240) [11], suggesting that, perhaps, this function is mediated by the proposed role of IKKα in histone modification (Chapter 5). Therefore, while NF-κB is critical to epidermal proliferation, differentiation, and function, IKKα has a distinct role in epidermal development that is independent of these pathways.

NF-κB has also been proposed to have an important role in the maintenance of other epithelial tissues. In particular, NF-κB is necessary for wound reconstitution using intestinal epithelial cells *in vitro* [12], while induction of NF-κB protects epithelial cells from radiation damage *in vivo* [13]. Despite contributing to inflammation, NF-κB is likely also important in repair of damaged lung epithelium [14].

6.3 DEVELOPMENT AND SURVIVAL OF INNATE IMMUNE CELLS

Many hematopoietic cells are subject to relatively high levels of turnover and consequently these populations are sensitive to changes in rates of apoptosis or proliferation. Consistent with its role as an antiapoptotic survival factor, NF-κB serves several key functions in the homeostasis of many hematopoetic lineages. There are two stages in the life of a leukocyte where cell survival decisions are particularly important — hematopoiesis and activation/maturation. Leukocyte development, like embryonic development, is marked by an abundance of both proliferation and apoptosis that shapes the resulting tissue/cell population. In addition, during activation and maturation into effector cells, many leukocytes undergo rapid proliferation that must also be resolved apoptosis, in this case a process termed activation induced cell death (AICD). In many regards our understanding of the contribution of NF-κB to the development and homeostasis of these cells has lagged relative to our knowledge of individual signaling events. However, as the pathways responsible for the development of natural killer (NK) cells, dendritic cells, macrophages, and granulocytes become better understood, it is likely that our appreciation for the role of NF-κB in their regulation will expand.

6.3.1 DEVELOPMENT OF INNATE IMMUNE CELLS

Hematopoietic components of the innate immune system include monocytes, macrophages, dendritic cells (myeloid and lymphoid), NK cells, granulocytes (basophils, eosinophils, and neutrophils), and mast cells (Figure 6.1). While many cells of the body contribute significantly to pathogen recognition, these marrow-derived cells mediate inflammation and activation of the adaptive immune system and are, there-

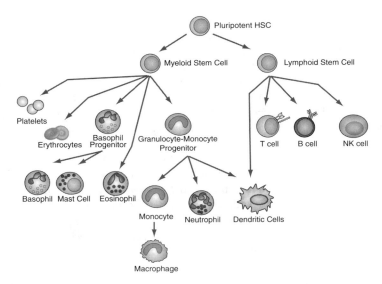

FIGURE 6.1 Basic hematopoietic schematic. Pluripotent hematopoietic stem cells (HSC) give rise to myeloid and lymphoid lineages. Myeloid stem cells differentiate to form erythrocytes, platelets, basophils, mast cells, eosinophils, monocytes, neutrophils, and myeloid dendritic cells. The lymphoid lineage gives rise to lymphoid dendritic cells, NK cells, and B and T cells whose development is discussed in detail in Chapter 7.

fore, the core constituents of the innate immune response. There is strong evidence that NF-κB plays a prosurvival role during the development of dendritic cells and NK cells and is also involved in granulopoiesis.

As mentioned above, IκBα knockouts display markedly abnormal nonlymphoid hematopoiesis. These mice are characterized by inflammatory infiltrates at multiple sites and exhibit robust granulocytosis [1]; however, the nature of the alteration in granulopoiesis has not been clearly defined. The most specific effect of IκB knockouts has been observed in NK cell development. Chimeras made with cells from *iκbα⁻/⁻iκbε⁻/⁻* mice show a moderate defect in both myelopoiesis and granulopoiesis [15]. More strikingly, cells from these mice have a pronounced alteration in NK development *in vivo*, consistent with the severe defect in development of lymphoid lineages [16]. Unexpectedly, the elevated levels of NF-κB activity in these cells is believed to exert a proapoptotic effect, a concept that will be discussed more extensively in the context of lymphopoiesis (Chapter 7).

Aspects of dendritic cell (DC) development are also dependent on NF-κB. Although knockouts of c-Rel, p50, and p65 do not have obvious effects, loss of RelB results in a striking defect in the development of DCs [17,18]. Dendritic cells may be of myeloid or lymphoid origin, with the latter being further divided into multiple subtypes, the most common distinction in mice being made on the basis of differential CD8α expression. RelB is specifically required in the development of CD8α⁻, but not CD8α⁺, DCs [19]. Conversely, classical p50/p65 complexes are required for the development of both CD8α⁺ and CD8α⁻ dendritic cells, despite normal development of other myeloid and lymphoid lineages under the same conditions [20]. As TNFα is thought to play an important role in DC development, it is not surprising that p65-/p50-deficient cells are susceptible to apoptosis induced by this stimulus [21]. Thus, it appears that DC development broadly requires classical NF-κB complexes, while CD8α⁻ DC development is selectively dependent on the alternative RelB-containing NF-κB species.

6.3.2 SURVIVAL OF CELLS OF THE INNATE IMMUNE SYSTEM

Hematopoietic cells of the innate immune system have a finite lifespan in the periphery. NF-κB plays a crucial role in protection of cells from apoptosis, and this role is particularly crucial during exposure to proinflammatory stimuli. Therefore, it is predicted that the cells of the innate immune system, which must function in the presence of these same stimuli, will require NF-κB to survive. Indeed, a large body of evidence, both genetic and otherwise, validates the hypothesis that NF-κB protects innate immune cells from the induction of apoptosis by proinflammatory cytokines. However, during the resolution phase of the inflammatory response, it is beneficial to rid the host of activated cells that would otherwise needlessly prolong the inflammatory process. As discussed in detail in Chapter 8, it has recently become apparent that NF-κB is also critical in mediating this process. Independent of inflammatory processes, however, ongoing maintenance of leukocyte homeostasis requires careful orchestration of cell survival and death.

Survival of DCs, which exhibit variable turnover rates depending on subtype, tends to be quite brief following activation. However, once stimulated, DC survival can be prolonged through CD40 signaling induced by interactions with T cells. DCs deficient in both p50 and c-Rel have decreased survival despite CD40 stimulation [20]. The importance of CD40-mediated NF-κB activation in the survival of activated DCs has been confirmed in human cells using expression of inhibitory IκBα [22]. Therefore, NF-κB functions to prolong the survival of DCs that are destined to undergo AICD. Whether NF-κB acts as a prosurvival factor in DC homeostasis, however, is not known.

The role of neutrophils in the inflammatory response is discussed in detail in Chapter 8. Nevertheless, it is relevant to mention here that NF-κB functions in regulating the survival of circulating neutrophils. Under normal circumstances, neutrophils undergo daily turnover and have been shown to rapidly apoptose *in vitro*. Despite their limited lifespan, there is evidence that NF-κB does enhance the survival of mature neutrophils. Inhibition of NF-κB in neutrophils results in accelerated apoptosis as well as sensitization to pro-apoptotic stimuli, including TNFα. It is interesting to note that neutrophils appear to lack p52 and RelB [23], which are crucial in the maintenance of long-lived lymphocytes. However, neutrophils are capable of activating NF-κB in response to many proinflammatory stimuli [24], and protection from apoptosis is likely to be important during the inflammatory response. Indeed, many TLR ligands increase neutrophil survival *in vitro,* and based on pharmacological inhibition, this effect is likely due to NF-κB mediated expression of antiapoptotic genes [25]. Thus, NF-κB promotes survival, differentiation, and proliferation during the development and homeostasis of leukocytes and other cellular components of the innate immune system.

6.4 NF-κB AND PRRs

In order to activate an appropriate immune response, the host must first recognize the presence of pathogen. This discrimination between self and nonself is an absolute requirement for the initiation of effector functions, such as the secretion of cytokines and antimicrobial peptides, carried out by the cells of the innate immune system. Consequently, pathogen recognition by germline encoded receptors of the innate immune system is an area of intense investigation in immunology. In addition to opening up a large field of inquiry with therapeutic potential, the discovery of Toll-like receptors (TLRs) and their ability to trigger NF-κB has highlighted the importance of NF-κB in innate immunity.

TLRs are evolutionarily conserved pattern recognition receptors that recognize unique, essential molecules characteristic of various classes of microbes. The function of TLRs as arbitrators of self/nonself discrimination highlights their central role in innate immunity as well as in the initiation of the adaptive immune response. The eleven characterized mammalian TLRs have varied tissue distribution and serve as recognition receptors for pathogen associated molecular patterns (PAMPs) present on bacteria, viruses, fungi, and parasites. Well-characterized PAMPs include LPS, dsRNA, nonmethylated CpG DNA, and peptidoglycan (Figure 6.2). Perhaps due to the multimeric nature of the TLR extracellular domain, which consists of multiple leucine rich repeats (LRRs), several receptors are capable of recognizing more than one microbial molecule. Heterodimerization of some TLRs and the use of coreceptors (e.g., CD14 and myeloid differentiation protein-2 [MD-2]) further expand the

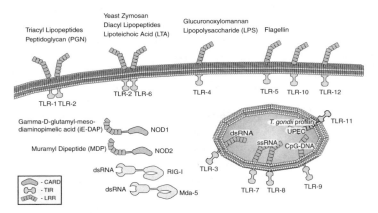

FIGURE 6.2 Pattern recognition receptors and their ligands. Various pattern recognition receptors signal to NF-κB. The most prominent among these are the Toll-like receptor (TLR) family (TLR1-13 in mouse), the nucleotide oligomerization domain (NOD)-LRR family, and RIG-I family. These receptors recognize pathogen associated molecular patterns through leucine-rich repeats (LRR; TLR; and NOD-LRR families) or RNA helicase domains (RIG-I family). Signaling to NF-κB, and other inducible transcription factors, is mediated by either cytoplasmic Toll/IL-1 receptor (TIR) domains or caspase activation and recruitment domains (CARDs), via homotypic interactions with downstream signaling molecules. Localization of TLRs to either the plasma membrane or endosomes is indicated, although in some cases localization is poorly established or cell type specific.

repertoire of PAMPs recognized. Significant progress has been made over the past few years in deciphering the relevant signaling pathways that operate downstream of TLRs (Chapter 3 and Section 6.4.1). The intracellular Toll/IL-1 receptor (TIR) domain of IL-1R and TLRs mediates interaction with downstream signaling adaptors that, in turn, lead to activation of three key families of transcription factors — NF-κB, activator protein-1 (AP-1), and IRF (interferon regulatory factor).

As stated above, pathogens recognized by TLRs can be categorized as bacterial, viral, or eukaryotic. In each of these categories, PAMPs have been described that more or less fit with existing hypotheses of how pathogen recognition by the innate immune system would occur [26]. The bacterial cell wall is a prominent source of PAMPs recognized by both TLRs and other PRRs. TLR2 mediates recognition of lipopeptides of mycoplasma and Gram-positive bacterial species, as well as peptidoglycan. NOD-LRRs are a non-Toll class of PRRs that also signal in response to bacterial cell wall components. In addition to cell wall components, other conserved structural components of bacteria are recognized by TLRs. TLR5 signals in response to bacterial flagellin, and binds a highly conserved, functionally important domain of the flagellin subunit [27]. TLR11 recognizes a protein component of uropathogenic bacteria [28]. Toll receptors are also capable of recognizing certain nucleic acids, for example, recognition of nonmethylated CpG DNA present in bacteria by TLR9 [29].

Nucleic acids are also key viral PAMPs, and are recognized by TLRs 3, 7, 8, and 9, as well as by cytoplasmic receptors of the RIG-I family discussed below. TLR3 recognizes dsRNA, a common viral replicative intermediate [30]. TLRs 7 and 8 signal in response to ssRNA [31–33] and, as mentioned, TLR9 recognizes nonmethylated cytosine-phosphate-guanosine (CpG) motifs. Recognition of cytoplasmic dsDNA leading to IRF, and perhaps NF-κB activation has also been reported, although the relevant receptor has not yet been identified [34,35].

MyD88 deficient cells have been used to demonstrate that many fungal species are capable of activating TLR pathways, although the receptors have not been identified in many cases. TLR4 has been shown to recognize *Cryptococcus neoformans* polysaccharides. TLR2 and TLR6 are required for recognition of yeast zymosan, while TLR4 is thought to recognize certain yeast mannans (for review see [36]). Parasite PAMPs have been more elusive, and, indeed, their existence has been somewhat controversial. However, TLR2 heterodimers reportedly recognize various parasite glycosylphosphatidylinositol (GPI)-anchored proteins, and TLR knockout mice have had variable defects in their ability to defend against various parasites (for review see [37]). Recently TLR9 has been reported to recognize the malarial pigment hemozoin, a byproduct of heme metabolism in infected erythrocytes [38] while TLR11 recognizes a profilin-like protein that is conserved in apicomplexan parasites including *Toxoplasma gondii* [39].

Thus members of the Toll-like receptor family are capable of recognizing a broad array of microbial markers. The diversity in PAMPs that act as TLR agonists is substantial, and highlights the dexterity of the innate immune system. As we shall see below, the ability of TLRs to distinguish pathogen types is translated into appropriate innate and adaptive responses through the selective activation of NF-κB and other inducible transcription factors.

6.4.1 TLR Signaling to NF-κB

Ligand binding to TLRs is just now beginning to be understood at the molecular level. Extracellular leucine rich repeats "LRRs" bind to ligand and, either through receptor oligomerization and/or induction of conformational change across the plasma membrane, induce the recruitment/activation of adaptor proteins through the intracellular TIR domain. These adaptors lead to the activation of classical IKKβ-dependent complexes; degradation of IκBα and IκBβ; and liberation of, primarily, p65 and c-Rel containing complexes. As discussed in Chapter 3, TLR signaling is divided into two pathways: those that are MyD88 dependent and those that are MyD88-independent/TRIF-dependent (Figure 6.3). We will base our discussion

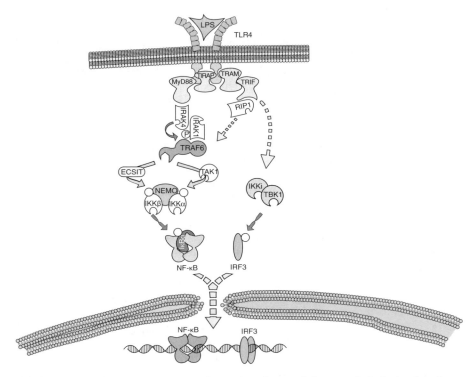

FIGURE 6.3 MyD88 and Toll-IL-1 resistant domain containing protein inducing interferon beta (TRIF)-dependent signaling to NF-κB. TLR4 utilizes both MyD88-dependent and TRIF-dependent pathways to signal to NF-κB. MyD88 is recruited to the TLR4 TIR domain by TIR domain-containing adaptor protein (TIRAP). MyD88, in turn, recruits and activates members of the IRAK family of kinases that interact with TRAF6. TRAF6 mediates IKK activation in a manner that requires TAK1 and ECSIT. TRIF is recruited to TLR4 by the TIR-containing adaptor protein TRAM. TRIF is thought to directly activate the atypical IKKs, IKKi, and TBK1, leading to IRF3 phosphorylation, dimerization, nuclear localization, and transactivation of target genes. Stimulation of NF-κB downstream of TRIF depends on RIP1 and TRAF6 and is likely mediated by TAK1, evolutionary conserved signaling intermediate in TIR pathways (ECSIT), and IKKβ-dependent phosphorylation of IκB proteins.

primarily on signaling events emanating from TLR4, which despite having the most complex downstream pathway is the most thoroughly studied TLR. Clear differences exist in signaling from other TLRs as noted throughout our discussion, and it is likely that further specializations will become apparent as individual TLR signaling pathways are investigated more thoroughly.

6.4.1.1 MyD88-Dependent Signaling

TLR4 signaling is relatively unique among TLRs in that the effector adaptors are one step removed from the receptor. For example, MyD88 recruitment to the receptor complex depends upon the TIR-domain containing adaptor protein Mal, also known as TIRAP [40–42]. TLR2 also requires Mal to bridge MyD88 to the receptor; however it is believed that other MyD88-utilizing TLRs directly recruit MyD88. Furthermore, the origin of the requirement for Mal in TLR4 and TLR2 signaling remains unclear. The N-terminal domain of MyD88 contains a death domain (DD) that recruits the DD-containing serine/threonine kinase interleukin-1 receptor-associated kinase (IRAK)-4. IRAK-4 and IRAK-1 form an active complex capable of recruiting TRAF6 (Chapter 3). The link between TRAF6 and the IKK complex remains somewhat enigmatic, although a few key players are known. As discussed in Chapter 3, transforming growth factor-beta-activated kinase 1 (TAK1) is required for NF-κB, as well as AP-1 and ERK, activation downstream of MyD88 [43,44]. While it is widely accepted that ubiquitination is a key switch at this crucial step of NF-κB activation, there remains considerable work to be done at the molecular level to understand how ubiquitination leads to activation (see Chapter 4).

In addition to TAK1, another protein, termed ECSIT (evolutionarily conserved signaling intermediate in toll pathways), has been reported to act as a bridge between TRAF6 and downstream signaling components. ECSIT binds to TRAF6 and is required for TLR and IL-1 signaling, but not TNF-signaling [45,46]. Although initial studies suggested ECSIT might function by recruiting and activating the kinase MEKK1 [45,46], knockouts of MEKK1 do not display an overt phenotype with respect to TLR or TNF signaling [47,48]. As discussed in Chapter 3, MEKK3 deficient cells do not transcribe IL-6 following TLR4 or IL-1R stimulation and exhibit delayed and weak NF-κB DNA binding following lipopolysaccharide (LPS) stimulation [49]. While TRAF6 and MEKK3 inducibly associate with each other in TLR4 signaling, the mechanisms that regulate this process are not known [49]. Therefore, it is possible, although not proven, that ECSIT exerts its role in TLR signaling by modulating the function of MEKK3.

The role of ECSIT is not limited to NF-κB signaling pathways. ECSIT also acts as a coactivator for SMAD transcription factors following Bmp/TGFβ stimulation [46]. Mice lacking ECSIT exhibit early embryonic lethality, preventing signaling studies; however, the *ecsit*[-/-] phenotype mimics that of Bmp receptor-1, a member of the TGFβ-receptor family that functions in early development. This surprising finding raises the possibility that ECSIT might regulate both TLR and TGF/Bmp pathways, and hence may explain the cross-repression that occurs between these pathways [50]. The TAK/TAB proteins were also initially identified and characterized as intermediates in TGFβ signaling; thus, it is possible that this link between adaptors in TLR and TGF signaling pathways may be more extensive than currently imagined. The link between inflammation and fibrosis is crucial to the pathophysiology of many diseases and communication between TGFβ and NF-κB pathways will therefore continue to be of interest.

6.4.1.2 TRIF-Dependent Signaling

As discussed in Chapter 3, some TLR signaling occurs independent of MyD88. LPS stimulation in *MyD88*$^{-/-}$ cells results in NF-κB activation with slower kinetics than WT cells, and leads to expression of only a subset of target genes [51]. When cells are stimulated through TLR3 and TLR4, TRIF (TICAM-1), a TIR-domain containing adaptor, mediates activation of NF-κB in the absence of MyD88 [52]. Indeed, in the case of TLR3, all downstream signaling is likely TRIF-dependent. Studies using cells from TRIF-deficient mice demonstrate that TRIF is required for late phase NF-κB responses and IRF3 responses to LPS, but not for jun N-terminal kinase (JNK) activation [53]. Reconstitution of *trif*$^{-/-}$ cells with a mutant TRIF lacking the TRAF-binding domain restores induction of IFNβ responsive genes, via activation of IRF3, but not NF-κB activation, indicating that TRIF is the point of divergence in signaling to NF-κB and IRF3. As discussed in Chapter 3, the mechanism by which TRIF activates IKK is quite unclear. Increasingly, however, it appears that these events share a common cast of characters with other NFκB pathways.

Recently it has been reported that TRIF binds receptor interacting protein (RIP)1 and RIP3, and that *rip1*$^{-/-}$ mouse embryo fibroblasts (MEFs) have decreased NF-κB activation following TLR3-poly(I:C) signaling [54]. Indeed, mutation of the RIP homotypic interaction motif (RHIM) abrogates signaling to NF-κB. Finally, another TIR-domain containing adaptor TRAM (TRIF-related adaptor molecule) functions upstream of TRIF in MyD88-independent signaling from TLR4. TRAM is required for IRF-3 activation and for the delayed phase of NF-κB activation following TLR4 engagement. TLR4-induced IRAK activation by MyD88, however, is unaffected by the absence of TRAM and TRAM does not function in TLR3 TRIF-dependent signaling pathways [55,56]. Therefore, it appears TRAM is only needed for TRIF signaling downstream of TLR4. Adding further complexity, two recent reports suggest that TLR4 TRIF-dependent NF-κB activation in MEFs is largely due to IRF3-induced expression of TNFα [57,58]. These results may be explained by difference in the recruitment of TRIF to the receptor, that is, by TRAM in the case of TLR4 vs. directly to TLR3, resulting in changes in the availability of TRIF's TRAF binding site and, consequently signaling to IKK.

Two divergent members of the IKK family, IKKi (IKKε) and TBK1 (T2K) act downstream of TRIF in signaling to IRF-3 [56,59], although neither is required for NF-κB activation by LPS or TNFα [60, 61, 62]. Instead these IKKs likely have cell-type specific roles in IRF activation downstream of TRIF. IKKi deficient cells have normal induction of IRF3 following stimulation with LPS (hence TLR4) while TBK1$^{-/-}$ cells do not [60]. IKKi expression is regulated by NF-κB, and it appears that IKKi is constitutively active once expressed [63]. IKKi facilitates CCAAA/enhancer-binding proteinδ (C/EBPδ) pathways that contribute to the expression of a subset of genes induced by IL-1, LPS, TNF, and phorbol myristate acetate (PMA) [64]. Therefore, TBK1 usually mediates TLR signaling to IRF-3 through direct phosphorylation of IRF-3. IKKi may have a more pronounced role in certain cell types or in propagating the response. In particular, because IKKi is a downstream target of NF-κB, this protein may provide positive feedback during interferon responses.

6.4.1.3 Negative Regulation of TLR Signaling

As we will see in Chapter 8, inflammatory responses are built upon waves of cytokine production and positive feedback mechanisms. As a result, tight control must be placed on the initiation and maintenance of these responses. The Toll-signaling pathway can also be negatively regulated by proteins that are induced or activated upon TLR signaling and therefore may help to limit signaling from these receptors. To date, the targets of many of these inhibitors are members of the IRAK family of proteins. The IRAK family member, IRAK-M (IRAK3) prevents the release of IRAK-1/4 from the TLR/MyD88 signaling complex and, therefore, inhibits activation of TRAF6; *irakm⁻/⁻* knockouts exhibit enhanced signaling to NF-κB [65]. Tollip, an adaptor protein constitutively associated with IRAK, is phosphorylated and dissociates following IRAK4 activation [66,67]. It has been suggested that high levels of Tollip in the intestinal epithelium help to prevent inflammation in response to commensal bacteria [68]. Tollip deficient cells demonstrate moderate defects in the production of NF-κB regulated cytokines [69]. SIGIRR (TIR8), a member of the IL-1R family binds to other Toll/IL-1 receptors and interacts with IRAK and TRAF6 [70,71]. In the absence of SIGIRR, Toll/IL-1 stimulation induces prolonged activation of NF-κB. SIGIRR may also function by inhibiting the association of IRAK with TLRs. Interestingly, like Tollip, SIGIRR is abundantly expressed in epithelial cells, suggesting that it too may suppress signaling at sites of constitutive microbial exposure. Finally, suppressor of cytokine signaling-1 (SOCS-1) has been reported to negatively regulate LPS signaling to NF-κB and SOCS-1 knockout mice exhibit an inflammatory phenotype [72,73]. Recent data suggest SOCS-1 may directly target Mal, thus downregulating TLR4 signaling through Mal/MyD88 [74].

In addition to these TLR specific regulators of signaling to NF-κB, other proteins function to control the extent and duration of NF-κB activation (Chapter 10). These factors both set thresholds for activation and help to prevent uncontrolled, and potentially deleterious, innate immune responses. The broad array of PAMPs recognized by the TLR system affords the host the ability to mount responses against many pathogens. Nevertheless, for some pathogens, TLRs alone are not sufficient, and some physical spaces, most notably the cytosol, are not effectively monitored by Toll-like receptors.

6.4.2 NF-κB and Caspase Activation and Recruitment Doman (CARD)-Carrying PRRs

PRRs that recognize bacterial PAMPs are expressed at the plasma membrane or with LRRs projecting into the lumen of vesicles that are topologically related to the extracellular space. However, in such a system, intracellular pathogens are uniquely protected from detection. Furthermore, viral infection and the resulting induction of interferon occur in many cells that do not express the full panoply of antiviral TLRs — suggesting that other PRRs must be at work. In fact, cells do have at their disposal families of cytoplasmic PRRs that, like TLRs, are capable of activating NF-κB and other transcriptional mediators of the innate immune response. Interestingly, many of the PRRs that have been discovered contain caspase activation and recruitment domains (CARDs) that are necessary for the activation of NF-κB following ligand binding. Here we provide a brief description of two classes of cytoplasmic PRRs — CARD-containing members of the CATEPILLAR and DExD/H-box helicase families.

6.4.2.1 CATEPILLER-NODs

Nucleotide oligomerization domain (NOD) proteins 1 and 2 are part of a large family termed the CATEPILLER family — CARD, transcription enhancer, R (purine)-binding, pyrin, lots of leucine repeats (Figure 6.2). Within this family exists the NOD-LRR subfamily, which is typified by the presence of leucine-rich repeats and nucleotide oligomerization domains. NOD1, NOD2, and IPAF have CARDs and can signal to NF-κB (for review see [75]). NOD proteins have been of particular interest due to the existence of mutations in the human population associated with inflammatory bowel disease [76]. NOD1 recognizes peptidoglycan containing *meso*-diaminopimelic acid (*meso*-DAP), with the γ-D-glutamyl-meso-DAP (iE-DAP) dipeptide being the minimum required portion of peptidoglycan. NOD1 induces signaling to NF-κB through a classical pathway that includes activation of IKKβ. NOD2 recognizes muramyl dipeptide (MDP), a component of peptidoglycan in nearly all bacterial cell walls. Relatively few signaling intermediates downstream of NOD-LRRs are known, however, there is growing evidence that these few components may be sufficient to activate NF-κB. The CARD-containing kinase RIP2 (RICK) is required for NF-κB activation, while the PYD-containing adaptor ASC mediates signaling to the caspase pathway. Intriguingly, the ATP binding cassette of both NOD1 and NOD2 is needed for signaling [77]. RIP2 is able to bind to NEMO and therefore is thought to directly mediate activation of the IKK complex [78]. In this model IKK activation results from induced proximity, mediated by the ligand-dependent oligomerization of NOD-LRRs, which is dependent on the ATP binding cassette (see Chapter 3 for a discussion of the induced proximity model of IKK activation).

6.4.2.2 RIG-I and MDA5

Two members of the DExD/H-box RNA helicase family are unique in that they have N-terminal CARDs. Retinoic acid inducible gene I (RIG-I) and melanoma differentiation-associated gene 5 (MDA5) are RNA helicase-containing cytoplasmic proteins. The RNA helicase domains of RIG-I and MDA5 bind directly to dsRNA and induce production of type I interferons [79,80,81]. Upon binding to dsRNA, representing either the viral genome or a replication intermediate, in the cytoplasm of virally infected cells, RIG-I and MDA5 induce the phosphorylation of IRF3 and activation of NF-κB. Interestingly, initiation of these signaling cascades is abrogated by point mutations in the Walker type ATP binding site, suggesting that ATPase activity is required [79]. The link between these two proteins and NF-κB/IRF3

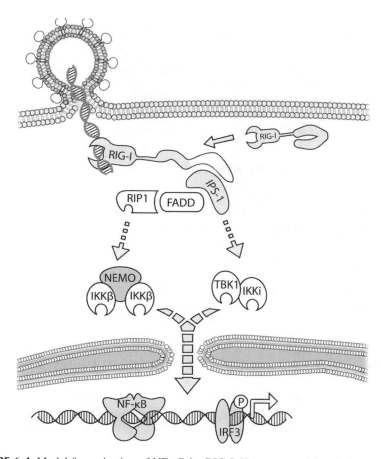

FIGURE 6.4 Model for activation of NF-κB by RIG-I. Upon entry of the viral genome into the cytoplasm dsRNA is recognized by the RNA helicase domain of RIG-I. It is thought that a conformational change following binding may allow homotypic interaction of the RIG-1 CARDs with the CARD-containing adaptor IPS-1 (MAVS, CARDIF, VISA). IPS-1 mediates signaling to NF-κB through RIP1 and IKK, and to IRF3 through TBK1 and IKKi. IPS-1 has also been proposed to require mitochondrial targeting in order to function in this signaling pathway, although the significance of this is not clear.

remains somewhat unclear (Figure 6.4), although, the overexpression of the N-terminal CARD domain alone is sufficient to induce signaling. Recently a CARD-containing protein, variably named CARDIF, IPS1, MAVS, and VISA, has been implicated downstream of RIG-I; however, the link between this protein and IKKβ leading to NF-κB, and TBK1/IKKi to IRF3, is unclear [82,83,84,85]. It appears, however, that there are similarities to TRIF mediated signaling, in that RIG-I induction of IRF3 depends on TBK1/IKKi, while NF-κB activation requires FADD and RIP-1 [86,87]. Currently the relative contributions of RIG-I and MDA5 *in vivo* are not clear. However, the lack of defects in responses of *TLR3-/-* mice is suggestive that these cytoplasmic receptors may have a crucial role. Indeed, the result of RIG-I deletion in mice suggests that these proteins may be crucial in type-I interferon responses to dsRNA viruses in many cell types while TLR recognition plays a more specialized role in plasmacytoid DCs [88]. Notably, RIG-I and MDA5 are, themselves, interferon inducible, and thus susceptible to positive feedback regulation, as well as upregulation in the context of type-I interferon production by cells of the innate immune system. A third member of the RIG-I DExD/H-box helicase family, LGP2, has been identified that lacks the N-terminal CARD, and may function as a negative regulator of RIG-I and MDA5 by competing for binding to dsRNA [79,87].

6.5   NF-κB AND THE INNATE RESPONSE TO PATHOGENS

The recognition of pathogens by PRRs initiates a complex series of events. The first is the mounting of immediate antimicrobial responses at the cellular level. This is an effective and evolutionarily conserved function of TLRs, and one in which NF-κB has an important role. PRR expression in epithelial cells is especially important in this early response, although it is believed that the variety of PRR expression in these cell types is somewhat limited. At the mRNA level, multiple TLRs are differentially expressed in epidermis, gut, pulmonary, urinary, and reproductive epithelium. However, in many cases it is thought that both TLR expression and responsiveness is tightly controlled in these cells. For example, keratinocytes upregulate TLRs expression and responsiveness following TGFα exposure [89]; renal epithelial cells increase expression of TLR2 and TLR4 in response to IFNγ or TNFα [90]; and intestinal epithelial cells have been shown to alter TLR expression under inflammatory conditions. However, sentinel cells of the innate immune system, particularly tissue resident dendritic cells and macrophages, express a more complete complement of PRRs, and thus are likely to bear the largest portion of the burden in the earliest events of pathogen recognition. The second part of the innate immune response, which may occur concomitantly with the first, is the elaboration of proinflammatory cytokines and chemokines. Aspects of this response, as they pertain to inflammation, are discussed in detail in Chapter 8.

The classical NF-κB pathway is crucial to the characteristic response that immediately follows TLR ligation. $p65^{-/-}/tnfr1^{-/-}$ mice have increased susceptibility to bacterial infection, highlighting the role of p65 in innate immune responses and the initiation of innate immune responses by nonhematopoietic cells [91]. In contrast to p65-deficient mice, p50-deficient mice do not show any developmental defects, however, B cells from $p50^{-/-}$ mice do not respond efficiently to LPS, emphasizing the importance of the classical p50 containing heterodimeric complexes, i.e., p65/p50 and c-Rel/p50, in TLR signaling [92]. TNFR/IKKβ double knockouts show a more pronounced defect in innate immune responses to bacterial infection, only surviving a few days after birth while the p65/TNFR1 knockouts can survive for many months. Therefore, while NF-κB family members may partially compensate for the loss of p65 in some aspects of the TLR transcriptional response, there is no compensation for the loss of IKKβ [93,94]. Likewise, MEFs from $nemo^{-/-}$ mice do not exhibit NF-κB activation by LPS or IL-1 [95]. Therefore, TLR-mediated activation of

NF-κB responsive genes progresses through the canonical pathway and exhibits the expected dependence on IKKβ and NEMO, but not IKKα. The genes regulated by this pathway contribute to each step of the innate response.

The liberation of products with direct antimicrobial activity occurs early at sites of pathogen entry. Toll ligation in Drosophila, as well as mammals, is at least partly responsible for the NF-κB-dependent expression of defensins — cationic peptides that exert direct bactericidal activity by inducing membrane permeabilization. Small intestine Paneth cells, for example, release large amounts of α-defensins into the intestinal lumen following exposure to a variety of bacteria/bacterial products [96]. The production of antimicrobial nitrogen and oxygen species, which are acutely toxic to a variety of microbes, augments the activity of antimicrobial peptides. Production of NOS is mediated in part by inducible nitric oxide synthase (iNOS), which is partially regulated by NF-κB. Consequently, NOS production results from TLR or NOD-LRR ligation by PAMPs.

The second event controlled by the innate immune response is the production of cytokines, which recruit and activate effector cells and induce tissue changes characteristic of inflammation (Chapter 8). A subset of cytokines also has more immediate antimicrobial functions. For example, recognition of viral dsRNA by TLR3 or RIG-I/MDA-5 induces the production of IFN-β through the concerted activation of NF-κB and IRF3. In turn, IFN-β induces host cell changes that are unfavorable to viral replication. RIG-I and MDA5 are specialized for the production of type-I interferons, and due to their wider tissue distribution may bear a greater share of this burden during *in vivo* antiviral responses than do the TLRs.

Plasmacytoid DCs are particularly adept at generating a large rapid burst of interferon following viral infection. The relatively selective expression of TLR7 and TLR9 in pDCs is consistent with a key role in the antiviral interferon response. However, pDCs lack TLR3, and therefore do not mount TLR-mediated responses to dsRNA. Nevertheless, both TLR3 and TLR9 are required for full protection against MCMV infection [97], similar to what is observed in mice with an inactivating TRIF mutation [98]. TLR9 signals only through MyD88, suggesting that this TLR, like TLRs 7 and 8, may induce IFNγ most likely through the production of IL-12 [99]. If TLR3 only signals through TRIF in response to MCMV infection, the more severe phenotype of the TRIF knockout mouse suggests that other PRRs that signal through TRIF are crucial. Therefore, RIG-I family members primarily mediate the interferon burst, while TLRs contribute indirectly through MyD88-dependent IL-12 production.

One effect of type-I interferons is the activation of NK cells. NK cells are responsible for killing pathogen-infected cells and are crucial in killing cells undergoing neoplastic changes. Evidence suggests two distinct roles for NF-κB in NK cell function. First, IL-2-responsive induction of perforin in NK cells is dependent on NF-κB activation [100]. However, this finding awaits confirmation in knockout models. Second, NK cells from *relb*-/- mice have defective cytolytic responses that are secondary to defective IFNγ production by NK cells [101]. It is not clear how RelB mediates IFNγ production in these cells. Finally, NK cells also express PRRs and respond directly to PAMPs.

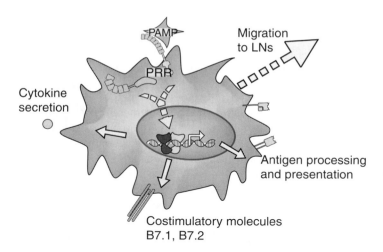

FIGURE 6.5 NF-κB facilitates initiation of the adaptive immune response. Binding of pathogen associated molecular patterns (PAMPs) to pattern recognition receptors (PRRs) leads to antigen presenting cell (APC) maturation. Induction of NF-κB dependent genes induces migration to the draining secondary lymphoid tissues where APCs will initiate and shape the adaptive immune response (Chapter 7). NF-κB-dependent upregulation of costimulatory molecules and cytokines is also required for the activation of naive lymphocytes. Finally, APCs also upregulate the expression of the genes responsible for antigen processing and presentation on MHC.

6.6 BRIDGING INNATE AND ADAPTIVE REPONSES

As discussed, a critical function of the innate immune system in the response to many pathogens is to alert the adaptive immune system to their presence. Thus, in addition to recognizing the pathogen, the innate system must convey this information to T and B lymphocytes. Activation of APCs following TLR ligation is chiefly responsible for this aspect of the response.

DC maturation mediated by pathogen recognition is crucial for the initiation of the adaptive immune response. To activate naïve T cells, DCs must undergo multiple changes (Figure 6.5). First, DCs must gain the ability to interact with T cells by changing their chemokine receptor expression and migrating into lymphoid tissues. Second, DCs must alter their antigen processing machinery to favor the presentation of pathogen epitopes on MHC. Third, APCs must upregulate the expression of costimulatory molecules B7.1 and B7.2, which ligate CD28 and provide the second signal necessary to induce T cell activation. Finally, as progress is made in exploring these events it is becoming increasingly clear that the responses to different pathogens are tailored-based on the distribution of PRRs in different cell types and the ability of different cell types to, in turn, interact with T cells in a biasing manner [102].

Maturation of DCs following viral infection depends on nucleic acid-binding PRRs. While it was initially thought the TLRs were responsible for this response, there is increasing evidence that RIG-I and MDA-5 are equally important. Indeed, DC maturation during viral infection occurs normally in the absence of MyD88 or TLR3 [103]. Bacterial responses are either mediated through other TLRs — DCs express TLRs 1, 2, 5, and 6 — or other classes of PRRs. Murine CD8α$^+$ DCs, which tend to induce T_H1 responses important in clearance of viral and parasitic infections, express TLR1, 2, 6, 9, and 11. In the absence of TLR11, for example, mice fail to mount a T_H1 response against *T. gondii*, due to the failure of CD8α$^+$ DCs to recognize the TLR11 ligand [39]. It remains unclear whether the ability of distinct TLR ligands to induce T_H1 vs. T_H2 responses is intrinsic to the specific TLR/ligand, dose of ligand, or the cell type within which this activation occurs. Evidence to date points toward the latter.

Stimulation with the NOD2 ligand MDP has been shown to induce dendritic cell maturation, although somewhat weakly. NOD1 is highly expressed in multiple tissues, while the expression of NOD2 is more limited. Thus, initial studies may have underestimated the contribution of this PRR, by examining MDP stimulation in the context of cells expressing little to no NOD2. There is evidence that TLR ligation may regulate NOD responsiveness; however, the subject of crosstalk between NOD-LRR and TLR families has been mired in rather confusing results, preventing a definitive consensus. While some results suggest negative regulation of TLRs by NOD1/2, in a manner somewhat analogous to LPS tolerance, other studies have found that NOD2 stimulation potentiates TLR responses. Furthermore, it has been shown that NOD2 is induced in an NF-κB dependent manner in response to TLR ligands or proinflammatory cytokines [104]. Furthermore, given that MDP is found as a contaminant in LPS, NOD2 responses may also be enhanced through TLR-dependent uptake of MDP. Conversely, however, it should be noted that the same authors have found that NOD2 upregulates the expression of MyD88. *In vivo* there has been little difference observed in responses in TLR ligands in mice lacking NOD2. The slight increase in resistance to LPS-induced shock in one report may be related to the expression of a mutant NOD2 instead of a true knockout [105]. However, NOD2 knockouts do show increased susceptibility to bacterial infection when challenged through the oral route [106]. In any case, these remain vital issues to clarify given the role of NOD2 in Crohn's disease, a relatively common (~10:100,000) and severely debilitating disease.

Finally, there is the question of the role of innate recognition of pathogens by lymphocytes themselves. Both B and T cells express subsets of TLRs, although exactly what complement of TLRs is expressed in these cells types is debatable. It has recently been shown that T-dependent B cell responses depend upon TLR ligands acting directly on B cells (Chapter 7). At the level of mRNA expression, human peripheral B cells express TLR1, 2, 4, 6, and 9 [107], thus leaving open the question of how B cell responses to pathogens lacking appropriate PAMPs might occur. Interestingly, it has been noted that T_H2 mediated IgE responses were not dependent on TLR ligation (MyD88).

6.6 CONCLUSIONS

The role of NF-κB in innate immunity begins prior to pathogen exposure and extends through resolution of the response. As we have seen, NF-κB performs vital functions in the many tissues that form the innate immune system. Indeed, even epithelia, which are often misconstrued as passive barriers, rely on NF-κB in their responses to physiological and nonphysiological stress. Likewise, the development and peripheral homeostasis of leukocytes is critically linked to NF-κB as a regulator of apoptosis. Research has tended to focus on NF-κB signaling in the effector functions of these hematopoietic components of the immune system and its role has thus been well established in innate responses to all classes of pathogen. In addition, there is now increasing evidence to support a broader model in which NF-κB also supports the resolution of inflammatory responses and, in situations that require it, initiation and coordination of the adaptive immune response.

REFERENCES

[1]   Beg, A.A., Sha, W.C., Bronson, R.T. et al., Constitutive NF-kappaB activation, enhanced granulopoiesis, and neonatal lethality in I kappaB alpha-deficient mice, *Genes Dev.,* 9, 2736, 1995.

[2]   Seitz, C.S., Lin, Q., Deng, H. et al., Alterations in NF-kappaB function in transgenic epithelial tissue demonstrate a growth inhibitory role for NF-kappaB, *Proc. Natl. Acad. Sci. USA,* 95, 2307, 1998.

[3]   Banno, T., Gazel, A., and Blumenberg, M., Pathway-specific profiling identifies the NF-kappaB-dependent tumor necrosis factor alpha-regulated genes in epidermal keratinocytes, *J. Biol. Chem.,* 280, 18973, 2005.

[4]   Pasparakis, M., Courtois, G., Hafner, M. et al., TNF-mediated inflammatory skin disease in mice with epidermis-specific deletion of IKK2, *Nature,* 417, 861, 2002.

[5]   Nenci, A., Huth, M., Funteh, A. et al., Skin lesion development in a mouse model of incontinentia pigmenti is triggered by NEMO deficiency in epidermal keratinocytes and requires TNF signaling, *Hum. Mol. Genet.,* 15, 531, 2006.

[6]   Li, Q., Lu, Q., Hwang, J.Y. et al., IKK1-deficient mice exhibit abnormal development of skin and skeleton, *Genes Dev.,* 13, 1322, 1999.

[7]   Takeda, K., Takeuchi, O., Tsujimura, T. et al., Limb and skin abnormalities in mice lacking IKKα, *Science,* 284, 313, 1999.

[8]   Hu, Y., Baud, V., Delhase, M. et al., Abnormal morphogenesis but intact IKK activation in mice lacking the IKKα subunit of IkappaB kinase, *Science,* 284, 316, 1999.

[9]   Hu, Y., Baud, V., Oga, T. et al., IKKalpha controls formation of the epidermis independently of NF-kappaB, *Nature,* 410, 710, 2001.

[10] Zandi, E., Chen, Y., and Karin, M., Direct phosphorylation of IkappaB by IKKalpha and IKKbeta: Discrimination between free and NF-kappaB-bound substrate, *Science,* 281, 1360, 1998.

[11] Sil, A.K., Maeda, S., Sano, Y. et al., IkappaB kinase-α acts in the epidermis to control skeletal and craniofacial morphogenesis, *Nature,* 428, 660, 2004.

[12] Egan, L.J., de Lecea, A., Lehrman, E.D. et al., Nuclear factor-kappaB activation promotes restitution of wounded intestinal epithelial monolayers, *Am. J. Physiol. Cell Physiol.,* 285, C1028, 2003.

[13] Egan, L.J., Eckmann, L., Greten, F.R. et al., IkappaB-kinaseβ-dependent NF-kappaB activation provides radioprotection to the intestinal epithelium, *Proc. Natl. Acad. Sci. USA,* 101, 2452, 2004.

[14] Jiang, D., Liang, J., Fan, J. et al., Regulation of lung injury and repair by Toll-like receptors and hyaluronan, *Nat. Med.,* 11, 1173, 2005.

[15]   Goudeau, B., Huetz, F., Samson, S. et al., IkappaBα/IkappaBε deficiency reveals that a critical NF-kappaB dosage is required for lymphocyte survival, *Proc. Natl. Acad. Sci. USA,* 100, 15800, 2003.

[16]   Samson, S.I., Memet, S., Vosshenrich, C.A. et al., Combined deficiency in IkappaBα and IkappaBε reveals a critical window of NF-kappaB activity in natural killer cell differentiation, *Blood,* 103, 4573, 2004.

[17]   Weih, F., Carrasco, D., Durham, S.K. et al., Multiorgan inflammation and hemato-poietic abnormalities in mice with a targeted disruption of RelB, a member of the NF-kappaB/Rel family, *Cell,* 80, 331, 1995.

[18]   Burkly, L., Hession, C., Ogata, L. et al., Expression of relB is required for the development of thymic medulla and dendritic cells, *Nature,* 373, 531, 1995.

[19]   Wu, L., DíAmico, A., Winkel, K.D et al., RelB is essential for the development of myeloid-related CD8alpha-dendritic cells but not of lymphoid-related CD8α+ den-dritic cells, *Immunity,* 9, 839, 1998.

[20]   Ouaaz, F., Arron, J., Zheng, Y. et al., Dendritic cell development and survival require distinct NF-kappaB subunits, *Immunity,* 16, 257, 2002.

[21]   Abe, K., Yarovinsky, F.O., Murakami, T. et al., Distinct contributions of TNF and LT cytokines to the development of dendritic cells *in vitro* and their recruitment *in vivo,* *Blood,* 101, 1477, 2003.

[22]   Kriehuber, E., Bauer, W., Charbonnier, A.S. et al., Balance between NF-kappaB and JNK/AP-1 activity controls dendritic cell life and death, *Blood,* 106, 175, 2005.

[23]   McDonald, P.P., Bald, A., and Cassatella, M.A., Activation of the NF-kappaB pathway by inflammatory stimuli in human neutrophils, *Blood,* 89, 3421, 1997.

[24]   McDonald, P.P., Transcriptional regulation in neutrophils: Teaching old cells new tricks, *Adv. Immunol.,* 82, 1, 2004.

[25]   Francois, S., El Benna, J., Dang, P.M. et al., Inhibition of neutrophil apoptosis by TLR agonists in whole blood: Involvement of the phosphoinositide 3-kinase/Akt and NF-kappaB signaling pathways, leading to increased levels of Mcl-1, A1, and phos-phorylated Bad, *J. Immunol.,* 174, 3633, 2005.

[26]   Janeway, C.A., Jr., Approaching the asymptote? Evolution and revolution in immu-nology, *Cold Spring Harb. Symp. Quant. Biol.,* 54 Pt 1, 1, 1989.

[27]   Hayashi, F., Smith, K.D., Ozinsky, A. et al., The innate immune response to bacterial flagellin is mediated by Toll-like receptor 5, *Nature,* 410, 1099, 2001.

[28]   Zhang, D., Zhang, G., Hayden, M.S. et al., A toll-like receptor that prevents infection by uropathogenic bacteria, *Science,* 303, 1522, 2004.

[29]   Hemmi, H., Takeuchi, O., Kawai, T. et al., A toll-like receptor recognizes bacterial DNA, *Nature,* 408, 740, 2000.

[30]   Alexopoulou, L., Holt, A.C., Medzhitov, R. et al., Recognition of double-stranded RNA and activation of NF-kappaB by toll-like receptor 3, *Nature,* 413, 732, 2001.

[31]   Lund, J.M., Alexopoulou, L., Sato, A. et al., Recognition of single-stranded RNA viruses by toll-like receptor 7, *Proc. Natl. Acad. Sci. USA,* 101, 5598, 2004.

[32]   Diebold, S.S., Kaisho, T., Hemmi, H. et al., Innate antiviral responses by means of TLR7-mediated recognition of single-stranded RNA, *Science,* 303, 1529, 2004.

[33]   Heil, F., Hemmi, H., Hochrein, H. et al., Species-specific recognition of single-stranded RNA via toll-like receptor 7 and 8, *Science,* 303, 1526, 2004.

[34]   Ishii, K.J., Coban, C., Kato, H. et al., A toll-like receptor-independent antiviral response induced by double-stranded B-form DNA, *Nat. Immunol.,* 7, 40, 2006.

[35]   Stetson, D.B. and Medzhitov, R., Recognition of cytosolic DNA activates an IRF3-dependent innate immune response, *Immunity,* 24, 93, 2006.

[36] Levitz, S.M., Interactions of toll-like receptors with fungi, *Microbes Infect.*, 6, 1351, 2004.

[37] Gazzinelli, R.T., Ropert, C., and Campos, M.A., Role of the toll/interleukin-1 receptor signaling pathway in host resistance and pathogenesis during infection with protozoan parasites, *Immunol. Rev.*, 201, 9, 2004.

[38] Coban, C., Ishii, K.J., Kawai, T. et al., Toll-like receptor 9 mediates innate immune activation by the malaria pigment hemozoin, *J. Exp. Med.*, 201, 19, 2005.

[39] Yarovinsky, F., Zhang, D., Andersen, J.F. et al., TLR11 activation of dendritic cells by a protozoan profilin-like protein, *Science*, 308, 1626, 2005.

[40] Horng, T., Barton, G.M., Flavell, R.A. et al., The adaptor molecule TIRAP provides signalling specificity for Toll-like receptors, *Nature*, 420, 329, 2002.

[41] Yamamoto, M., Sato, S., Mori, K. et al., Cutting edge: A novel Toll/IL-1 receptor domain-containing adapter that preferentially activates the IFN-beta promoter in the Toll-like receptor signaling, *J. Immunol.*, 169, 6668, 2002.

[42] Fitzgerald, K.A., Palsson-McDermott, E.M., Bowie, A.G. et al., Mal (MyD88-adapter-like) is required for Toll-like receptor-4 signal transduction, *Nature*, 413, 78, 2001.

[43] Shim, J.H., Xiao, C., Paschal, A.E. et al., TAK1, but not TAB1 or TAB2, plays an essential role in multiple signaling pathways *in vivo*, *Genes Dev.*, 19, 2668, 2005.

[44] Sato, S., Sanjo, H., Takeda, K. et al., Essential function for the kinase TAK1 in innate and adaptive immune responses, *Nat. Immunol.*, 6, 1087, 2005.

[45] Kopp, E., Medzhitov, R., Carothers, J. et al., ECSIT is an evolutionarily conserved intermediate in the Toll/IL-1 signal transduction pathway, *Genes Dev.*, 13, 2059, 1999.

[46] Xiao, C., Shim, J.H., Kluppel, M. et al., Ecsit is required for Bmp signaling and mesoderm formation during mouse embryogenesis, *Genes Dev.*, 17, 2933, 2003.

[47] Xia, Y., Makris, C., Su, B. et al., MEK kinase 1 is critically required for c-Jun N-terminal kinase activation by proinflammatory stimuli and growth factor-induced cell migration, *Proc. Natl. Acad. Sci. USA*, 97, 5243, 2000.

[48] Yujiri, T., Ware, M., Widmann, C. et al., MEK kinase 1 gene disruption alters cell migration and c-Jun NH2- terminal kinase regulation but does not cause a measurable defect in NF-kappaB activation, *Proc. Natl. Acad. Sci. USA*, 97, 7272, 2000.

[49] Huang, Q., Yang, J., Lin, Y. et al., Differential regulation of interleukin 1 receptor and Toll-like receptor signaling by MEKK3, *Nat. Immunol.*, 5, 98, 2004.

[50] Moustakas, A. and Heldin, C.H., Ecsit-ement on the crossroads of Toll and Bmp signal transduction, *Genes Dev.*, 17, 2855, 2003.

[51] Kawai, T., Adachi, O., Ogawa, T. et al., Unresponsiveness of MyD88-deficient mice to endotoxin, *Immunity*, 11, 115, 1999.

[52] Oshiumi, H., Matsumoto, M., Funami, K. et al., TICAM-1, an adaptor molecule that participates in Toll-like receptor 3-mediated interferon-beta induction, *Nat. Immunol.*, 4, 161, 2003.

[53] Yamamoto, M., Sato, S., Hemmi, H. et al., Role of adaptor TRIF in the MyD88-independent toll-like receptor signaling pathway, *Science*, 301, 640, 2003.

[54] Meylan, E., Burns, K., Hofmann, K. et al., RIP1 is an essential mediator of Toll-like receptor 3-induced NF-kappaB activation, *Nat. Immunol.*, 5, 503, 2004.

[55] Yamamoto, M., Sato, S., Hemmi, H. et al., TRAM is specifically involved in the Toll-like receptor 4-mediated MyD88-independent signaling pathway, *Nat. Immunol.*, 4, 1144, 2003.

[56] Fitzgerald, K.A., Rowe, D.C., Barnes, B.J. et al., LPS-TLR4 signaling to IRF-3/7 and NF-kappaB involves the toll adapters TRAM and TRIF, *J. Exp. Med.*, 198, 1043, 2003.

[57] Covert, M.W., Leung, T.H., Gaston, J.E. et al., Achieving stability of lipopolysaccha-
      ride-induced NF-kappaB activation, *Science,* 309, 1854, 2005.

[58] Werner, S.L., Barken, D., and Hoffmann, A., Stimulus specificity of gene expression
      programs determined by temporal control of IKK activity, *Science,* 309, 1857, 2005.

[59] Sharma, S., tenOever, B.R., Grandvaux, N. et al., Triggering the interferon antiviral
      response through an IKK-related pathway, *Science,* 300, 1148, 2003.

[60] Hemmi, H., Takeuchi, O., Sato, S. et al., The roles of two IkappaB kinase-related
      kinases in lipopolysaccharide and double stranded RNA signaling and viral infection,
      J. Exp. Med., 199, 1641, 2004.

[61] Akira, S., Toll-ike receptor signaling, in *Keystone Symposium on NF-kB: Biology and
      Pathology,* Ulrich K. Siebenlist, A.R., Sankar Ghosh Keystone Symposia, Snowbird
      Resort, Snowbird, Utah, 2004.

[62] McWhirter, S.M., Fitzgerald, K.A., Rosains, J. et al., IFN-regulatory factor 3-depen-
      dent gene expression is defective in TBK1-deficient mouse embryonic fibroblasts,
      Proc. Natl. Acad. Sci. USA, 101, 233, 2004.

[63] Mercurio, F., IKK-related kinases as a target for drug therapy, in *Keystone Symposium
      on NF-kB: Biology and Pathology,* Ulrich K. Siebenlist, A.R., Sankar Ghosh Keystone
      Symposia, Snowbird Resort, Snowbird, Utah, 2004.

[64] Kravchenko, V.V., Mathison, J.C., Schwamborn, K. et al., IKKi/IKKepsilon plays a
      key role in integrating signals induced by pro-inflammatory stimuli, *J. Biol. Chem.,*
      278, 26612, 2003.

[65] Kobayashi, K., Hernandez, L.D., Galan, J.E. et al., IRAK-M is a negative regulator
      of Toll-like receptor signaling, *Cell,* 110, 191, 2002.

[66] Burns, K., Clatworthy, J., Martin, L. et al., Tollip, a new component of the IL-1RI
      pathway, links IRAK to the IL-1 receptor, *Nat. Cell. Biol.,* 2, 346, 2000.

[67] Zhang, G. and Ghosh, S., Negative regulation of toll-like receptor-mediated signaling
      by Tollip, *J. Biol. Chem.,* 277, 7059, 2002.

[68] Melmed, G., Thomas, L.S., Lee, N. et al., Human intestinal epithelial cells are broadly
      unresponsive to Toll-like receptor 2-dependent bacterial ligands: implications for
      host-microbial interactions in the gut, *J. Immunol.,* 170, 1406, 2003.

[69] Didierlaurent, A., Brissoni, B., Velin, D. et al., Tollip regulates proinflammatory
      responses to interleukin-1 and lipopolysaccharide, *Mol. Cell. Biol.,* 26, 735, 2006.

[70] Thomassen, E., Renshaw, B.R., and Sims, J.E., Identification and characterization of
      SIGIRR, a molecule representing a novel subtype of the IL-1R superfamily, *Cytokine,*
      11, 389, 1999.

[71] Wald, D., Qin, J., Zhao, Z. et al., SIGIRR, a negative regulator of Toll-like receptor-
      interleukin 1 receptor signaling, *Nat. Immunol.,* 4, 920, 2003.

[72] Kinjyo, I., Hanada, T., Inagaki-Ohara, K. et al., SOCS1/JAB is a negative regulator
      of LPS-induced macrophage activation, *Immunity,* 17, 583, 2002.

[73] Nakagawa, R., Naka, T., Tsutsui, H. et al., SOCS-1 participates in negative regulation
      of LPS responses, *Immunity,* 17, 677, 2002.

[74] Mansell, A., Smith, R., Doyle, S.L. et al., Suppressor of cytokine signaling 1 nega-
      tively regulates Toll-like receptor signaling by mediating Mal degradation, *Nat. Immu-
      nol.,* 7, 148, 2006.

[75] Inohara, N. and Nunez, G., NODs: Intracellular proteins involved in inflammation
      and apoptosis, *Nat Rev. Immunol.,* 3, 371, 2003.

[76] Vermeire, S. and Rutgeerts, P., Current status of genetics research in inflammatory
      bowel disease, *Genes Immun.,* 6, 637, 2005.

[77] Tanabe, T., Chamaillard, M., Ogura, Y. et al., Regulatory regions and critical residues
      of NOD2 involved in muramyl dipeptide recognition, *Embo J.,* 23, 1587, 2004.

[78] Inohara, N., Koseki, T., Lin, J. et al., An induced proximity model for NF-kappaB activation in the Nod1/RICK and RIP signaling pathways, *J. Biol. Chem.,* 275, 27823, 2000.

[79] Yoneyama, M., Kikuchi, M., Natsukawa, T. et al., The RNA helicase RIG-I has an essential function in double-stranded RNA-induced innate antiviral responses, *Nat. Immunol.,* 5, 730, 2004.

[80] Andrejeva, J., Childs, K.S., Young, D.F. et al., The V proteins of paramyxoviruses bind the IFN-inducible RNA helicase, MDA-5, and inhibit its activation of the IFN-beta promoter, *Proc. Natl. Acad. Sci. USA,* 101, 17264, 2004.

[81] Kang, D.C., Gopalkrishnan, R.V., Wu, Q. et al., mda-5: An interferon-inducible putative RNA helicase with double-stranded RNA-dependent ATPase activity and melanoma growth-suppressive properties, *Proc. Natl. Acad. Sci. USA,* 99, 637, 2002.

[82] Meylan, E., Curran, J., Hofmann, K. et al., Cardif is an adaptor protein in the RIG-I antiviral pathway and is targeted by hepatitis C virus, *Nature,* 437, 1167, 2005.

[83] Seth, R.B., Sun, L., Ea, C.K. et al., Identification and characterization of MAVS, a mitochondrial antiviral signaling protein that activates NF-kappaB and IRF 3, *Cell,* 122, 669, 2005.

[84] Kawai, T., Takahashi, K., Sato, S. et al., IPS-1, an adaptor triggering RIG-I- and Mda5-mediated type I interferon induction, *Nat. Immunol.,* 6, 981, 2005.

[85] Xu, L.G., Wang, Y.Y., Han, K.J. et al., VISA is an adapter protein required for virus-triggered IFN-β signaling, *Mol. Cell,* 19, 727, 2005.

[86] Balachandran, S., Thomas, E., and Barber, G.N., A FADD-dependent innate immune mechanism in mammalian cells, *Nature,* 432, 401, 2004.

[87] Yoneyama, M., Kikuchi, M., Matsumoto, K. et al., Shared and unique functions of the DExD/H-box helicases RIG-I, MDA5, and LGP2 in antiviral innate immunity, *J. Immunol.,* 175, 2851, 2005.

[88] Kato, H., Sato, S., Yoneyama, M. et al., Cell type-specific involvement of RIG-I in antiviral response, *Immunity,* 23, 19, 2005.

[89] Miller, L.S., Sorensen, O.E., Liu, P.T. et al., TGF-alpha regulates TLR expression and function on epidermal keratinocytes, *J. Immunol.,* 174, 6137, 2005.

[90] Wolfs, T.G., Buurman, W.A., van Schadewijk, A. et al., *In vivo* expression of Toll-like receptor 2 and 4 by renal epithelial cells: IFN-γ and TNF-α mediated up-regulation during inflammation, *J. Immunol.,* 168, 1286, 2002.

[91] Alcamo, E., Mizgerd, J.P., Horwitz, B.H. et al., Targeted mutation of TNF receptor I rescues the RelA-deficient mouse and reveals a critical role for NF-kappaB in leukocyte recruitment, *J. Immunol.,* 167, 1592, 2001.

[92] Sha, W.C., Liou, H.C., Tuomanen, E.I. et al., Targeted disruption of the p50 subunit of NF-kappaB leads to multifocal defects in immune responses, *Cell,* 80, 321, 1995.

[93] Li, Z.W., Chu, W., Hu, Y. et al., The IKKbeta subunit of IkappaB kinase (IKK) is essential for nuclear factor κB activation and prevention of apoptosis, *J. Exp. Med.,* 189, 1839, 1999.

[94] Senftleben, U., Li, Z.W., Baud, V. et al., IKKβ is essential for protecting T cells from TNFα-induced apoptosis, *Immunity,* 14, 217, 2001.

[95] Rudolph, D., Yeh, W.C., Wakeham, A. et al., Severe liver degeneration and lack of NF-kappaB activation in NEMO/IKKγ-deficient mice, *Genes Dev.,* 14, 854, 2000.

[96] Ayabe, T., Satchell, D.P., Wilson, C.L. et al., Secretion of microbicidal α-defensins by intestinal Paneth cells in response to bacteria, *Nat. Immunol.,* 1, 113, 2000.

[97] Tabeta, K., Georgel, P., Janssen, E. et al., Toll-like receptors 9 and 3 as essential components of innate immune defense against mouse cytomegalovirus infection, *Proc. Natl. Acad. Sci. USA,* 101, 3516, 2004.

[98]   Hoebe, K., Du, X., Georgel, P. et al., Identification of LPS2 as a key transducer of MyD88-independent TIR signalling, *Nature,* 424, 743, 2003.

[99]   Hart, O.M., Athie-Morales, V., OíConnor, G.M et al., TLR7/8-mediated activation of human NK cells results in accessory cell-dependent IFN-γ production, *J. Immunol.,* 175, 1636, 2005.

[100] Zhou, J., Zhang, J., Lichtenheld, M.G. et al., A role for NF-kappaB activation in perforin expression of NK cells upon IL-2 receptor signaling, *J. Immunol.,* 169, 1319, 2002.

[101] Caamano, J., Alexander, J., Craig, L. et al., The NF-kappaB family member RelB is required for innate and adaptive immunity to *Toxoplasma gondii, J. Immunol.,* 163, 4453, 1999.

[102] Iwasaki, A. and Medzhitov, R., Toll-like receptor control of the adaptive immune responses, *Nat. Immunol.,* 5, 987, 2004.

[103] Lopez, C.B., Garcia-Sastre, A., Williams, B.R. et al., Type I interferon induction pathway, but not released interferon, participates in the maturation of dendritic cells induced by negative-strand RNA viruses, *J. Infect. Dis.,* 187, 1126, 2003.

[104] Gutierrez, O., Pipaon, C., Inohara, N. et al., Induction of Nod2 in myelomonocytic and intestinal epithelial cells via nuclear factor-kappaB activation, *J. Biol. Chem.,* 277, 41701, 2002.

[105] Maeda, S., Hsu, L.C., Liu, H. et al., Nod2 mutation in Crohn's disease potentiates NF-kappaB activity and IL-1β processing, *Science,* 307, 734, 2005.

[106] Kobayashi, K.S., Chamaillard, M., Ogura, Y. et al., NOD2-dependent regulation of innate and adaptive immunity in the intestinal tract, *Science,* 307, 731, 2005.

[107] Hornung, V., Rothenfusser, S., Britsch, S. et al., Quantitative expression of toll-like receptor 1-10 mRNA in cellular subsets of human peripheral blood mononuclear cells and sensitivity to CpG oligodeoxynucleotides, *J. Immunol.,* 168, 4531, 2002.

7 NF-κB in the Adaptive Immune System: Lymphocyte Survival and Function

Matthew S. Hayden and Sankar Ghosh

CONTENTS

7.1 INTRODUCTION

The innate immune response is, in many cases, sufficient for the prevention of microbial invasion and dissemination. For some pathogens, however, innate responses alone are inadequate for clearance and instead play the crucial role of alerting the adaptive immune system to the threat of a microbial pathogen. In Chapter 6, we discussed aspects of NF-κB function that shape the innate immune response and how these lead to the initiation of the adaptive response. In this chapter we will focus on the biological role of NF-κB family members in the adaptive immune response. Many of the functions of NF-κB in adaptive immunity are analogous to the functions of NF-κB in the innate immune system. NF-κB is necessary for the hematopoietic development of lymphocytes (T and B cells); lymphocyte survival and maturation; and lymphocyte effector functions. Like the innate immune system, the adaptive immune system requires continual replenishment of its cellular components and careful regulation of these cells during and after the immune response. As was alluded to in Chapter 6, adaptive responses rely on signaling events that occur within carefully organized primary and secondary lymphoid organs that require NF-κB for their development and maintenance.

Of its many functions, one of the most critical roles of NF-κB in the adaptive immune system is the regulation of antiapoptotic genes. This supposition is based on genetic studies demonstrating that deficits in adaptive immune responses in NF-κB knockout mice can be rescued by either removal of proapoptotic stimuli or forced expression of antiapoptotic genes. However, as is often the case, careful examination reveals unexpected complexities — NF-κB may protect from or facilitate apoptosis depending on the timing, target, and context of a given signaling event. NF-κB-regulated cytokines are also critical to the adaptive response, while the regulation of organogenic chemokines by NF-κB is crucial for lymphoid organogenesis. Consequently, analyzing the role of NF-κB in adaptive immunity requires not only dissecting overlapping functions in lymphocyte development and activation, but also looking at the organization and formation of the tissues that support these processes.

7.2 LYMPHOID ORGANOGENESIS

We begin our look at NF-κB function in the adaptive immune system with a brief discussion of the development and function of tissues that are central to lymphocyte biology. These tissues can be broadly divided into two categories: primary and secondary lymphoid organs. Primary (central) lymphoid organs include the bone marrow and thymus, where B cell and T cell development occur, respectively. B and T cells develop continuously over the mammalian lifespan. The bone marrow remains active throughout life, while thymic activity dwindles with the onset of adulthood. Secondary (peripheral) lymphoid organs include the lymph nodes, Peyer's patches, mucosal-associated lymphoid tissue (MALT), and spleen. These tissues are associated with the maintenance and activation of mature lymphocytes, and in some cases these tissues also can contribute to lymphocyte development. As discussed in Chapters 6 and 8, the peripheral lymphoid organs provide an environment within which the interaction of lymphocytes and other leukocytes can be carefully orchestrated. Deletions of NF-κB family genes have revealed a multitude of defects in lymphoid organogenesis, although in some instances it has remained unclear whether the deficit is in the development of the tissue itself or in the hematopoietic cells that populate the fully differentiated tissue.

7.2.1 NF-κB in Primary Lymphoid Organs

NF-κB function is critical in the development and maintenance of the central lymphoid organs, bone marrow, and thymus. Although, there has been relatively little effort to characterize the role of NF-κB in the development of primary lymphoid tissues, one NF-κB family member has a clear role in thymic development — RelB. RelB is responsible for much of the constitutive NF-κB activity observed in various tissues, in addition to being the primary mediator of gene expression resulting from activation of the alternative pathway. Deletion of *relB* results in decreased basal NF-κB activity in the thymus and spleen as well as severe deficits in adaptive immunity [1]. Furthermore, there is a specific defect in thymic medullary epithelial cells and a loss of functional thymic dendritic cells in mice lacking *relB* [2]. Deletion of p52 and p50 results in thymic atrophy, perhaps also due to a decrease in RelB-dependent signaling via RelB:p52 containing complexes. Although the role of RelB in the thymus is not known definitively, there is one promising RelB target. In the thymus RelB regulates the expression of autoimmue regulator (AIRE), a transcription factor that has a role in organizing medullary thymic stroma [3], and there is some evidence that suggests RelB does so in response to LTβR signaling. LTβR knockouts share phenotypic similarities with AIRE knockouts, in terms of autoimmune tissue infiltration, and have decreased upregulation of AIRE in the thymic medullary epithelial cells [4]. Given that lymphotoxin knockouts do not have thymic medullary defects [5], it is likely that RelB acts both upstream and downstream of LTβR signaling and that AIRE expression requires stimuli in addition to that mediated by LTβR. Consequently, regulation of AIRE expression following LTβR stimulation is but one of the important functions carried out by RelB in thymic organogenesis.

Bone marrow is the central repository for all hematopoietic lineages, is the site of the initial stages of hematopoietic development and differentiation, and has a role in lymphocyte homeostasis and function. A subset of NF-κB pathway knockouts display phenotypes in bone that are worth discussing here, as they are not discussed elsewhere in the text (for review see [6]). Osteoclasts are responsible for the absorption of bone and develop from monocytes, a process mediated primarily by activation of receptor activator of NF-κB (RANK, also known as TNFR super family member 11A) by RANKL (also known as TRANCE, ODF, OPGL and TNFSF11) expressed on stromal cells [7]. RANKL is also important in the activation of mature osteoclasts in normal bone remodeling, as well as inflammation-induced bone destruction. RANK mediates NF-κB activation through TRAF6, and TRAF6-deficient animals show a similar defect in osteoclastogenesis to that observed in RANK knockouts [7,8,9,10,11]. The NF-κB p50 and p52 dKO, which interferes with classical and alternative pathways, leads to a loss of osteoclasts and consequent osteopetrosis, a disease in which bones are overly dense [12,13].

While these data clearly show the importance of NF-κB in osteoclastogenesis, the relative contribution of the alternative and classical pathways is somewhat unclear. NIK deficient mice have normal bone development, but display defects in late stages of osteoclastogenesis when analyzed *in vitro* [14,15]. IKKα deficient embryos have decreased multinucleate osteoclast formation *in vivo*, although when IKKα knockouts are complemented *in vivo* with defective IKKα(AA) no osteopetrosis is observed in adult mice [16,17]. The same study that failed to observe defects in IKKα(AA) mice reported defects in the formation of multinucleated giant cells, the final stage of osteoclastogenesis, in IKKβ/TNFR1 dKO mice. This suggests that the classical pathway is more important for osteoclastogenesis *in vivo*.

IKKβ deficient mice are also resistant to the induction of inflammatory arthritis, which is dependent on osteoclast activation. Likewise, inhibition of NF-κB in the adult animal inhibits osteoclast activity and diminishes the extent of arthritis and inflammation-induced bone loss [18,19]. Taken together, these results suggest that both alternative and classical pathways have a role in the generation of fully competent osteoclasts. However, the results of inhibiting in NF-κB signaling in the adult animal suggest that the classical pathway may have a more prominent role in basal osteoclastogenesis and osteoclast activation and survival during inflammatory bone loss. The alternative pathway is involved in osteoclast development *in vitro* and, therefore, may be required for development under pathological conditions. In terms of the adaptive immune system, it remains to be seen whether the defects in bone structure that occur in the absence of NF-κB affect the ability of the bone marrow to support hematopoiesis. In the setting of inflammatory arthritis, however, NF-κB has a clear role in the activation of osteoclasts by cytokines generated by the immune response.

7.2.2 NF-κB in Secondary Lymphoid Organs

The secondary lymphoid organs have highly characteristic structural features that are crucial to the development and activation of lymphocytes. The most basic view of lymphoid organogenesis involves the association between an LT$\alpha_1\beta_2$ expressing hematopoietic cell and VCAM1 expressing stromal cell — for review, see [20] (Figure 7.1). The interaction of these two key cell types initiates a signaling loop whereby LTβR signaling activates stromal cells to produce organogenic chemokines and VCAM1, resulting in increased recruitment of hematopoietic cells. Expression of RANKL results in differentiation of these hematopoietic cells and upregulation of LT$\alpha_1\beta_2$, thus initiating a positive feedback loop leading to organogenesis. Many reports implicate NF-κB in lymphoid organogenesis. For example, it has long been known that NF-κB family members are selectively expressed in certain cell populations of the developing and adult spleen [21]. Cytokines that are known to induce NF-κB activation have been implicated in lymphoid organogenesis: e.g., LT$\alpha_1\beta_2$, RANKL (TRANCE), and TNFα. In turn, mediators of lymphoid organogenesis and homeostasis, such as the adhesion molecules ICAM, VCAM, PNAd, GlyCAM-1, and MadCAM, cytokines including TNFα, and organogenic chemokines such as CXCL12 (GRO/MIP-2), CXCL13 (BLC), CCL19 (ELC), and CCL21 (SLC), are regulated by NF-κB. Finally, multiple gene targeting experiments have confirmed that NF-κB has an important role in lymphoid organogenesis.

Lymphoid organogenesis exhibits distinct requirements for the alternative and classical NF-κB pathways. Signaling through TNFR, LTβR, and RANK activates classical p65 containing complexes and, consequently, it is not surprising that *p65$^{-/-}$/tnfr1$^{-/-}$* mice lack Peyer's patches and lymph nodes, and exhibits a disorganized [22]. The requirement for p65 in development of these tissues lies with the stromal cells (Figure 7.1). The requirement for p65-containing NF-κB signaling complexes is likely due to a combination of effects: regulation of apoptosis (e.g., that induced by TNF); regulation of organogenic gene expression including VCAM and LT$\alpha_1\beta_2$; and enhancement of the alternative pathway through upregulation of p100.

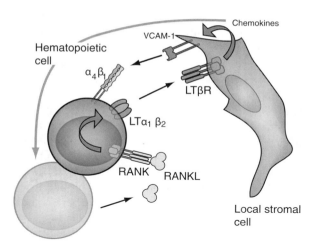

FIGURE 7.1 NF-κB in early events of lymphoid organogenesis. The early events of lymphoid organogenesis involve the association of hematopoietic lineage cells and VCAM1 expressing stromal cell. This interaction initiates an NF-κB-dependent signaling loop whereby LTα₁β₂ ligates LTβR on stromal cells. Activation of LTβR induces NF-κB-dependent upregulation of organogenic chemokines and VCAM1, which results in increased recruitment of hematopoietic cells. Increasing expression of RANKL results in NF-κB activation via RANK, which is also expressed on hematopoietic cells. The resulting differentiation of these hematopoietic cells, which includes increased expression of LTα₁β₂, initiates a positive feedback loop leading to organogenesis. Large arrows indicate NF-κB-dependent gene regulation.

Alternative pathway signaling downstream of LTβR is crucial during secondary lymphoid organogenesis. Indeed, the alymphoplasia (*aly*) mouse, which lacks multiple secondary lymphoid organs, has an inactivating point mutation in NIK [23,24,25]. RelB/p52 heterodimers, liberated by processing of p100 to p52, are thought to be the primary transcriptional mediator of several key organogenic factors – CXCL12, CXCL13, CCL19, CCL21, and MadCAM-1 [26]. The p52 knockout lacks normal B cell follicles, germinal centers, and Peyer's patch development [27, 28, 29]; RelB is also required for Peyer's patch development [26]. Although lymph node development occurs in RelB knockouts, the nodes are small at birth and are resorbed perinatally. Many of these defects are shared in Lymphotoxin, NIK, and IKKα knockout animals [20,30]. In addition to LTβR, knockouts of RANK also lack peripheral lymph nodes, suggesting that more than one alternative pathway stimulus may be involved in LN development [7].

The spleen is a particularly important secondary lymphoid organ because it has a vital role in the final steps of B cell development as well as the initiation and maturation of B cell responses. Splenic architecture must, therefore, be maintained to mount a normal adaptive response. The spleen is divided histologically into areas of white and red pulp. The red pulp is the site where erythrocytes that have reached the end of their lifespan are phagocytosed by macrophages. The white pulp is populated by splenic lymphocytes and consists of B cell follicles and T cell zones. Splenic architecture allows for dynamic changes, most notably the formation of germinal centers, during the initiation and maturation of a B cell response. Multiple NF-κB knockouts exhibit defects in some aspect of splenic architecture; however, because TNFR knockouts are used to rescue the embryonic lethality associated with some NF-κB knockouts, the analysis of splenic architecture has been complicated by defects that occur in the TNFR knockout. Nevertheless, there has been a considerable progress in deciphering the role of NF-κB family members in development and maintenance of splenic architecture. Mice in which p65 has been targeted for deletion exhibit disrupted segregation of B and T cell areas and defects in one particular macrophage population — metallophilic marginal zone macrophages. In addition, $p65^{-/-}/tnfr1^{-/-}$ spleens have a more pronounced defect in GC generation following immunization than do $tnfr1^{-/-}$ mice [22]. However, it is worth emphasizing that defects observed in $tnfr1^{-/-}$ animals may, in fact, be due to changes in p65-dependent responses. Thus, the role of p65/classical pathway in the spleen is likely underappreciated.

Mice in which alternative pathway components have been targeted have severe defects in splenic architecture, consistent with that seen in $ltbr^{-/-}$ spleens. These defects, seen in RelB, NIK, or IKKα deficient animals, are due to deficiencies in splenic radiation-resistant stromal cells, rather than hematopoietic cells [23,24]. Alternative pathway deficient mice lack segregation of B cell–T cell zones and FDC networks, and they fail to form germinal centers following immunization. Marginal zone macrophages, which line the border between red and white pulp areas, are absent or found dispersed throughout the spleen in RelB, p52, NIK, or IKKα knockouts [29,31]. While localization of macrophages to the marginal zone depends on alternative pathway components in stromal cells, the absence of metallophilic marginal zone macrophages in p52 knockouts is a hematopoietic defect [12]. Finally, knockout of the atypical IκB family member BCL-3 also leads to alterations in lymphoid architecture. These mice lack spleen germinal centers, and although they exhibit normal serum antibody levels, they fail to develop antigen specific humoral responses [12,32]. The BCL-3 knockout phenotype is partially shared by p52 knockout mice, providing genetic evidence in support of BCL-3 and p52 forming a transcriptionally active complex [12,32,33,34].

In summary, both alternative and classical NF-κB pathways are required for the development of most secondary lymphoid organs. However, the role of the alternative pathway, as assessed by examining mice deficient for IKKα, p52, NIK, or RelB, is especially important both during organogenesis and maintenance of splenic architecture. However, it is important to note that appreciation of the role of the classical pathway in these events has been hindered by embryonic lethality and complicated by the existing deficits in $tnfr^{-/-}$ animals. Nevertheless, our understanding of alternative pathway function in secondary lymphoid organs is consistent with the ability of RelB-containing complexes to regulate key organogenic chemokines and adhesion molecules that direct leukocyte trafficking. The functional consequences of defects in these processes are severe. Loss of secondary lymphoid organs has direct ramifications for the host's ability to mount a robust immune response. Alterations in lymphoid architecture likewise impede the initiation of the adaptive response as well as the fine-tuning of this response through processes such as B cell affinity maturation.

7.3 NF-κB AND LYMPHOPOIESIS

NF-κB also has a vital role in the development and function of the effector cells of the adaptive immune system. The selection process leading to the production of functional lymphocytes is characterized by a high rate of apoptosis. Thus, the antiapoptotic effect of NF-κB is important in many aspects of lymphopoiesis. For example, during T cell development, most of the requirements for signaling to NF-κB, i.e., through the T cell receptor, can be overcome by transgenic expression of the antiapoptotic protein Bcl-2 [35]. In some cases, the embryonic lethality of NF-κB knockouts precedes the development of the mature hematopoietic system and, as such, impedes analysis of the hematopoietic pathways of the adult organism. Furthermore, the defects in both primary and secondary lymphoid tissues make analyses of mice that do survive to adulthood quite difficult. Nevertheless, efforts to circumvent these obstacles through adoptive transfer of lymphocyte precursors and, more recently, use of conditional genetic alterations have supported the crucial role of NF-κB in lymphopoiesis. Finally, there are examples from human diseases that further support the importance of NF-κB in lymphopoiesis. The gene encoding NEMO is located on the X-chromosome and is therefore subject to random inactivation. In female patients who are heterozygous for a mutant version of NEMO all peripheral lymphocytes possess an intact NEMO gene (see Chapter 9). Thus, in the absence of NF-κB signaling, the key cells of the adaptive immune response fail to develop.

Lymphopoiesis is, in and of itself, an area of intense research, and a thorough analysis of this topic is well beyond the scope of this chapter. Instead, the level of detail given in Figure 7.2 is sufficient to allow an uninitiated reader a basic understanding of B and T cell development with which to ascertain the importance of NF-κB in these processes. Both B and T cells develop from hematopoietic stem cells in the bone marrow and from there proceed through partially analogous developmental paths.

T cell development occurs primarily in the thymus (Figure 7.2). Lymphoid precursors leave the bone marrow and undergo proliferation following immigration to the thymus and interaction with cortical epithelial cells therein. Thymocyte maturation occurs along a spatial path from the cortical to medullary space of the thymus. Thymocytes are typically classified by their expression of CD4 and CD8 — markers of mature helper (T_H) and cytotoxic (T_C) T cells, respectively. Initially cells lack both CD4 and CD8, and are referred to as double negatives (DNs). Thymocytes undergo recombination of the TCR locus to produce the β-chain of the T cell receptor (TCR) during the DN stage. Successful β-chain rearrangement results in expression of the pre-TCR consisting of recombined β-chain and germline encoded pre-Tα. Signaling through the pre-TCR halts β-chain rearrangement and induces proliferation and expression of CD4 and CD8. Double positive (DP) thymocytes then reexpress the recombination machinery required for rearrangement of the TCR-α chain, resulting in expression of a mature α/βTCR. DP thymocytes undergo positive selection through interaction of the α/βTCR with self-peptide: MHC expressed on thymic stroma. Thymocytes that lack TCRs capable of initiating signaling will undergo apoptosis — a failure of positive selection termed death by neglect. Thymocytes

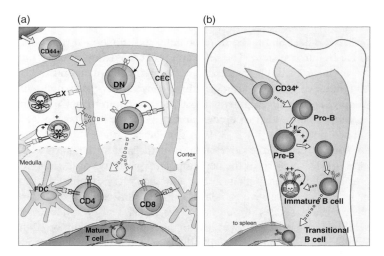

FIGURE 7.2 Basic stages of lymphopoiesis. T cell development occurs in the thymus (a). CD44+ T cell progenitors enter the thymus and undergo rapid proliferation and migration into the cortex followed by rearrangement of the TCR β-chain. These double negative (DN) thymocytes receive a survival signal when a functionally recombined β-chain pairs with the germline encoded pre-TCRα chain to form the pre-TCR. These thymocytes upregulate the expression CD4 and CD8, becoming double positive (DP) cells and undergo rearrangement of the TCR α chain locus. DP thymocytes expressing mature TCRα/β undergo one of several fates depending on the strength of TCR binding to MHC expressed on cortical epithelial cells (CEC): TCR that is unable to bind MHC (X) results in death by neglect; high affinity interaction (++) induces negative selection; while a balanced stimulus results in survival and migration. Within the thymic medulla differentiation into naïve CD4 or CD8 single positive (SP) cells occurs through interactions with follicular dendritic cells (FDC). B cell development occurs within the bone marrow and spleen (b). CD34+ hematopoietic progenitors differentiate into prepro and then pro-B cells, which undergo heavy chain (Igμ) rearrangement. Recombined heavy chain pairs with germline encoded VpreB and Igλ5 light chains resulting in expression of the pre-BCR. These cells begin to proliferate and are called large pre-B cells. A survival signal is then imparted by expression of the pre-BCR. As they cease proliferating, pre-B cells downregulate the pre-BCR, and are called small pre-B cells. Rearrangement of the light chain loci occurs in these cells, resulting in expression of a mature BCR. Immature B cells expressing the recombined BCR are subject to negative selection if they recognize antigen at this stage. Immature B cells the express a mature BCR but are not self-reactive become transitional B cells, which exit the bone marrow and complete B cell development in the spleen.

with TCRs of high affinity for self-peptide: MHC are negatively selected on thymic antigen presenting cells located further into the cortico-medullary boundary. Cells surviving negative selection become single positive (SP) thymocytes and emigrate from the thymus as mature naïve T_H (CD4) or T_C (CD8) cells.

The comparable events of B cell development mainly occur within the bone marrow (Figure 7.2). Signals from bone marrow stroma induce recombination of the immunoglobulin heavy chain in pro-B cells. B cell precursors expressing a functionally recombined heavy chain pair it with a surrogate light chain molecule, resulting in signaling though the pre-B cell receptor (pre-BCR) and the formation of proliferating large pre-B cells. As these precursors cease dividing, they become small-pre-B cells, recombine the immunoglobulin light chain, and, if successful, express a competent immunoglobulin receptor and exit from the bone marrow as transitional B cells. Negative selection of B cells is analogous to that of thymocytes. B cells expressing high affinity receptors for self-antigen undergo apoptosis or, alternatively, may continue to rearrange their immunoglobulin loci in an attempt to generate a more suitable BCR. Following exit from the bone marrow, transitional B cells complete development within the spleen.

7.3.1 NF-κB in Early T and B Cell Development

NEMO inactivation in both mice and humans underscores the need for NF-κB in lymphopoiesis even if, in many cases, the where and why of NF-κB function remain obscure. Transfer of p65/p50 dKO fetal liver cells demonstrates that there is a requirement for classical NF-κB signaling early in lymphopoiesis, prior to expression of the preantigen receptors (pre-TCR or pre-BCR) [36]. Transfer of cells lacking both c-Rel and p65 exhibit a similar developmental failure [37]. It is likely, though unproven, that NF-κB is involved in the expression of antiapoptotic factors required for cell survival downstream of proapoptotic stimuli, such as TNFα, to which lymphoid progenitors are exposed (Figures 7.3 and 7.4). In fact, early CD34+ bone marrow cells do respond to various NF-κB stimuli, including TNFα, and in these cells NF-κB acts as a prosurvival factor allowing colony formation during the early stages of hematopoiesis [38]. Although these progenitor cells do clearly respond to NF-κB stimuli, it is hypothesized that in a genetically intact hematopoietic system NF-κB may play a more important role in suppressing the production of TNFα to which the cells in the knockout systems are susceptible. Evidence for this comes from cotransfer experiments in which it has been shown that inclusion of normal fetal liver cells during the transfer of NF-κB deficient cells rescues the ability of the knockout cells to generate lymphocytes.

7.3.1.1 NF-κB in Preantigen Receptor Signaling

The expression of the precursor antigen receptor during lymphocyte development produces survival signals that depend, at least in part, on NF-κB. Pre-TCR expression in DN thymocytes coincides with high levels of NF-κB activity, and NF-κB activity at this stage is necessary for DN survival and progression to the DP stage (Figure 7.3). Indeed, overexpression of constitutively active IKK allows DN lymphocytes to survive in the absence of recombination (TCR signaling), while inhibition of NF-κB by expression of IκBα-SR decreases DN thymocyte maturation and induces apoptosis in these cells [39]. Likewise, signaling through the pre-BCR likely also induces antiapoptotic signals through NF-κB (Figure 7.4). Retroviral transduction of BM with an IBα-SR results in a decrease in pre-B cells that can be rescued by overexpression of antiapoptotic Bcl-2 family member Bcl-XL [40,41]. Despite these data it is unclear how NF-κB is activated downstream of the pre-AgR. Multiple knockouts that block the TCR-induced activation of NF-κB in mature lymphocytes do not affect lymphopoiesis. Thus, there is a clear requirement for NF-κB during pre-TCR expression although the signaling pathways involved are not yet delineated.

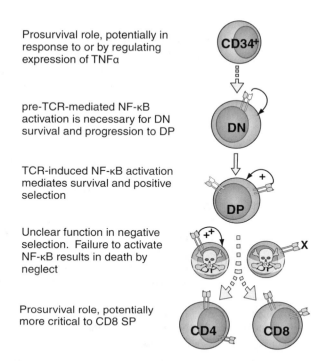

Prosurvival role, potentially in response to or by regulating expression of TNFα

pre-TCR-mediated NF-κB activation is necessary for DN survival and progression to DP

TCR-induced NF-κB activation mediates survival and positive selection

Unclear function in negative selection. Failure to activate NF-κB results in death by neglect

Prosurvival role, potentially more critical to CD8 SP

FIGURE 7.3 Role of NF-κB in T cell development. See text for detailed discussion.

7.3.1.2 Positive and Negative Selection of Thymocytes

DP thymocytes are selected on the basis of the ability of their TCR to recognize peptide:MHC complexes. Cells that express TCRs that are unable to bind to MHC die in a process termed death by neglect, while those that can bind peptide: MHC may be either positively or negatively selected depending on the strength of this interaction. Cells that bind with very high affinity are those that would be self-reactive and are, therefore, negatively selected. Thus only DP thymocytes that recognize self-peptide:MHC with an affinity that falls within a defined range will be positively selected to become SP T cells. Somewhat counterintuitively, it appears that NF-κB functions in both positive and negative selection of thymocytes.

Using agonistic anti-CD3 antibodies to mimic the high affinity TCR binding responsible for negative selection, it has been shown that inhibition of NF-κB blocks the induction of apoptosis [42]. Especially intriguing is the finding that DP thymocytes display the expected sensitivity to other cell stresses, i.e., inhibition of NF-κB sensitizes DP thymocytes to proapoptotic stimuli except TCR ligation. Furthermore, thymocytes in male mice expressing a TCR transgene for H-Y, the so-called male antigen, that normally would undergo negative selection, is rescued by inhibiting NF-κB [43]. In the context of the best-known roles for NF-κB, this function is somewhat difficult to explain; however, other situations in which NF-κB appears to foster apoptosis, though few in number, have been described. Various proapoptotic genes, e.g., Fas and FasL, are under the control of NF-κB. Late,

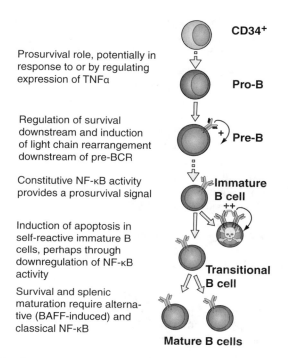

Prosurvival role, potentially in response to or by regulating expression of TNFα

Regulation of survival downstream and induction of light chain rearrangement downstream of pre-BCR

Constitutive NF-κB activity provides a prosurvival signal

Induction of apoptosis in self-reactive immature B cells, perhaps through downregulation of NF-κB activity

Survival and splenic maturation require alterna-tive (BAFF-induced) and classical NF-κB

CD34+

Pro-B

Pre-B

Immature B cell

Transitional B cell

Mature B cells

FIGURE 7.4 Role of NF-κB in B cell development. See text for detailed discussion.

medullary negative selection of TCR transgenic thymocytes may be Fas-dependent at high antigen doses [44], suggesting NF-κB could mediate negative selection by sensitizing thymocytes to FasL expressed in the thymus [45]. Although negative selection appears normal in Fas-deficient mice, suggesting this is not the target of NF-κB in negative selection; this example provides a model for conceptualizing how NF-κB might promote negative selection. Arguably, this example may be more akin to clonal deletion of peripheral T cells than to negative selection of thymocytes; however, the underlying theme of robust TCR activation inducing NF-κB-dependent, pro-apoptotic genes is germane.

While the role of NF-κB in positive selection of thymocytes is more in keeping with widely held views of NF-κB, e.g., as an inducer of antiapoptosis genes, the details remain to be fully understood. Studies of transgenic mice expressing constitutively active IκBα mutants suggest that NF-κB may be preferentially required for the positive selection of those cells destined to become CD8+ SP cells. It is thought that NF-κB acts in this capacity by regulating the expression of prosurvival Bcl-2 family members. The molecular mechanisms by which strength of signaling through the TCR determines whether NF-κB mediates positive or negative selection, however, remain unclear.

7.3.1.3 Negative Selection of B Cells

As in thymocytes, negative selection of immature B cells is mediated by strength of signaling through the antigen receptor; that is, high affinity BCR binding to self-antigen induces apoptosis. Unlike thymocytes, however, NF-κB functions as a pro-survival factor in B cell negative selection. Immature B cells display constitutive NF-κB activity. It is thought that BCR ligation in immature B cells induces a stable downregulation of basal NF-κB activity, sensitizing these cells to what are likely preexisting proapoptotic signals. Alternatively, subsequent BCR stimulation itself may provide a proapoptotic stimulus in the setting of inhibited NF-κB. Evidence for this model comes primarily from *in vitro* experiments in which it was shown that blocking basal NF-κB activity induces apoptosis in an *immature-like* B cell line that otherwise expresses constitutive NF-κB activity [46]. Similar to T cells, some signaling components that are required for NF-κB activation in mature B cells can be genetically targeted without disrupting B cell development. Therefore, there are differences in the NF-κB signaling pathway in developing B cells that may facilitate the somewhat anomalous response to BCR ligation that is observed in these cells.

7.3.2 NF-κB in Late Stages of Lymphocyte Development

Part of the difficulty in ascertaining the role of NF-κB in the late events of lymphocyte development is the lack of appropriate experimental tools. Genetic alterations that affect survival frequently affect earlier stages of lymphocyte development, while promoters that have been used in the conditional deletion of genes tend to act gradually, incompletely, and over multiple stages of thymocytes and B cell maturation. Although no single NF-κB knockout has a severe phenotype with regards to the generation of mature lymphocytes, double knockouts demonstrate that the antiapoptotic function of NF-κB is important to the maturation and survival of lymphocytes.

Following positive and negative selection, DP thymocytes must make a lineage commitment and become single positive (SP) thymocytes (CD4$^+$CD8$^-$ or CD4$^-$CD8$^+$), which shortly thereafter emigrate from the thymus. Whether NF-κB has a role in the CD4/CD8 decision is not yet entirely clear, however, there is evidence that NF-κB is required for survival of these cells once they reach the SP stage. Targeted deletion of NEMO through *cd4*-promoter driven Cre recombinase expression, or overexpression of kinase dead IKKβ, results in loss of mature peripheral T cells [47]. Of note, this deficit in thymocyte survival and emigration is most likely thymocyte-intrinsic as it occurs in the presence of wild-type cells that fail to delete NEMO and survive to populate the periphery. These data strongly suggest that NF-κB activation is required in late stages of T cell development, however, *ikk$^{-/-}$*, *tnfr1$^{-/-}$* double knockouts, *IKKβ$^{-/-}$* chimeras, or *cd4*-Cre IKKβ conditional knockouts do not have a defect in the production of naïve T cells [47,48]. These data are rather intriguing, as they represent a rare instance in which there is a requirement for NEMO but not IKK.

Analysis of B-luciferase transgenic mice, in which luciferase expression is driven by canonical κB binding sites, have shown that CD8 SP thymocytes have significantly higher levels of NF-κB activity than CD4 SP thymocytes [39]. Conversely, the anti-apoptotic factor Bcl-2 is more highly expressed in CD4 cells than CD8 cells, suggesting that CD8 SP thymocytes may be more dependent on NF-κB for survival. Therefore, it may be that the primary function of NF-κB in SP thymocytes is to provide a pro-survival signal, upon which CD8$^+$ thymocytes are particularly dependent.

Following completion of the selection process, immature B cells emigrate from the bone marrow to complete development in the spleen. Postbone marrow B cells, which are referred to as transitional B cells, mature into either follicular or marginal zone B cells. Both p50/p52 and p65/c-Rel dKO progenitor cells are defective in their ability to generate mature B cells following adoptive transfer [12,37]. These and other studies demonstrate that NF-κB regulation of the expression of prosurvival factors, likely Bcl-2 family members, is crucial during these final steps of B cell development [49]. Interestingly, it appears that both alternative and classical pathways have a role in late B cell maturation; perhaps a partial requirement for both pathways explains why deletion of p50 and p52 produces a more complete block in B cell development than loss p65 and c-Rel. Furthermore, deficiency in NEMO, IKKα, or IKKβ, likewise decreases the numbers of mature B cells [50,51,52]. While these data suggest that both alternative and classical pathways are important for B cell maturation, recent work [53,54,55] has persuasively established the importance of BAFF in activation of the alternative pathway and in promoting the expression of antiapoptotic Bcl-2 family members in transitional B cells (see [56]). Indeed BAFF knockout mice exhibit a complete failure of transitional B cell maturation, which mirrors that of Bcl-X_L knockout mice [53,57,58]. Nevertheless, in terms of NF-κB's function, only those knockouts that target both alternative and classical pathways have an effect of the magnitude of BAFF or Bcl-X_L. The inducing stimulus that mandates this requirement for the classical NF-κB pathway in transitional B cells is unknown.

7.4 LYMPHOCYTE ACTIVATION

7.4.1 B Cell and T Cell Receptor Signaling to NF-κB

Signaling from B cell and T cell receptors is the central event of the adaptive immune response, as it is this event that is responsible for conferring antigen specificity. Activation of NF-κB downstream of BCR and TCR ligation facilitates antigen specific proliferation and maturation of lymphocytes into effector cells. Signaling through these two antigen receptors is functionally analogous, although the molecular details differ. We will therefore discuss the pathway leading to NF-κB from antigen receptors by focusing on the TCR, and highlight those aspects in which the pathways differ.

The T cell receptor complex consists of α/β subunits which are associated with the CD3 protein heterodimers γ/ε, δ/ε, and either η/η, η/ζ, or ζ/ζ, for a total of eight membrane proteins. The BCR is, likewise, a multiprotein complex consisting of the surface immunoglobulin receptor associated with a heterodimer of Igα and Igβ. These AgR complexes interact with Src family tyrosine kinases (SFKs), Lck, and Fyn in T cells and Lyn in B cells, as well as the phosphatase CD45. CD45 facilitates activation of SFKs by removing inhibitory phosphorylation. Following receptor ligation, SFKs are activated and induce in the phosphorylation of immunoreceptor tyrosine activation motifs (ITAMs) on CD3 and Igα/ chains. The cytoplasmic tyrosine kinases ZAP70 or Syk are then recruited via SH2 domains to the phosphorylated ITAMs of the TCR or BCR complexes, respectively. ZAP70 and Syk, in turn, initiate activation of the IP$_3$ and Ras family pathways. However, the pathways that lead to NF-κB from these receptor proximal tyrosine kinases had remained relatively elusive.

Recent studies have implicated a number of potential signaling intermediates that appear to comprise a novel signaling pathway that includes PKCθ, CARMA1/ CARD11, BCL10, and MALT1 (see Chapter 3). Knockouts of each of these genes led to a specific block in NF-κB activation in response to antigen-receptor signaling, and therefore current efforts are focused on determining how these proteins are linked to the IKK complex [59,60,61,62]. The IKK complex is rapidly recruited to the immunological synapse and can be colocalized to the TCR by confocal immunofluorescence analysis and coimmunoprecipitated with the TCRζ chain [63,64]. In ZAP-70 deficient cells, NF-κB activation is rescued by overexpression of a NEMO-(SH2)$_2$ chimera, which directly targets NEMO to the immunological synapse, suggesting that signaling downstream of ZAP-70 may only be necessary for IKK recruitment [63].

It is clear from knockout studies that PKCθ is essential for activation of NF-κB via T cell stimulation [65]. PKCθ is specifically recruited to the immunological synapse, although how specificity for PKCθ is achieved remains unclear since T cells express most of the other PKC isoforms. PKCθ is capable of directly interacting with the IKK complex in primary T cells [64]. PKCθ might function by bringing IKK to the receptor complex and, thus, into proximity with other essential components in this pathway, namely BCL10, and MALT1 (collectively known as the CBM complex). Recently we have provided evidence, using RNAi-mediated knockdown experiments, that the protein kinase PDK1 recruits PKCθ and the IKK complex to lipid rafts. In addition, we have found that PDK1 simultaneously recruits the CBM complex by binding to CARMA1 [66]. Interestingly, T cell-specific PDK1 conditional deletion results in a defect in T cell development, preventing the production of peripheral T cells. On the other hand, PCKθ knockouts do not display defects in thymocyte development.

The CBM complex is essential to both T cell and B cell antigen receptor signaling. What is surprising, however, is the lack of role for this complex in developing lymphocytes, and by extrapolation signaling through the pre-AgRs. The MAGUK family protein CARMA1 is required for activation of NF-κB in T cells following TCR ligation, but its loss has no effect on the development of thymocyte [60,67,68]. Similarly, BCL10 is critical for NF-κB activation via the BCR and TCR, yet normal numbers of peripheral T cells are seen in BCL10 knockouts, and no clear defects in B cell development are observed [62]. BCL10 interacts with CARMA1 leading to BCL10 phosphorylation, although CARMA-1 lacks kinase activity [69,70].

Interestingly genetic evidence of a role for RIP2 has recently been reported in T cell signaling [71]. RIP2 associates with BCL10 and is necessary for TCR-induced BCL10 phosphorylation and IKK activation. It is not yet clear how upstream mediators of TCR signaling may regulate RIP2 and, in turn, how RIP2 might affect IKK activity. BCL10 oligomerization has been implicated in IKK activation through a process that involves ubiquitination of NEMO [72]. This ubiquitination event appears to be mediated by MALT1, and possibly CARMA1. Perhaps this TCR-induced ubiquitination of NEMO may depend on TRAF6, although no T cell deficits have been reported in TRAF6 deficient mice [10,73]. If this is so, one might have predicted that this effect would be through the kinase TAK1, which functions in IKK activation downstream of TRAF6 in other pathways. However, B cells in which TAK1 has been conditionally knocked out have defective NF-κB signaling from LPS but normal BCR signaling to NF-κB. More work must be done with TAK1 conditional knockouts in both the BCR and TCR pathways to address the role of this kinase.

BCR and TCR signaling are analogous in many aspects, in particular with respect to the activation of NF-κB (Figure 7.5). In both cells, NF-κB activation is critically dependent on the CBM complex and on PKC family members for the recruitment and activation of IKK. In the case of BCR signaling, the conventional PKCβ, instead of PKCθ, is the important PKC family member. This difference underscores another likely difference in signaling in these lymphocytes as conventional PKC isoforms

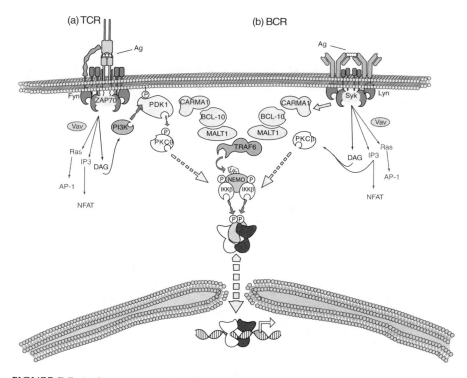

FIGURE 7.5 Antigen receptor signaling to NF-κB. Signaling initiates with the interaction of the TCR complex and MHC/antigen (Ag; a) or the BCR complex with antigen alone (b), and occurs through highly analogous pathways to NF-κB activation. Key differences between the pathways include the receptor complex itself, the receptor proximal kinases, and PKC isoform utilization. It is likely that PDK1 plays a similar role in both BCR and TCR signaling, although this remains to be shown. Not shown are costimulatory signaling events required for activation of naïve lymphocytes by antigens other than superantigens (T cells) and thymus independent antigens (B cells). Many receptor-proximal signaling components, as well as the details of the NFAT and AP-1 pathways, have been left out for clarity.

require both DAG and Ca^{2+} for activation. Nevertheless, PKCβ is required for recruitment of the IKK complex to lipid rafts and signaling to NF-κB in response to BCR ligation [74]. Thus, differences in PKC isoform utilization between BCR and TCR pathways do not appear to correlate with functional differences in the mechanism of NF-κB activation.

7.5 NF-κB IN LYMPHOCYTE RESPONSES

Signaling through antigen receptors is accompanied by costimulatory signals that modulate the resulting transcriptional response. In order to respond to antigen, lymphocytes must undergo both proliferative (clonal expansion) as well as differentiative processes. In addition, activation of antigen receptors may also induce apoptosis, a process termed "activation induced cell death" (AICD). Thus, as in the response of innate immune system (Chapter 6), the adaptive system must regulate NF-κB in both the initiation and resolution of the effector response.

To become activated, naïve T cells must receive two distinct signals: antigen specific and costimulatory. Antigen-specific activation signals emanate from the binding of the TCR to cognate antigen expressed in the binding cleft of MHC. Costimulatory signaling is provided through ligation of T cell CD28 by B7 molecules expressed on activated APCs (Figure 7.6). Stimulation of naïve T cells results

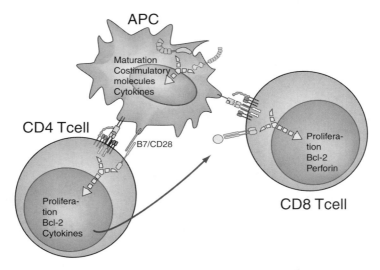

FIGURE 7.6 NF-κB in the activation of naïve T cells. Naïve CD4$^+$ T cells are activated following TCR recognition of antigen/MHC-II as well as costimualtion through binding of CD28 to B7 expressed on mature antigen presenting cells (APCs). Activation of NF-κB-regulated genes mediates proliferation, protection from apoptosis and cytokine production in CD4$^+$ T cells. Production of cytokines by T$_H$1 cells provides a costimulatory signal to naïve CD8$^+$ T cells that recognize their cognate antigen displayed on MHC-I. Costimulation in CD8$^+$ T cells results in NF-κB-dependent proliferation, protection from apoptosis, and the production of perforin and other mediators of effector function.

in the production of IL-2, which is necessary for proliferation and survival. Activated naïve T cells proliferate rapidly and simultaneously undergo differentiation into effector cells. In the case of T_H cells, proliferation leads to differentiation into immature effector cells, T_H0, which subsequently differentiate into T_H1 or T_H2 cells depending on the predominant cytokine milieu. CD8 T cells are likewise activated by professional APCs, although they may receive secondary signals from activated T_H1 cells. The lack of CD8 conditional knockouts, and selective loss of this population due to inhibition of NF-κB, has resulted in a paucity of data on NF-κB function in these cells.

Nevertheless, it is clear that NF-κB has an important role in activation and function of T cells. In particular, rapidly proliferating activated T cells rely on NF-κB activity for protection from apoptosis as well as the production of cytokines supporting proliferation and differentiation. As we might expect, inhibition of NF-κB in activated T cells facilitates progression towards AICD or apoptosis [75,76]. Indeed, stimulation of p65-deficient naïve T cells does induce cell death [77]. Peripheral T cells lacking c-Rel do not undergo apoptosis, but, nevertheless, fail to proliferate in response to typical mitogenic stimuli [78] and both p65 and c-Rel containing complexes accumulate in the nucleus following TCR/CD28 stimulation [79]. Interestingly, c-Rel deficient T cells appear to have a defect in T_H1 proliferation and production of IFNγ, indicating a selective role for NF-κB family members in T_H1/T_H2 differentiation, independent of that mediated by the innate response (Chapter 6).

Multiple transcriptional activators and repressors regulate expression of IL-2. Among these, members of the NF-κB family play multiple roles. In naïve T cells, which are not expressing IL-2, repressive p50 homodimers are found associated with the IL-2 promoter [80]. Failure of T cell proliferative responses in c-Rel knockouts is attributable to a failure to produce IL-2 [78]. In naïve T cells, c-Rel is responsible for mediating chromatin remodeling across the IL-2 locus following CD3/CD28 costimulation [81]. Naïve T cells can be primed by exposure to inflammatory cytokines such that they generate a more robust response to CD3/CD28 costimulation. The use of dn-IκBα overexpression suggested that NF-κB is required for this priming event in T cells [82]. More recent data indicate that c-Rel is necessary for naïve helper T cell priming by proinflammatory cytokines elicited following stimulation with TLR ligands [83]. NF-κB p65 containing complexes, on the other hand, appear to function more traditionally in mediating transactivation of IL-2 gene expression, and overexpression of p65 with c-Jun can overcome the requirement for costimulation in naïve T cells [84]. However, as discussed below, these complexes may also be the target of negative regulation following T cell differentiation.

Recent work in T_H1/T_H2 differentiation has focused on the induction of specific transcription factors in these two effector cell types — T-bet and GATA3 respectively. Interestingly, mice lacking p50 are unable to mount an asthma-like airway T_H2 response, and do not induce GATA-3 expression during T cell stimulation under T_H2 differentiating conditions [85]. Consistent with this finding, BCL-3 deficient T cells also fail to undergo T_H2 differentiation. Furthermore, BCL-3 can induce expression of a reporter gene from a *gata-3* promoter, suggesting that p50/BCL-3 complexes are crucial for T_H2 differentiation [86]. Conversely, the same authors found that RelB deficient T cells are deficient in T_H1 differentiation and IFNγ production, and

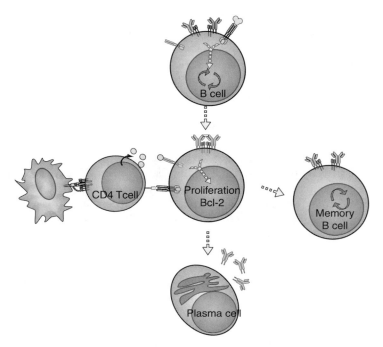

FIGURE 7.7 NF-κB in the activation of B cells. B cells require NF-κB for baseline homeostasis, perhaps downstream of BAFF-R, and tonic BCR signaling (top). B cells that recognize
their cognate antigen (center) through surface IgM, and also costimulatory cytokines from CD4⁺
T cells (left), undergo NF-κB dependent proliferation and maturation into antibody secreting
plasma cells. NF-κB is also required for maintenance and survival of memory B cells (right).

show decreased expression of T-bet; likely through a failure to upregulate STAT4,
which functions in signaling from IFN to T-bet induction. Therefore, it appears that
NF-κB activation during TCR stimulation may render cells competent for both
proliferative and differentiating stimuli.

As T_H cells differentiate in to T_H1 or T_H2 they decrease their expression of IL-2
and, instead, become dependent on T_H1 and T_H2 cytokines (e.g., IFNγ and IL-4).
As a correlate, NF-κB transactivation of the IL-2 gene is repressed. Direct binding
of T-bet to p65 that is associated with the IL-2 gene enhancer may mediate the
repression of IL-2 production in T_H1 cells [87]. Alternatively, in T_H2 cells the lack
of IL-2 transcription may be directly attributed the observation that there is decreased
constitutive p65 activation in T_H2 cells [88].

B cell responses are either of two forms — thymus-dependent (TD) or -inde-
pendent (TI) (Figure 7.7). In responses to T-dependent antigens, B cells require
costimulatory signaling from T_H cells expressing CD40L and cytokines, such as
IL-4. B cells from individuals with a mutation in CD40L are unable to undergo class
switch recombination in response to T-dependent antigens [89]. Signaling through
CD40 activates both canonical and alternative NF-κB pathways, although it is
unclear which is operative in the response to T-dependent antigens. For example,

while B cells from *p52-/-* mice mount inadequate humoral responses to various T-dependent antigens, they exhibit a normal response following adoptive transfer into *rag-1-/-* mice, indicating that this deficit is not intrinsic to B cells [29]. Furthermore, B cells from *RelB-/-* mice, although crippled in their proliferative response, undergo normal IgM secretion and class switching in response to various stimuli [90]. Therefore, it is likely that alternative pathway activation downstream of CD40 is not required for class switching during T-dependent antigen responses.

Analysis of the classical pathway is complicated by more generalized defects in lymphocyte responses due to the requirement for this pathway in AgR signaling. Nevertheless, whether downstream of CD40 or other stimuli, evidence supports classical pathway activation in the process of class switch recombination. Following adoptive transfer, B cells from *p65-/-* mice exhibit markedly diminished class switching, despite a less severe loss of lymphocyte proliferation following various stimuli [91]. Likewise, c-Rel deficient mice, or mutants lacking the c-Rel TAD, fail to generate a productive humoral immune response suggesting a requirement for c-Rel in class switch recombination [78,92,93]. B cells from *p50-/-* mice exhibit decreased proliferation in response to mitogenic stimulation, and p65/p50 dKO B cells exhibit greater defects in proliferation and class switching [90,94]. Therefore, analyses of knockout animals suggest that the classical NF-κB pathway likely has a role in maturation of the B cell response in addition to directly mediating proliferative responses following BCR ligation.

As discussed in Chapter 6, signaling through TLRs has an important role in the initiation of the adaptive immune response via antigen presenting cells of the innate immune system. In recent years, however, there has also been increasing interest in the ability of TLR signaling to directly modulate the adaptive response. For example, it has been observed that homeostatic polyclonal activation of B cells, which results in so called serological memory, i.e., detectable antibody to antigens that are no longer present in the host, can be induced/maintained by TLR ligation [95]. Analogously, TLR2 is upregulated in CD4+ T cells following TCR stimulation, and TLR2 ligands may thus provide an activation/maintenance signal in these cells [96]. More recently, it has been suggested that TLR signaling in B cells may also be required for optimal response to TD antigens [97]. That signaling through TLRs in these aspects of B cell responses requires NF-κB seems likely, but has yet to be demonstrated.

Thymus independent antigens have an intrinsic ability to activate B cell responses in the absence of T cell help by acting as B cell mitogens, TLR ligands, for example, or by having high avidity binding to the BCR through repetitive structural features. In such cases it is expected that B cell responses are more dependent on members of the classical pathway that have well-documented roles in TLR signaling (Chapter 6) or BCR signaling (as discussed above). For example, c-Rel-deficient B cells are highly sensitive to apoptosis following BCR cross-linking [98,99,100]. As mentioned above, p50 and p50/p65 dKO B cells are deficient in responses to TI stimulation. Likewise, IKKβ-deficient B cells fail to mount TI or TD responses [101]. These IKKβ-deficient B cells also exhibit increased spontaneous apoptosis, suggesting that NF-κB is important in survival of B cells.

7.6 LYMPHOCYTE MAINTENANCE

Lymphocyte homeostasis is dependent on the survival of mature lymphocytes in addition to replenishment of the peripheral lymphocyte pool through lymphopoiesis. Consequently, there is increasing interest in the possible role of NF-κB in the survival of mature lymphocytes. It is widely accepted that lymphocyte survival is mediated through tonic stimulation downstream of the AgR, as well as certain cytokine receptors. As discussed above, genetic targeting experiments support an important role for NF-κB family members in lymphocyte survival. Naïve T cells require continued contact with MHC: self-peptides, most likely expressed on lymphoid DCs, to generate the tonic TCR signal that is essential for continued survival. Survival of memory cells, on the other hand, is independent of continued contact with self-peptide:MHC complexes.

B cells are formed at a far higher rate than T cells, but they also undergo a significantly higher rate of turnover in the periphery. Nonetheless, they too require maintenance signals to achieve peripheral homeostasis. The antigen receptor on B cells most likely provides a basal level of signaling, albeit independent of the presence of antigen, which is required for maintenance of mature B cells. Not surprisingly, B cells from $p65^{-/-}$, $p100^{-/-}$, $p105^{-/-}$, and $c\text{-}Rel^{-/-}$ mice display increased sensitivity to apoptosis and/or decreased survival *ex vivo* [54,99,102].

In large part, these defects appear to be due to a loss of BCR signaling, as demonstrated in an elegant study that demonstrated that deletion of the BCR from mature B cells led to a complete loss of the peripheral B cell pool [103]. Most likely this was due to the loss of signaling to NF-κB in these cells because loss of IKKβ, NEMO, or components of the CBM complex in mature B cells also results in a complete loss of peripheral B cells [50,101,104].

The alternative pathway is also relevant to B cell survival, as the loss of IKKα results in striking defects in B cell survival [51,52]. However, rather than acting downstream of tonic BCR signaling, recent studies have implicated signaling from BAFF-R in this aspect of the B lymphocyte survival (reviewed in [56]). Together these data suggest that a subset of Bcl-2 family members, e.g., the antiapoptotic factor A1, are regulated by RelB/p52 containing complexes and are necessary for the maintenance of mature B cells.

7.7 CONCLUSIONS

NF-κB was originally described as a regulator in the adaptive immune response and in many respects this continues to be an area of extremely productive research. Investigations of mice with targeted deletions have shed considerable light on the process of lymphoid organogenesis and, for example, through characterization of the *aly/aly* mouse, studies of lymphoid organogenesis have driven the discovery of the alternative NF-κB pathway and various receptors that signal to NF-κB. Progress continues to be made in understanding antigen receptor signaling to NF-κB, and this work has led to the appreciation of regulatory ubiquitination events in IKK activation (see Chapter 4). The specificity of the requirement for the CBM complex and PKCθ/β in signaling by T cell and B cell receptors opens the door to the development of highly specific NF-κB inhibitors. However, there remain fundamental gaps in our understanding of how these components mediate IKK activation and what the role of regulatory ubiquitination is in this process (Chapter 4). To date we have little more than extrapolation, based largely on our understanding of phosphorylation, to explain how these events might function at the molecular level. While a considerable amount of traditional biochemistry remains to be done, there is also optimism that a crystal structure of the IKK complex will allow rapid progress in this particular area of NF-κB research.

The role of NF-κB in hematopoiesis, as discussed in this and the preceding chapter, remains poorly defined, although there is little doubt of its importance. Genetic targeting has allowed enumeration of multiple steps at which various NF-κB pathway components are required. However, a cohesive picture of how NF-κB functions in these pathways has remained elusive. For example, while we know that classical NF-κB activity is required in early lymphopoiesis, it is unclear whether this is in the regulation of TNFα production or in mediating a survival signal in the developing lymphocyte. NF-κB is required in the process of positive and negative selection, but again, its mechanism(s) of action remain poorly defined. Progress using *in vitro* systems for studying lymphocyte development may allow a more rigorous assessment of NF-κB function in these processes. Likewise, advances in conditional gene targeting approaches and the generation of animals in which defective versions of NF-κB genes have been knocked-in may help to overcome the problems of embryonic lethality and functional redundancy that have made existing studies difficult to interpret.

T and B cell responses require NF-κB as a prosurvival factor as well as for the regulation of genes involved in differentiation to effector cells. Most recently, a limited number of studies have suggested a role for pattern recognition receptors in lymphocyte activation, and the role of NF-κB in this process remains to be elucidated. The role of NF-κB in the differentiation of T_H cells is partially understood, but requires further clarification. CD8+ activation, differentiation, and function are, by comparison, quite poorly understood; progress in this area awaits development of better tools for genetically manipulating this cell population. Upon successful clearance of pathogen, regulation of NF-κB allows resolution of the response and facilitates the process of memory. To date, relatively little is known about NF-κB in memory cells; although recent progress in identifying and characterizing memory precursors bodes well for future progress in this area.

In summary, research to date has highlighted the importance of NF-κB in regulating genes that prevent apoptosis and promote differentiation and development in cells of the adaptive immune system. A large body of work has elucidated many of the molecular mechanisms governing regulation of NF-κB by lymphocyte antigen specific receptors. In turn, these data provide a trove of information that will hopefully prove useful in attempts to manipulate the immune system to prevent and treat disease.

REFERENCES

[1] Weih, F., Carrasco, D., Durham, S.K. et al., Multiorgan inflammation and hematopoietic abnormalities in mice with a targeted disruption of RelB, a member of the NF-kappaB/Rel family, *Cell,* 80, 331, 1995.

[2] Burkly, L., Hession, C., Ogata, L. et al., Expression of relB is required for the development of thymic medulla and dendritic cells, *Nature,* 373, 531, 1995.

[3] Zuklys, S., Balciunaite, G., Agarwal, A. et al., Normal thymic architecture and negative selection are associated with AIRE expression, the gene defective in the autoimmune-polyendocrinopathy-candidiasis-ectodermal dystrophy (APECED), *J. Immunol.,* 165, 1976, 2000.

[4] Chin, R.K., Lo, J.C., Kim, O. et al., Lymphotoxin pathway directs thymic AIRE expression, *Nat. Immunol.,* 4, 1121, 2003.

[5] Grech, A.P., Riminton, D.S., Gabor, M.J. et al., Increased thymic B cells but maintenance of thymic structure, T cell differentiation and negative selection in lymphotoxin-alpha and TNF gene-targeted mice, *Dev. Immunol.,* 8, 61, 2000.

[6] Jimi, E. and Ghosh, S., Role of nuclear factor-kappaB in the immune system and bone, *Immunol. Rev.,* 208, 80, 2005.

[7] Dougall, W.C., Glaccum, M., Charrier, K. et al., RANK is essential for osteoclast and lymph node development, *Genes Dev.,* 13, 2412, 1999.

[8] Naito, A., Azuma, S., Tanaka, S. et al., Severe osteopetrosis, defective interleukin-1 signalling and lymph node organogenesis in TRAF6-deficient mice, *Genes Cells,* 4, 353, 1999.

[9] Wong, B.R., Josien, R., Lee, S.Y. et al., The TRAF family of signal transducers mediates NF-kappaB activation by the TRANCE receptor, *J. Biol. Chem.,* 273, 28355, 1998.

[10] Lomaga, M.A., Yeh, W.C., Sarosi, I. et al., TRAF6 deficiency results in osteopetrosis and defective interleukin-1, CD40, and LPS signaling, *Genes Dev.,* 13, 1015, 1999.

[11] Darnay, B.G., Haridas, V., Ni, J. et al., Characterization of the intracellular domain of receptor activator of NF-kappaB (RANK). Interaction with tumor necrosis factor receptor-associated factors and activation of NF-kappaB and c-Jun N-terminal kinase, *J. Biol. Chem.,* 273, 20551, 1998.

[12] Franzoso, G., Carlson, L., Xing, L. et al., Requirement for NF-kappaB in osteoclast and B-cell development, *Genes Dev.,* 11, 3482, 1997.

[13] Iotsova, V., Caamano, J., Loy, J. et al., Osteopetrosis in mice lacking NF-kappaB1 and NF-kappaB2, *Nat. Med.,* 3, 1285, 1997.

[14] Aya, K., Alhawagri, M., Hagen-Stapleton, A. et al., NF-(kappa)B-inducing kinase controls lymphocyte and osteoclast activities in inflammatory arthritis, *J. Clin. Invest.,* 115, 1848, 2005.

[15] Novack, D.V., Yin, L., Hagen-Stapleton, A. et al., The IkappaB function of NF-kappaB2 p100 controls stimulated osteoclastogenesis, *J. Exp. Med.,* 198, 771, 2003.

[16] Chaisson, M.L., Branstetter, D.G., Derry, J.M. et al., Osteoclast differentiation is impaired in the absence of inhibitor of kappa B kinase alpha, *J. Biol. Chem.,* 279, 54841, 2004.

[17] Ruocco, M.G., Maeda, S., Park, J.M. et al., I{kappa}B kinase (IKK){beta}, but not IKK{alpha}, is a critical mediator of osteoclast survival and is required for inflammation-induced bone loss, *J. Exp. Med.,* 201, 1677, 2005.

[18] Jimi, E., Aoki, K., Saito, H. et al., Selective inhibition of NF-kappa B blocks osteoclastogenesis and prevents inflammatory bone destruction in vivo, *Nat. Med.,* 10, 617, 2004.

[19] Dai, S., Hirayama, T., Abbas, S. et al., The IkappaB kinase (IKK) inhibitor, NEMO-binding domain peptide, blocks osteoclastogenesis and bone erosion in inflammatory arthritis, *J. Biol. Chem.,* 279, 37219, 2004.

[20] Mebius, R.E., Organogenesis of lymphoid tissues, *Nat. Rev. Immunol.,* 3, 292, 2003.

[21] Weih, F., Carrasco, D., and Bravo, R., Constitutive and inducible Rel/NF-kappaB activities in mouse thymus and spleen, *Oncogene*, 9, 3289, 1994.

[22] Alcamo, E., Hacohen, N., Schulte, L.C. et al., Requirement for the NF-kappaB family member RelA in the development of secondary lymphoid organs, *J. Exp. Med.*, 195, 233, 2002.

[23] Miyawaki, S., Nakamura, Y., Suzuka, H. et al., A new mutation, aly, that induces a generalized lack of lymph nodes accompanied by immunodeficiency in mice, *Eur. J. Immunol.*, 24, 429, 1994.

[24] Koike, R., Nishimura, T., Yasumizu, R. et al., The splenic marginal zone is absent in alymphoplastic aly mutant mice, *Eur. J. Immunol.*, 26, 669, 1996.

[25] Shinkura, R., Kitada, K., Matsuda, F. et al., Alymphoplasia is caused by a point mutation in the mouse gene encoding NF-kappaB-inducing kinase, *Nat. Genet.*, 22, 74, 1999.

[26] Yilmaz, Z.B., Weih, D.S., Sivakumar, V. et al., RelB is required for Peyer's patch development: Differential regulation of p52-RelB by lymphotoxin and TNF, *Embo J.*, 22, 121, 2003.

[27] Paxian, S., Merkle, H., Riemann, M. et al., Abnormal organogenesis of Peyer's patches in mice deficient for NF-kappaB1, NF-kappaB2, and Bcl-3, *Gastroenterol.*, 122, 1853, 2002.

[28] Caamano, J.H., Rizzo, C.A., Durham, S.K. et al., Nuclear factor (NF)-kappaB2 (p100/p52) is required for normal splenic microarchitecture and B cell-mediated immune responses, *J. Exp. Med.*, 187, 185, 1998.

[29] Franzoso, G., Carlson, L., Poljak, L. et al., Mice deficient in nuclear factor (NF)-kappaB/p52 present with defects in humoral responses, germinal center reactions, and splenic microarchitecture, *J. Exp. Med.*, 187, 147, 1998.

[30] Bonizzi, G. and Karin, M., The two NF-kappaB activation pathways and their role in innate and adaptive immunity, *Trends Immunol.*, 25, 280, 2004.

[31] Weih, D.S., Yilmaz, Z.B., and Weih, F., Essential role of RelB in germinal center and marginal zone formation and proper expression of homing chemokines, *J. Immunol.*, 167, 1909, 2001.

[32] Schwarz, E.M., Krimpenfort, P., Berns, A. et al., Immunological defects in mice with a targeted disruption in Bcl-3, *Genes Dev.*, 11, 187, 1997.

[33] Caamano, J.H., Perez, P., Lira, S.A. et al., Constitutive expression of Bcl-3 in thymocytes increases the DNA binding of NF-kappaB1 (p50) homodimers in vivo, *Mol. Cell. Biol.*, 16, 1342, 1996.

[34] Sha, W.C., Liou, H.C., Tuomanen, E.I. et al., Targeted disruption of the p50 subunit of NF-kappaB leads to multifocal defects in immune responses, *Cell*, 80, 321, 1995.

[35] Sentman, C.L., Shutter, J.R., Hockenbery, D. et al., bcl-2 inhibits multiple forms of apoptosis but not negative selection in thymocytes, *Cell*, 67, 879, 1991.

[36] Horwitz, B.H., Scott, M.L., Cherry, S.R. et al., Failure of lymphopoiesis after adoptive transfer of NF-kappaB-deficient fetal liver cells, *Immunity*, 6, 765, 1997.

[37] Grossmann, M., Metcalf, D., Merryfull, J. et al., The combined absence of the transcription factors Rel and RelA leads to multiple hemopoietic cell defects, *Proc. Natl. Acad. Sci. USA*, 96, 11848, 1999.

[38] Pyatt, D.W., Stillman, W.S., Yang, Y. et al., An essential role for NF-kappaB in human CD34(+) bone marrow cell survival, *Blood*, 93, 3302, 1999.

[39] Voll, R.E., Jimi, E., Phillips, R.J. et al., NF-kappaB activation by the pre-T cell receptor serves as a selective survival signal in T lymphocyte development, *Immunity*, 13, 677, 2000.

[40] Jimi, E., Phillips, R.J., Rincon, M. et al., Activation of NF-kappaB promotes the transition of large, CD43+ pre-B cells to small, CD43- pre-B cells, *Int. Immunol.,* 17, 815, 2005.

[41] Feng, B., Cheng, S., Hsia, C.Y. et al., NF-kappaB inducible genes BCL-X and cyclin E promote immature B-cell proliferation and survival, *Cell Immunol.,* 232, 9, 2004.

[42] Hettmann, T., DiDonato, J., Karin, M. et al., An essential role for nuclear factor kappaB in promoting double positive thymocyte apoptosis, *J. Exp. Med.,* 189, 145, 1999.

[43] Mora, A.L., Stanley, S., Armistead, W. et al., Inefficient ZAP-70 phosphorylation and decreased thymic selection *in vivo* result from inhibition of NF-kappaB/Rel, *J. Immunol.,* 167, 5628, 2001.

[44] Kishimoto, H., Surh, C.D., and Sprent, J., A role for Fas in negative selection of thymocytes *in vivo, J. Exp. Med.,* 187, 1427, 1998.

[45] French, L.E., Hahne, M., Viard, I. et al., Fas and Fas ligand in embryos and adult mice: Ligand expression in several immune-privileged tissues and coexpression in adult tissues characterized by apoptotic cell turnover, *J. Cell. Biol.,* 133, 335, 1996.

[46] Wu, M., Lee, H., Bellas, R.E. et al., Inhibition of NF-kappaB/Rel induces apoptosis of murine B cells, *Embo J.,* 15, 4682, 1996.

[47] Schmidt-Supprian, M., Tian, J., Ji, H. et al., IkappaB kinase 2 deficiency in T cells leads to defects in priming, B cell help, germinal center reactions, and homeostatic expansion, *J. Immunol.,* 173, 1612, 2004.

[48] Senftleben, U., Li, Z.W., Baud, V. et al., IKKbeta is essential for protecting T cells from TNFα-induced apoptosis, *Immunity,* 14, 217, 2001.

[49] Grossmann, M., O'Reilly, L.A., Gugasyan, R. et al., The anti-apoptotic activities of Rel and RelA required during B-cell maturation involve the regulation of Bcl-2 expression, *Embo J.,* 19, 6351, 2000.

[50] Pasparakis, M., Schmidt-Supprian, M., and Rajewsky, K., IkappaB kinase signaling is essential for maintenance of mature B cells, *J. Exp. Med.,* 196, 743, 2002.

[51] Senftleben, U., Cao, Y., Xiao, G. et al., Activation by IKKalpha of a second, evolutionary conserved, NF-kappaB signaling pathway, *Science,* 293, 1495, 2001.

[52] Kaisho, T., Takeda, K., Tsujimura, T. et al., IkappaB kinase alpha is essential for mature B cell development and function, *J. Exp. Med.,* 193, 417, 2001.

[53] Schiemann, B., Gommerman, J.L., Vora, K. et al., An essential role for BAFF in the normal development of B cells through a BCMA-independent pathway, *Science,* 293, 2111, 2001.

[54] Claudio, E., Brown, K., Park, S. et al., BAFF-induced NEMO-independent processing of NF-kappaB2 in maturing B cells, *Nat. Immunol.,* 3, 958, 2002.

[55] Batten, M., Groom, J., Cachero, T.G. et al., BAFF mediates survival of peripheral immature B lymphocytes, *J. Exp. Med.,* 192, 1453, 2000.

[56] Mackay, F., Schneider, P., Rennert, P. et al., BAFF AND APRIL: A tutorial on B cell survival, *Annu. Rev. Immunol.,* 21, 231, 2003.

[57] Gross, J.A., Dillon, S.R., Mudri, S. et al., TACI-Ig neutralizes molecules critical for B cell development and autoimmune disease: Impaired B cell maturation in mice lacking BLyS, *Immunity,* 15, 289, 2001.

[58] Motoyama, N., Wang, F., Roth, K.A. et al., Massive cell death of immature hematopoietic cells and neurons in Bcl-x-deficient mice, *Science,* 267, 1506, 1995.

[59] Ruefli-Brasse, A.A., French, D.M., and Dixit, V.M., Regulation of NF-kappaB-dependent lymphocyte activation and development by paracaspase, *Science,* 302, 1581, 2003.

[60] Hara, H., Wada, T., Bakal, C. et al., The MAGUK family protein CARD11 is essential for lymphocyte activation, *Immunity,* 18, 763, 2003.

[61] Ruland, J., Duncan, G.S., Wakeham, A. et al., Differential requirement for MALT1 in T and B cell antigen receptor signaling, *Immunity*, 19, 749, 2003.

[62] Ruland, J., Duncan, G.S., Elia, A. et al., Bcl10 is a positive regulator of antigen receptor-induced activation of NF-kappaB and neural tube closure, *Cell*, 104, 33, 2001.

[63] Weil, R., Schwamborn, K., Alcover, A. et al., Induction of the NF-kappaB cascade by recruitment of the scaffold molecule NEMO to the T cell receptor, *Immunity*, 18, 13, 2003.

[64] Khoshnan, A., Bae, D., Tindell, C.A. et al., The physical association of protein kinase C theta with a lipid raft-associated inhibitor of kappaB factor kinase (IKK) complex plays a role in the activation of the NF-kappaB cascade by TCR and CD28, *J. Immunol.*, 165, 6933, 2000.

[65] Sun, Z., Arendt, C.W., Ellmeier, W. et al., PKC-theta is required for TCR-induced NF-kappaB activation in mature but not immature T lymphocytes, *Nature*, 404, 402, 2000.

[66] Lee, K.Y., D'Acquisto, F., Hayden, M.S. et al., PDK1 nucleates T cell receptor-induced signaling complex for NF-kappaB activation, *Science*, 308, 114, 2005.

[67] Egawa, T., Albrecht, B., Favier, B. et al., Requirement for CARMA1 in antigen receptor-induced NF-kappaB activation and lymphocyte proliferation, *Curr. Biol.*, 13, 1252, 2003.

[68] Gaide, O., Favier, B., Legler, D.F. et al., CARMA1 is a critical lipid raft-associated regulator of TCR-induced NF-kappaB activation, *Nat. Immunol.*, 3, 836, 2002.

[69] Bertin, J., Wang, L., Guo, Y. et al., CARD11 and CARD14 are novel caspase recruitment domain (CARD)/membrane-associated guanylate kinase (MAGUK) family members that interact with BCL10 and activate NF-kappaB, *J. Biol. Chem.*, 276, 11877, 2001.

[70] Gaide, O., Martinon, F., Micheau, O. et al., Carma1, a CARD-containing binding partner of Bcl10, induces Bcl10 phosphorylation and NF-kappaB activation, *FEBS Lett.*, 496, 121, 2001.

[71] Ruefli-Brasse, A.A., Lee, W.P., Hurst, S. et al., Rip2 participates in Bcl10 signaling and T-cell receptor-mediated NF-kappaB activation, *J. Biol. Chem.*, 279, 1570, 2004.

[72] Zhou, H., Wertz, I., O'Rourke, K. et al., Bcl10 activates the NF-kappaB pathway through ubiquitination of NEMO, *Nature*, 427, 167, 2004.

[73] Sun, L., Deng, L., Ea, C.K. et al., The TRAF6 ubiquitin ligase and TAK1 kinase mediate IKK activation by BCL10 and MALT1 in T lymphocytes, *Mol. Cell*, 14, 289, 2004.

[74] Su, T.T., Guo, B., Kawakami, Y. et al., PKC-beta controls I kappa B kinase lipid raft recruitment and activation in response to BCR signaling, *Nat. Immunol.*, 3, 780, 2002.

[75] Ivanov, V.N. and Nikolic-Zugic, J., Transcription factor activation during signal-induced apoptosis of immature CD4(+)CD8(+) thymocytes: A protective role of c-Fos, *J. Biol. Chem.*, 272, 8558, 1997.

[76] Jeremias, I., Kupatt, C., Baumann, B. et al., Inhibition of nuclear factor kappaB activation attenuates apoptosis resistance in lymphoid cells, *Blood*, 91, 4624, 1998.

[77] Wan, Y.Y. and DeGregori, J., The survival of antigen-stimulated T cells requires NFkappaB-mediated inhibition of p73 expression, *Immunity*, 18, 331, 2003.

[78] Kontgen, F., Grumont, R.J., Strasser, A. et al., Mice lacking the c-rel proto-oncogene exhibit defects in lymphocyte proliferation, humoral immunity, and interleukin-2 expression, *Genes Dev.*, 9, 1965, 1995.

[79] Ghosh, P., Tan, T.H., Rice, N.R. et al., The interleukin 2 CD28-responsive complex contains at least three members of the NF-kappaB family: c-Rel, p50, and p65, *Proc. Natl. Acad. Sci. USA*, 90, 1696, 1993.

[80] Grundstrom, S., Anderson, P., Scheipers, P. et al., Bcl-3 and NFkappaB p50-p50 homodimers act as transcriptional repressors in tolerant CD4+ T cells, *J. Biol. Chem.,* 279, 8460, 2004.

[81] Rao, S., Gerondakis, S., Woltring, D. et al., c-Rel is required for chromatin remodeling across the IL-2 gene promoter, *J. Immunol.,* 170, 3724, 2003.

[82] Mora, A., Youn, J., Keegan, A. et al., NF-kappa B/Rel participation in the lymphokine-dependent proliferation of T lymphoid cells, *J. Immunol.,* 166, 2218, 2001.

[83] Banerjee, D., Liou, H.C., and Sen, R., c-Rel-dependent priming of naïve T cells by inflammatory cytokines, *Immunity,* 23, 445, 2005.

[84] Parra, E., McGuire, K., Hedlund, G. et al., Overexpression of p65 and c-Jun substitutes for B7-1 costimulation by targeting the CD28RE within the IL-2 promoter, *J. Immunol.,* 160, 5374, 1998.

[85] Das, J., Chen, C.H., Yang, L. et al., A critical role for NF-kappaB in GATA3 expression and TH2 differentiation in allergic airway inflammation, *Nat. Immunol.,* 2, 45, 2001.

[86] Corn, R.A., Hunter, C., Liou, H.C. et al., Opposing roles for RelB and Bcl-3 in regulation of T-box expressed in T cells, GATA-3, and Th effector differentiation, *J. Immunol.,* 175, 2102, 2005.

[87] Hwang, E.S., Hong, J.H., and Glimcher, L.H., IL-2 production in developing Th1 cells is regulated by heterodimerization of RelA and T-bet and requires T-bet serine residue 508, *J. Exp. Med.,* 202, 1289, 2005.

[88] Lederer, J.A., Liou, J.S., Todd, M.D. et al., Regulation of cytokine gene expression in T helper cell subsets, *J. Immunol.,* 152, 77, 1994.

[89] Aruffo, A., Farrington, M., Hollenbaugh, D. et al., The CD40 ligand, gp39, is defective in activated T cells from patients with X-linked hyper-IgM syndrome, *Cell,* 72, 291, 1993.

[90] Snapper, C.M., Zelazowski, P., Rosas, F.R. et al., B cells from p50/NF-kappa B knockout mice have selective defects in proliferation, differentiation, germ-line CH transcription, and Ig class switching, *J. Immunol.,* 156, 183, 1996.

[91] Doi, T.S., Takahashi, T., Taguchi, O. et al., NF-kappa B RelA-deficient lymphocytes: normal development of T cells and B cells, impaired production of IgA and IgG1 and reduced proliferative responses, *J. Exp. Med.,* 185, 953, 1997.

[92] Carrasco, D., Cheng, J., Lewin, A. et al., Multiple hemopoietic defects and lymphoid hyperplasia in mice lacking the transcriptional activation domain of the c-Rel protein, *J. Exp. Med.,* 187, 973, 1998.

[93] Zelazowski, P., Carrasco, D., Rosas, F.R. et al., B cells genetically deficient in the c-Rel transactivation domain have selective defects in germline CH transcription and Ig class switching, *J. Immunol.,* 159, 3133, 1997.

[94] Horwitz, B.H., Zelazowski, P., Shen, Y. et al., The p65 subunit of NF-kappa B is redundant with p50 during B cell proliferative responses, and is required for germline CH transcription and class switching to IgG3, *J. Immunol.,* 162, 1941, 1999.

[95] Bernasconi, N.L., Traggiai, E., and Lanzavecchia, A., Maintenance of serological memory by polyclonal activation of human memory B cells, *Science,* 298, 2199, 2002.

[96] Komai-Koma, M., Jones, L., Ogg, G.S. et al., TLR2 is expressed on activated T cells as a costimulatory receptor, *Proc. Natl. Acad. Sci. USA,* 101, 3029, 2004.

[97] Pasare, C. and Medzhitov, R., Control of B-cell responses by Toll-like receptors, *Nature,* 438, 364, 2005.

[98] Grumont, R.J., Rourke, I.J., and Gerondakis, S., Rel-dependent induction of A1 transcription is required to protect B cells from antigen receptor ligation-induced apoptosis, *Genes Dev.,* 13, 400, 1999.

[99] Grumont, R.J., Rourke, I.J., O'Reilly, L.A. et al., B lymphocytes differentially use the Rel and nuclear factor kappaB1 (NF-kappaB1) transcription factors to regulate cell cycle progression and apoptosis in quiescent and mitogen-activated cells, *J. Exp. Med.*, 187, 663, 1998.

[100] Owyang, A.M., Tumang, J.R., Schram, B.R. et al., c-Rel is required for the protection of B cells from antigen receptor-mediated, but not Fas-mediated, apoptosis, *J. Immunol.*, 167, 4948, 2001.

[101] Li, Z.W., Omori, S.A., Labuda, T. et al., IKK beta is required for peripheral B cell survival and proliferation, *J. Immunol.*, 170, 4630, 2003.

[102] Prendes, M., Zheng, Y., and Beg, A.A., Regulation of developing B cell survival by RelA-containing NF-kappaB complexes, *J. Immunol.*, 171, 3963, 2003.

[103] Kraus, M., Alimzhanov, M.B., Rajewsky, N. et al., Survival of resting mature B lymphocytes depends on BCR signaling via the Igalpha/beta heterodimer, *Cell*, 117, 787, 2004.

[104] Thome, M., CARMA1, BCL-10 and MALT1 in lymphocyte development and activation, *Nat. Rev. Immunol.*, 4, 348, 2004.

8 NF-κB and IKK as Key Mediators of Inflammation and Oncogenic Progression

Albert S. Baldwin

CONTENTS

8.1 INTRODUCTION

As a transcription factor that is inducible and transiently active downstream of physiologically important and stress-induced pathways, it is not surprising that NF-κB dysregulation is associated with many disease mechanisms. This chapter focuses on the role of NF-κB in promoting and resolving inflammation, in controlling inflammatory diseases, and in promoting cancer. Additionally, I discuss relatively new data regarding an important link between NF-κB activation and the ability of inflammation to promote oncogenesis.

8.1.1 Process of Inflammation

Inflammation is a generally beneficial response to infection or to tissue injury that results in the restoration of homeostasis and repair, at both the cellular and tissue levels, and requires both innate and adaptive immune responses (see Chapters 6 and 7). It is clear that unresolved inflammation is the underlying cause of the pathogenesis of a variety of diseases such as arthritis, asthma, inflammatory bowel disease, diabetes, and cancer.

Inflammation is characterized by increased blood flow to the affected tissue, which then increases temperature, swelling, redness, and pain — calor, tumor, rubor, and dolor, as described by Celsus in the first century A.D. Numerous cells are involved in the inflammatory process, including mast cells, monocytes/macrophages, eosinophils, neutrophils, dendritic cells, and B and T cells. These inflammatory cells are essential to the promotion as well as the resolution phases of the inflammatory response. For example, resident mast cells (Figure 8.1) appear to be key players in the initiation of inflammation through their ability to bind pathogen-associated molecules such as bacterial lipopolysaccaride (LPS) through toll-like receptors (TLRs) [1]. Once activated by recognition of foreign molecules, mast cells release potent mediators of the inflammatory response, including cytokines such as tumor

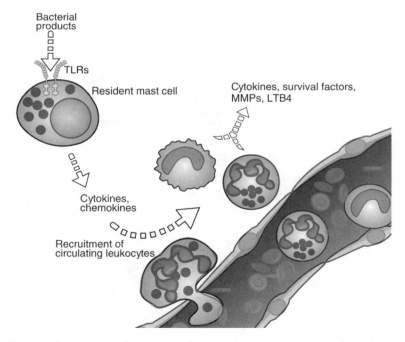

FIGURE 8.1 Mast cells are involved in the initiation of inflammation through an innate immune response.

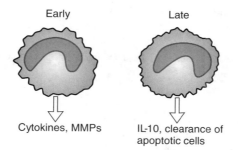

FIGURE 8.2 Macrophages are involved in pro- and antiinflammatory responses.

necrosis factor (TNF) and chemokines such as IL-8 [1]. Release of these inflammatory mediators is then involved in the recruitment of circulating leukocytes, mediated by changes in vascular endothelial cells, which become activated at the site of inflammation and release additional inflammatory mediators [1] (see Figure 8.1).

Within the complex interplay of cells and molecules that function during inflammation, a key early event is the recruitment of neutrophils, the cells that act as first line effectors of the innate immune defense against bacterial and fungal infection. Neutrophil migration and activation during an inflammatory response results from several events, including the release of chemoattractants by tissue resident cells [2]. Thus, it has been shown that macrophages, mast cells, and leukocytes control neutrophil recruitment during inflammation through the release of cytokines, chemokines, and LTB4 (see Figure 8.1 and [2]).

Macrophage infiltration is also one of the hallmarks of severe or progressive inflammatory disease. Macrophages function through the release of inflammatory cytokines, chemokines, and nitric oxide (NO) and by directly phagocytosing bacteria [3]. Studies from several groups indicate that these cells cause tissue injury in progressive inflammatory disease. Severity of injury is often directly correlated with the level of the macrophage infiltrate in human disease and in animal models [4]. Studies show that the infiltrating macrophages express inducible nitric oxide synthase (iNOS), as well as other known markers of the proinflammatory phenotype (see Figures 8.1 and 8.2). Inducible NO production is controlled by iNOS, which is transcriptionally controlled by NF-κB downstream of cell stimulation by cytokines or LPS [5]. Demonstration of a pathogenic role for macrophages in disease is provided by depletion/adoptive transfer experiments in rodent models of acute nephritis (see [6]).

CC chemokines and their receptors play a fundamental role in trafficking and activating leukocytes at inflammation sites by facilitating chemoattraction and migration of leukocytes from the circulation [7]. A key regulatory factor in the process of inflammation is the CC chemokine MCP-1 (monocyte chemoattractant protein 1). In addition to monocytes, this protein is also important in recruiting and activating T cells, mast cells, and basophils during inflammation [8]. MCP-1 is produced by fibroblasts, endothelial cells, monocytes, and macrophages. MCP-1 and other chemokines such as RANTES/CCL5 and macrophage inflammatory protein (MIP)-1α/CCL3 are strongly implicated in the process of inflammation; for example, blockade of chemokine receptors CCR2, CCR5, and CXCR3 protects mice from experimental colitis [9]. Importantly, MCP-1 has been shown to regulate the balance between pro- and antiinflammatory cytokines in a murine model of septic shock [10], which is consistent with a role for macrophages in mediating as well as resolving inflammation.

In vivo, local increases in COX-2 gene expression are associated with inflammation and inflammatory diseases [11]. COX-1, which is constitutively expressed, and COX-2, which is inducibly expressed downstream of inflammatory mediators, are responsible for prostaglandin H_2 biosynthesis from arachidonic acid. COX products, mainly PGE_2, modulate the classic signs of inflammation [11].

The traditional proinflammatory cytokines TNFα and IL-1 are elevated in inflammatory lesions. TNFα is a key mediator in the host's response against bacteria and other pathogens [12]. Mice deficient for TNF receptor 1 are resistant to lethal doses of LPS but are severely impaired in their ability to clear *L. monocytogenes* and easily succumb to infection [13]. IL-1 induces tissue factor production, thus triggering the clotting cascade, and induces systemic hormonal effects that induce fever [14]. IL-1 mediates inflammation by recruitment of neutrophils, activation of macrophages, and stimulation of T and B cell proliferation and differentiation [15]. Other potent inflammatory mediators include the leukotrienes, which are produced by the 5-lipoxygenase pathway downstream of arachidonic acid production [16].

Treatment for inflammation typically involves the utilization of steroidal compounds, such as hydrocortisone or dexamethasone (or derivatives), or nonsteroidal antiinflammatory drugs (NSAIDs) [11,17]. Given that TNFα is a crucial cytokine in the establishment and maintenance of inflammation, it was logical to target TNF as a therapeutic approach for several inflammatory diseases. Infliximab and etanercept, two injectable biologic TNF-blocking drugs, have shown efficacy in rheumatoid arthritis and Crohn's disease [18], and trials are underway for the use of these drugs in other inflammatory diseases. β2-adrenergic agonists have antiinflammatory effects based on the fact that neutrophils, monocytes, and macrophages express β2-receptors [19]. Additionally, statin compounds function as antiinflammatory compounds independent of their ability to modulate cholesterol levels [20, 21].

8.1.2 RESOLUTION OF INFLAMMATION

Except under conditions of chronic inflammation and disease, inflammation resolves through a series of recently identified mechanisms. For example, evidence has been presented that COX-2, which has proinflammatory functions related to the production to prostaglandin PGE_2, also has antiinflammatory functions (see Figure 8.3). In these studies, COX 2 expression occurred at 2 hours following inflammatory stimulus, and this coincided with elevated PGE_2 and a proinflammatory response [22]. COX-2 expression reoccurred at 48 hours, coinciding with minimal PGE_2 but elevated cyclopentonone prostaglandin 15d-PGJ_2; 15d-PGJ_2 is now known to be the endogenous ligand for the nuclear receptor PPARγ [23]. Interestingly, cyclopentonone prostaglandins appear to serve as antiinflammatory mediators and to promote resolution of inflammation *in vivo* [22]. Thus, 15d-PGJ_2 is proposed to be an antiinflammatory prostaglandin through its ability to signal

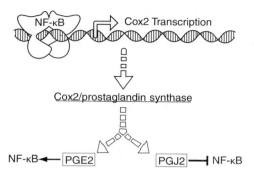

FIGURE 8.3 NF-κB-dependent transcription of COX-2 can have pro- and antiinflammatory outcomes.

through PGD$_2$ receptors and through interaction with certain cellular targets [24] (see Figure 8.3, and see Section 8.2.2).

While macrophages clearly are mediators of the proinflammatory phenotype, there is evidence that these cells contribute to resolution of inflammation (see Figure 8.2). Suspicion that macrophages might function in inflammatory resolution originated from the observation that the degree of macrophage infiltration at sites of inflammation does not always correlate with severity of injury [25]. Macrophages secrete high levels of the antiinflammatory cytokine IL-10 late in the inflammatory response (see Figure 8.2 and [26]). Adoptive transfer of macrophages expressing antiinflammatory molecules such as IL-4 and IL-10 into the kidney reduces renal inflammation [27]. Consistent with this finding, IL-10 was shown to be involved in both the onset and the resolution of inflammation in a mouse model of asthma [28]. In addition, observations of inflammation in skin and lung also demonstrate the importance of macrophages in the resolution of injury and the repair of tissue [29,30].

Apoptosis of inflammatory cells is crucial in the resolution of inflammation because of the high levels of recruitment and expansion of leukocyte populations that occurs during the inflammatory process [3,31]. Thus, the resulting clearance of apoptosed leukocytes by macrophages is important for the resolution of inflammation [31]. Consistent with this, and as discussed above, the ability of organisms to resolve inflammation appears to be significantly controlled by the production of IL-10 and TGFβ, derived from epithelial and regulatory T cells [32,33]. In this regard, it has been postulated that defects in inflammatory cell apoptosis contribute to the pathogenesis of inflammatory diseases.

Endogenous glucocorticoids are well-known antiinflammatory mediators, functioning through their ability to bind glucocorticoid receptors and to modulate a variety of pathways (reviewed in [34]). As discussed in Section 8.2.1, activated glucocorticoid receptors can function to suppress NF-κB activity. Additionally, glucocorticoids can induce the expression of genes encoding proteins with antiinflammatory functions [34]. These gene targets include annexin 1/lipocortin 1, which can inhibit the production of prostaglandins (see [35]), and lipoxins, which suppress neutrophil and eosinophil chemotaxis [35,36]. Lipoxins also promote the ability of macrophages to clear apoptotic cells [37].

Nitric oxide is a cell-derived radical with diverse roles in physiological and immunological functions [see 3]. NO exerts repressive effects on several cell types, including mast cells and T cells. For example, macrophage-derived NO inhibits Ag-induced mast cell degranulation [38], T lymphocyte proliferation [39], and T cell responses during cognate Ag recognition [40]. Recent evidence indicates that NO is involved in the resolution of inflammation, partly through its ability to inhibit IgE-dependent cytokine production in mast cells [41].

8.1.3 INFLAMMATION AND DISEASE

Many human diseases have an inflammatory basis. Examples include arthritis, inflammatory bowel disease/colitis, atherosclerosis, psoriasis, cystic fibrosis, type 2 diabetes, chronic obstructive pulmonary disease, and gastritis. Rheumatoid arthritis is characterized by joint inflammation and connective tissue destruction associated with elevated proinflammatory cytokines and lipid mediators [15]. Macrophages/monocytes are prominent cells in chronic inflammatory conditions such as arthritis. A significant role for the cytokine IL-1 in mediating arthritis and tissue destruction is now well established [15]. Destruction of connective tissue is a major contributing factor in arthritis-related pathology and is thought to be the result of the activation of matrix metalloproteinases (MMPs) — collagenases, gelatinases, stromelysin, metalloelastase, and membrane-type MMPs [42]. Thus, it is thought that these enzymes are involved in the connective tissue loss associated with chronic inflammatory diseases. Similarly, although neutrophils have a generally protective effect during acute inflammation, these cells can mediate tissue damage observed in inflammatory diseases such as rheumatoid arthritis, glomerulonephritis, and inflammatory bowel disease (see [43]).

The human chronic inflammatory bowel diseases (IBDs), Crohn's disease and ulcerative colitis, affect more than one million people in the United States. Research has identified contributing factors for these diseases that include defects in the barrier mechanism of the intestinal epithelium and a poorly regulated immune response against the normal resident microbial flora [44]. Crohn's disease is known to be based on both environmental and genetic determinants [45]. One genetic component of Crohn's disease is the nucleotide oligomerization domain (NOD2) locus, which encodes a protein involved in the intracellular recognition of bacterially derived muramyl depeptide (MDP) (see Chapter 6 and [46]). Mutations in NOD2 strongly increase susceptibility to Crohn's disease [47]. One key aspect of Crohn's disease is the presence within the intestinal lamina propria of activated macrophages that secrete high levels of proinflammatory cytokines such IL-1β and TNFα [45].

Gastritis, gastric ulcers, and gastric adenocarcinoma can be promoted by *Helicobacter pylori* infection. *H. pylori*-associated gastritis is characterized by infiltration of neutrophils and mononuclear cells in the gastric mucosa [48]. *H. pylori* adheres to gastric epithelial and stimulates IL-8 production by these cells. IL-8 expression is thought to be particularly important in the recruitment of neutrophils associated with *H. pylori*-associated gastritis [49,50]. Signaling induced by *H. pylori* LPS is likely through TLR4 and the associated factor MD-2 [51]. NOD1 has also been implicated in mediating this inflammatory process in response to the *H. pylori* peptidoglycan [52].

Type 2 diabetes is characterized by insulin resistance, impaired insulin secretion, and elevated hepatic glucose production. In addition, a strong relationship between obesity and type 2 diabetes has been established [53,54]. Based on the fact that obese individuals have elevated serum levels of inflammatory cytokines such as TNF and IL-1β, it has been suggested that diabetes associated with obesity is an inflammatory disease [53]. In fact, mouse models linking obesity-induced type 2 diabetes and inflammation have been established [54].

8.1.4 CANCER MECHANISMS

Cancer is an extremely complex disease based on multiple etiologies, cell targets, and developmental stages. Furthermore, most advanced cancer cells exhibit genomic instability, leading to further complexity in the nature of the disease. However, as described below, cancer cells exhibit common features with defects in regulatory pathways that govern normal proliferation and homeostasis. Many of these alterations are intrinsic to the cancer cells themselves, and others are manifested from signals that emanate from the surrounding tumor microenvironment.

Hanahan and Weinberg [55] have proposed six alterations that are shared by advanced cancer cells and that drive malignant growth: (1) self-sufficiency in growth generating signals; (2) profound resistance to growth inhibitory signals; (3) strong resistance to apoptosis; (4) extended replication potential; (5) the potential to induce angiogenesis; and (6) the ability to invade local tissue and to metastasize to distant sites.

Self-sufficiency in growth can be provided by dysregulated expression of growth factors or growth factor receptors leading to uncontrolled cell division. For example, many cancers exhibit upregulation of expression of members of the epidermal growth factor receptor (e.g., EGF receptor or Her2/ErbB2). Furthermore, cancer cells (e.g., glioblastoma and sarcomas) produce growth factors such as PDGF and TGFα (reviewed in [55]). Mutations in proteins that can drive proliferation are also relatively common in cancers. For example, mutations in Ras alleles drive proliferation through chronic stimulation of signal transduction pathways [56]. Paracrine mechanisms involving normal bystanders or recruited inflammatory cells can also promote growth. Resistance to growth-inhibitory signals can be achieved through mutations in tumor suppressor genes such as p53, Rb, ARF, and antigen presenting cell (APC), or in receptors such as that for TGFβ. Additionally, upregulation of expression of cyclin D1 or c-myc, or activating mutations in transcription factors such as β-catenin, can promote cell proliferation or cell growth (see [55]).

A key process in the ability of tumor cells to expand locally and to metastasize is the ability of the cell to resist apoptosis [57]. Resistance to apoptosis can involve the activation of expression of antiapoptotic factors, such as Bcl-2 or Bcl-x_L, or the loss of expression or mutation of proapoptotic factors, such as p53. Alternatively, mutations in tumor suppressors such as PTEN can lead to the activation of intracellular signaling pathways (in this case, the PI3 kinase/Akt pathway) that suppress apoptosis [57]. An additional mechanism of suppression of apoptosis is the induction of cytokine release from the tumor stroma. Cancer cells also are able to proliferate beyond the normal regenerative capacity of the original untransformed cell type. Extended replication potential is controlled in part through telomere maintenance, a property associated with virtually all cancer cells [58]. This mechanism is commonly controlled in cancer through upregulation of expression of the telomerase enzyme.

Cancer cells also exhibit properties associated with inducing and sustaining angiogenesis, a process that appears to be required for tumor progression. Animal models of tumorigenesis indicate that angiogenesis is a midstage property, occurring prior to the appearance of malignancy (see [55]). Angiogenesis is mediated through a complex interplay of regulatory factors, including vascular endothelial growth factor (VEGF). In fact, many tumors exhibit transcriptional upregulation of VEGF. The appearance of angiogenesis is often associated with the ability of tumor cells to invade adjacent tissues and subsequently migrate systemically to preferred sites (metastasis). Local invasion is mediated by changes in the expression of cell adhesion molecules and integrins and in changes in the expression of extracellular proteases such as MMP-2 and MMP-9. In some situations, these matrix-degrading proteases are produced by tumor-associated stromal and inflammatory cells. Some cancer cells, for example, induce urokinase expression in stromal cells, which then binds the urokinase receptor to the cancer cells [59].

Another important mechanism in the progression of certain cancers is the epithelial to mesenchymal transition (EMT) [60,61]. The differentiated epithelial phenotype is typically characterized by the polarization of the cell surface into apical and basolateral domains and by a junctional complex that controls intercellular adhesion. The occurrence of EMT in cancer is associated with the downregulation of the expression of E-cadherin, a member of the classic cadherin family, which controls cell polarity and tight junctions [61].

8.1.5 PROMOTION OF CANCER BY INFLAMMATION

Scientists have long suspected that chronic inflammation is intimately associated with the promotion of cancer. In this regard, evidence for the involvement of inflammation in cancer was provided by clinical studies correlating tumor infiltration of immune cells with poor clinical outcome [62]. A causal role for inflammation and cancer is suggested by studies reporting reduced cancer incidence in patients undergoing treatment with antiinflammatory drugs [63]. Since many antiinflammatory compounds inhibit cyclooxygenase-2 activity, this enzyme has been suspected as the primary target for prevention. Proteolytic enzymes, the transcription factor NF-κB, and cytokines such as TNF are also believed to be functionally important in potentiating inflammation-based cancer [62,64,65].

Although there are a number of examples whereby inflammation is implicated or suspected in cancer progression (e.g., pancreatitis with pancreatic cancer and gastritis with gastric cancer), probably the best studied link is between inflammatory bowel disease and colorectal cancer. Colitis is associated with an approximate ten-fold increase in the risk of developing colorectal cancer, and this risk

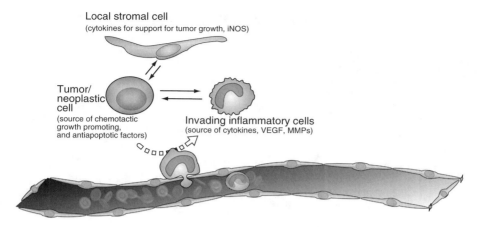

FIGURE 8.4 Cells involved in tumor growth and progression/invasion and processes controlled by NF-κB.

increases significantly with the duration and extent of disease (reviewed in [66,67]). Recent evidence suggests that the development of colitis-associated cancer is manifested by inflammation of the intestinal submucosa directed by contact with the intestinal microflora, which promotes tumor outgrowth in the overlying epithelium (see [66]).

The role of inflammation in cancer has been unclear. Tumor development and progression is controlled by reciprocal interactions between neoplastically initiated cells, vascular cells, fibroblasts, and immune cells (see [68] and Figure 8.4). In this regard, recent evidence indicates a significant role for innate immune cell involvement in cancer development [69]. Obviously, the local production of cytokines and growth factors would favor cancer cell survival and local invasive properties, but these effects would not, on their own, be expected to generate oncogenic events. However, it is accepted that the local production of reactive oxygen species and metabolites from inflammation pose a mutagenic threat to DNA. Thus, it is still unclear whether inflammation simply promotes the outgrowth and maintenance of independently generated neoplastic cells or whether inflammation itself can generate *de novo* oncogenesis.

As described, IL-8 is a key factor in the initiation and progression of inflammation. Interestingly, Bar-Sagi and colleagues [70] demonstrate that oncogenic Ras expression leads to the upregulation of IL-8, which then functions to promote tumor-associated inflammation, angiogenesis, and tumor growth. The authors showed that inhibition of IL-8 with a neutralizing antibody inhibited the growth of tumor cells that express IL-8. Furthermore, the antibodies reduced recruitment of neutrophils and macrophages in the derived tumors and reduced recruitment of endothelial cells and resulting angiogenesis. These studies provide evidence of a regulatory cascade initiated by oncoprotein expression and leading to upregulation of a proinflammatory cytokine that recruits key inflammatory cells that promote tumorigenesis.

Proinflammatory Functions of NF-κB Include Transcription of:

-iNOS*
-Cytokines
-Chemokines
-MMPs
-Antiapoptotic factors
-Cox2*

Antiinflammatory Functions of NF-κB Include Transcription of:

-Cox2*
-iNOS*

*Cox2 and iNOS gene products can function in both proinflammatory and antiinflammatory pathways.

FIGURE 8.5 Pro- and antiinflammatory effects of NF-κB.

8.2 ROLE OF NF-κB IN MEDIATING PRO- AND ANTIINFLAMMATORY MECHANISMS

8.2.1 NF-κB IS A KEY MEDIATOR OF THE INFLAMMATORY RESPONSE AND OF INFLAMMATORY DISEASES

Previous studies of mice genetically deficient in components of the NF-κB pathway demonstrate that these proteins play key roles in promoting the inflammatory response (Chapters 6 and 7). For example, Alcamo and colleagues [71] showed that p65/RelA is critical for leukocyte recruitment during inflammation. c-Rel has a role in promoting hematopoietic cell survival and production of mediators that maintain the inflammatory response. Gerondakis and colleagues [72] showed that c-Rel has both positive and negative regulatory functions in different macrophage populations, and other studies indicate that c-Rel and p65 serve redundant functions in controlling the differentiation and survival of committed progenitors in multiple hemopoietic cell lineages [73]. NF-κB activation in models of inflammation indicates a role for this transcription factor in promoting inflammation through upregulation of COX-2 and iNOS and subsequent production of PGE_2 and NO (see [74]). NF-κB inhibitors given early in the rat carrageenin pleurisy model suppressed the inflammatory response [74]. However, when given later in the model, these inhibitors led to sustained inflammation (see comments in Section 8.2.2 regarding this point). The role of NF-κB in promoting the transcription of key inflammatory genes such as IL-8, COX-2, iNOS, several chemokines, and MCP-1 is consistent with the role of this transcription factor in regulating the cellular hierarchy that promotes the inflammatory response (see Figure 8.5).

The production of certain MMPs in monocytes, which promote migration of these inflammatory cells as well as disease-related connective tissue destruction, is controlled through NF-κB activation [75]. This same group has provided evidence that the production of MMP-9 in activated monocytes is controlled through a PI3K/Akt/IKKα/ NF-κB pathway [76]. Prostaglandins such as PGE_2 and leukotrienes — molecules associated with inflammation — are known to stimulate NF-κB transcriptional activity [77,78], apparently at a level secondary to the regulation of IKK [77].

An essential role for NF-κB in promoting inflammation may be to promote survival of invading inflammatory cells. Thus, Cheah and colleagues [79] found NF-κB to be activated in inflammatory neutrophils from premature infants. This activation was inversely correlated with proapoptotic caspase-3 activation. Additionally, others have shown that the resolution of sepsis-associated neutrophil-mediated inflammation occurs through apoptosis, and suppression of NF-κB activation in these cells promotes apoptosis [80]. NF-κB inhibitors increase apoptosis in neutrophils and eosinophils, causing these cells to undergo TNF-induced apoptosis. In these cells, evidence has been presented that the production of the prostaglandin PGD2 suppresses NF-κB activation in late stages of inflammation [81]. Thus, NF-κB functions early in inflammation to promote inflammation, partly through its ability to suppress apoptosis. Later, in the resolution phase, NF-κB may function to induce its own inhibition through the production of distinct prostaglandins.

The gut maintains a physiological level of inflammation in response to endogenous bacterial flora. The presence of pathogenic bacteria is typically sufficient to activate an inflammatory response that appears to be mediated by NF-κB. For example, flagellin, the structural component of bacterial flagella, activates NF-κB through TLR5 [82]. These innate mucosal events are essential for combating gut infections. Recent evidence indicates that certain nonpathogenic bacteria can block NF-κB activation; for example, nonvirulent Salmonella strains inhibit NF-κB activation in intestinal epithelial cells through a mechanism that prevents IκB ubiquitination and subsequent degradation [83]. Another group [84] showed that a commensal bacterium, *B. thetaiotamicron*, suppresses inflammation by promoting an association between PPARγ and p65, leading to p65/RelA export from the nucleus. These reports suggest that prokaryotic determinants are responsible for tolerance of the gastrointestinal mucosa to inflammatory signals. Additionally, it suggests that NF-κB is a major mediator of the inflammatory response.

Naturally occurring antiinflammatory molecules have been shown to inhibit NF-κB. In this way, the antiinflammatory cyclopentonone prostaglandin 15d-PGJ$_2$, the natural ligand for PPARγ, can block NF-κB activation. This prostaglandin appears to suppress NF-κB activation both through the induced upregulation of IκBα and through direct inhibition of IKK [85,86]. We reported previously that the antiinflammatory protein IL-10 can suppress NF-κB through a mechanism associated with reduced or delayed IKK activity [87]. Additionally, antiinflammatory mechanisms associated with NO production (see above) have been attributed to the ability of NO to induce S-nitrosylation of IKK [88].

Inhibition of NF-κB in bone marrow-derived macrophages by expression of superrepressor IκBα converted these cells from highly proinflammatory to strongly antiinflammatory [89]. Macrophages were reoriented from expression of iNOS, TNF, and IL-12 to express high levels of IL-10. These cells, when injected in association with a glomerulonephritis model, strongly reduced renal injury and suppressed the inflammatory milieu. These data indicate that NF-κB promotes the proinflammatory phenotype in macrophages and actively suppresses the antiinflammatory phenotype.

Experiments from Karin and colleagues [90] indicate a key role for the NF-κB pathway in promoting inflammation-mediated, obesity-based insulin resistance. Mice with IKKβ knocked out in the liver retain insulin sensitivity in the liver in response to a high-fat diet or obesity, but exhibit muscle and fat insulin resistance. In contrast, mice with IKKβ knocked out in myeloid cells exhibit global insulin sensitivity. Similarly, Cai and colleagues [91] have demonstrated that a high-fat diet and obesity activate NF-κB in the liver. Expression of IKKβ in the liver of mice mimics a type 2 diabetes phenotype — presumably through the production of proinflammatory cytokines that mediate the disease. Inhibition of NF-κB in the liver through superrepressor IκB expression inhibited the inflammatory effect observed in the IKKβ transgenic animals. In other studies, Chen and colleagues [92] used the IKKβ conditional knockout to explore the role of the classical NF-κB pathway in controlling systemic inflammation using a model of gut ischemia-reperfusion. Knockout of IKKβ in intestinal enterocytes blocked the induced systemic inflammatory response, but sensitized these cells to apoptosis in response to reperfusion.

Evidence suggests that NF-κB activation underlies much of the pathology associated with *H. pylori* infection and associated gastritis, duodenal ulceration, and potentially gastric adenocarcinoma [49]. *H. pylori* infection induces IL-8 gene expression from gastric epithelial cells through the activation of NF-κB [50]. Data suggest the involvement of MD2/TLR4 in *H. pylori*-dependent signaling leading to NF-κB activation and IL-8 gene expression [51]. However, another group [52] has identified NOD1, a key intracellular protein involved in innate immune responses, as a key effector of the inflammatory response to *H. pylori* infection, although there was no connection linking NOD1 engagement with NF-κB activation.

Early experiments indicated a role for NF-κB in mediating inflammatory bowel disease. More recently, Maeda and colleagues [93] demonstrated a link between NOD2 and enhanced NF-κB activation. Maeda and colleagues knocked in a variant of NOD2 and showed elevated NF-κB in macrophages. This alteration was associated with enhanced susceptibility to intestinal inflammation induced by DSS. In rheumatoid arthritis, it has been proposed that NF-κB functions as a master switch controlling expression of cytokines, small molecule mediators, and metalloproteinases involved in the disease [94]. Consistent with this finding, inhibition of NF-κB activity through the use of *in vivo*-delivered DNA decoys suppressed arthritis in an experimental model [95]. More recently, inhibition of NF-κB activation by IκB blocked local cytokine production and the influx of leukocytes to the synovium (see [94]). The NEMO binding domain peptide developed by Ghosh and colleagues suppressed several inflammatory disorders in animal models, including inflammatory bone resorption and arthritis [96]. Similarly, delivery of a dominant negative IKKβ to lung epithelium blocked proinflammatory mechanisms in a model for asthma [97].

Consistent with a key role for NF-κB in promoting inflammation are the observations that many antiinflammatory compounds function to block NF-κB activation (discussed in detail in Chapter 10). Glucocorticoids are known to modulate NF-κB functional activity through distinct mechanisms [17]. Several NSAIDs, including aspirin and celecoxib, have been shown to inhibit NF-κB activation through inhibition of IKK activity [98–100]. Furthermore, the immunomodulatory activity of statins has been partly attributed to the ability of these compounds to block NF-κB activation [21]. Finally, the antiinflammatory activity of β2-adrenergic agonists has been attributed to their ability to block NF-κB activation [101]. Therefore, NF-κB appears to be a highly relevant target of most antiinflammatory compounds.

8.2.2 NF-κB FUNCTION IN RESOLUTION OF INFLAMMATION

As discussed in Section 8.1.2, mechanisms associated with resolution of inflammation have been elaborated in recent years. Thus, COX-2 expression appears to mediate the inflammatory response through PGE_2 production; however, at later time points, COX-2 induces production of 15d-PGJ_2 [22]. Given that NF-κB controls COX-2 expression, it was speculated that NF-κB may drive both inflammatory and antiinflammatory mechanisms (see Figure 8.3). In fact, Lawrence and colleagues [74] presented evidence that NF-κB is critically involved in resolving inflammation. These studies utilized a carrageenin-induced model for pleurisy. Inhibition of NF-κB after the initiation of inflammation resulted in the inability to resolve inflammation. However, it should be noted that the role of specific NF-κB subunits in this model was not identified; the inhibitor used in these studies was a proteasome inhibitor that can block both activation of p50-p65 heterodimers as well as p50 homodimers. Interestingly, the authors suggest that p50 homodimers are the dominant complex in the resolution phase [74]. The ability of NF-κB to regulate iNOS gene expression, along with evidence that NO can function both in pro- and antiinflammatory mechanisms, could partly explain the ability of NF-κB to induce and resolve inflammation (see Figure 8.5). Lawrence and colleagues [102] have provided evidence that IKKα promotes resolution of inflammation by suppressing the NF-κB response in macrophages. The data indicate that IKKα contributes to the suppression of inflammation by accelerating the turnover of p65 and c-Rel and by functioning to remove these NF-κB subunits from the promoters of genes encoding proinflammatory mediators.

8.3 ROLE OF NF-κB IN PROMOTING BASIC CANCER MECHANISMS

8.3.1 EVIDENCE FOR INVOLVEMENT OF NF-κB IN CANCER

A potential role for NF-κB in cancer was suggested by the cloning of the NF-κB p50/p105 subunit, which revealed homology to the cellular homologue (c-Rel) of the oncogene (v-Rel) for the avian reticuloendotheliosis virus. Subsequently, the p52 NF-κB subunit was shown to be encoded by a gene that undergoes translocations in certain B cell lymphomas (the lyt-10 translocation) [103,104]. Extensive evidence has now emerged indicating a critical role for NF-κB in promoting oncogenic conversion and in facilitating later-stage tumor properties such as metastasis [104].

Various oncoproteins have been shown to induce NF-κB activation as measured either through reporter assays or through analysis of nuclear levels of NF-κB [103,104]. For example, oncogenic Ras (see [103]) and Her-2/Neu (ErbB2) have been demonstrated to activate NF-κB [105]. In murine fibroblasts, p65 and c-Rel are required for efficient cellular transformation induced by oncogenic Ras [106]. BCR-Abl was shown to activate an NF-κB-dependent reporter, and inhibition of NF-κB with IκBα-SR expression blocked BCR-Abl-induced tumor growth [107].

The mechanisms of stimulation of NF-κB functional activity by oncoproteins often lags behind the significantly clearer understanding of pathways associated with cytokine and LPS-induction of NF-κB. One of the better understood mechanisms for the activation of NF-κB by an oncoprotein is the mechanism whereby the human T cell leukaemia virus type I (HTLV-I) Tax protein activates NF-κB. This protein directly binds and activates the IKK complex [104]. The MALT1/c-IAP2 fusion found in certain lymphomas leads to ubiquitination of NEMO and subsequent NF-κB activation [108]. In addition to the known responses of oncoproteins in activating NF-κB, evidence has been presented that certain tumor suppressors can block NF-κB activation. For example, the tumor suppressor ARF has been reported to inhibit NF-κB through a mechanism involving phosphorylation of Thr505 on p65 downstream of ATR and Chk1 (Chapter 5) [109].

Studies showing that NF-κB is activated (i.e., nuclear) in a number of tumors are consistent with a role for NF-κB in cancer, although some tumor cell lines exhibit NF-κB activity without significant nuclear accumulation. Adding further complexity, the p50 subunit of NF-κB appears to facilitate both the positive and negative regulation of the expression of the KAI1 tumor suppressor gene. Expression of β-catenin converts the p50 transcriptional complex to a repressive complex through loss of Tip60 coactivator association and recruitment of transcriptional corepressors, leading to loss of expression of KAI1 [110].

Given the important role that growth factors play in promoting oncogenesis, it is not surprising that several growth factors have been shown to activate NF-κB. For example, Biswas and colleagues [111] have shown that EGF can activate NF-κB in certain cell types. We have provided evidence that EGF can induce recruitment of p65 to the EAAT2 promoter through a mechanism independent of IκB degradation [112]. Finally, we and others have shown that PDGF can induce NF-κB activation and promote c-myc transcription [103].

While there is strong evidence that NF-κB promotes oncogenesis in a variety of tumors, evidence has been presented that inhibition of NF-κB in skin promotes oncogenic potential and potentiates Ras-induced transformation (see [113]). One mechanism to explain this concept is that inhibition of NF-κB in the skin leads to JNK activation [114], a finding consistent with several reports that NF-κB activation suppresses the phosphorylation and activation of JNK. Similar questions regarding the oncogenic potential of NF-κB arise with the consideration that NF-κB appears to function downstream of the tumor suppressor p53. Vousden and colleagues [115] have presented evidence that NF-κB activation is required downstream of p53 in order for this tumor suppressor to induce apoptosis. Hung and colleagues [116] have reported that the oncoprotein β-catenin can block NF-κB activation (also see [110]). These findings suggest that NF-κB can, under certain conditions, also function as a tumor suppressor [117].

8.3.2 MODES OF ACTION OF NF-κB IN CANCER AND CANCER THERAPY RESISTANCE

Clearly, NF-κB functions as a transcriptional regulator in a variety of cancer cells as evidenced through the identification of cancer-specific cancer-relevant gene targets and processes controlled by NF-κB (see Figure 8.4). For example, gene profiling has identified a subset of diffuse large B cell lymphoma that requires NF-κB for growth and survival. Inhibition of NF-κB activation blocks the expression of key genes associated with this type of lymphoma [118]. We have identified a set of approximately 25 genes, which are regulated by NF-κB in a manner dependent on oncogenic Ras expression in murine fibroblasts [106].

One of the key properties associated with transformed cells is their ability to resist apoptosis. Experiments have revealed that induction of RasV12 in immortalized Rat1 fibroblasts leads to cellular transformation but not to apoptosis. If NF-κB is inhibited in these cells by expression of IκBα-superrepressor (SR), then the induction of RasV12 expression induces high levels of apoptosis [119]. Consistent with this point, inhibition of NF-κB in certain tumor cell lines leads to apoptotic cell death [103]. Hodgkin's lymphoma has proven to be a cancer that is strongly controlled by NF-κB activation. Proliferation and survival of Hodgkin/Reed-Sternberg cells is blocked when NF-κB is inhibited by IκBα expression [120]. Genes regulated by NF-κB that suppress apoptosis, such as Bcl-2 and Bcl-x_L, are often expressed in human cancers, and inhibition of NF-κB in Hodgkin/Reed-Sternberg cells led to the loss of expression of antiapoptotic effectors A1/Bfl-1, c-IAP2, TRAF1, and Bcl-x_L [121]. Relating to a role for the NF-κB pathway in preventing apoptosis, Hu and colleagues [122] have shown that IKKβ activation in breast cancer cells leads to the direct phosphorylation and degradation of the proapoptotic factor Foxo3a, suppressing apoptotic potential in certain breast cancer cells and promoting cell proliferation. Consistent with these findings, NF-κB has been shown to be activated by certain chemotherapies and radiation, and this response is generally antiapoptotic [103]. Inhibition of NF-κB in these models results in enhanced chemotherapy-induced apoptosis [103,123,124]. However, it has been reported that NF-κB can function in some cancer cell types in a proapoptotic manner downstream of certain chemotherapies [125]. Thus, as we have seen in inflammation and hematopoiesis (Chapter 7), NF-κB does not function uniformly in inhibiting apoptosis in all cancer cells.

NF-κB activation also appears to promote cellular proliferation, which is consistent with a role in promoting growth sufficiency of cancer cells. Evidence has been presented that NF-κB can bind and activate the cyclin D1 promoter, promoting Rb hyperphosphorylation (see [103]). Additionally, the IκB homologue Bcl-3, in association with p52 homodimers, has also been found to potently activate transcription of the cyclin D1 gene. Interestingly, IKKα has been proposed to play a role in cyclin D1 transcription through the Tcf site via its ability to control β-catenin phosphorylation [126]. Consistent with a role for IKKα in promoting cyclin D1 transcription, Karin and colleagues found that IKKα is required for RANK signaling and cyclin D1 expression in mammary gland development [127]. Other mechanisms whereby NF-κB may potentiate oncogenic conversion and maintenance are through the upregulation of HIF-1α [128] and the regulation of c-myc transcription [129].

NF-κB activation in tumor cells and in tumor-associated stromal and endothelial cells likely plays a role in tumor progression. In this regard, NF-κB has been reported to promote both angiogenesis and metastasis in certain tumor models, potentially through regulation of VEGF and MMPs [103,104]. A role for NF-κB in invading myeloid cells has been suggested from IKKβ ablation studies in these cells (see discussion in Section 8.4 and [64]). In this study [64], NF-κB was shown to control tumor growth-promoting cytokines in myeloid cells. NF-κB appears to promote gastric cancer cell invasion in a manner dependent on its ability to control expression of COX-2 [130]. In an interesting model, NF-κB was shown to mediate an IL-1/nitric oxide paracrine growth loop involving stromal fibroblasts and pancreatic neoplastic cells [131]. The potential of NF-κB to promote tissue homing and successful metastatic colonization is suggested by the fact that RANTES and its receptor CCR5 are important for melanoma metastasis and that NF-κB is known to control the expression of these genes. NF-κB has been shown to mediate the epithelial-mesenchymal transition [132], which is also consistent with a role in metastasis.

8.4 NF-κB IN INFLAMMATION-BASED ONCOGENESIS

As described in the previous section, there is compelling evidence that inflammation promotes cancer progression. Given the key role that NF-κB plays in both innate and acquired immunity, it would not be surprising that NF-κB plays an essential role in providing the link between inflammation and cancer progression (reviewed in [133]). In fact, recently published data directly implicate NF-κB activation as a key component in inflammation-based cancer progression. Karin and colleagues [64] creatively utilized their IKKβ conditional knockout to test the role of the NF-κB activation pathway in controlling tumorigenesis in a colitis-associated cancer model. Deletion of IKKβ in intestinal epithelial cells dramatically reduced tumor number but not tumor size in this model. The reduction in tumor number was explained by strongly enhanced apoptosis in the DNA-damaged intestinal target cells, consistent with a role for NF-κB in suppressing apoptotic potential. Importantly, tumor-associated inflammation was not reduced in this component of the model. These data argue that NF-κB activation is important in the early stages of DNA-damaged induced tumorigenesis (as an antiapoptotic mediator) but not in the prominent growth phase of tumorigenesis. Karin and colleagues [64] went on to show that deletion of IKKβ in myeloid cells, however, leads to a significant reduction in tumor size but not tumor number, and to a reduction in tumor-associated proinflammatory cytokine levels that are likely to serve as tumor growth factors. This result is consistent with the importance of NF-κB in promoting myeloid cell recruitment and inflammatory gene expression as part of the inflammatory phase of oncogenesis (see Figure 8.4). In another related study, Maeda and colleagues [93] showed that loss of IKKβ in hepatocytes actually promoted chemical-induced hepatocarcinogenesis through a mechanism involving enhanced ROS production and JNK activation with associated cell death, leading to a compensatory response in surviving hepatocytes. Knockout of IKKβ in Kupffer cells suppressed hepatocarcinogenesis, indicating an inflammatory role for the hemopoietic-derived cells in this model.

In another recently described model, Ben-Neriah and colleagues [65] utilized a mouse knockout for the Mdr2 gene, which develops spontaneous hepatitis followed by hepatocellular carcinoma. This serves as a straightforward model for inflammation-based oncogenesis. Inhibition of NF-κB through IκBα-SR expression did not block hepatitis or the earliest stages of neoplasia. However, inhibition of NF-κB at later stages (either through IκBα expression or through TNF antibodies) blocked progression to hepatocellular carcinoma at least partly through the induction of apoptosis. In this model, NF-κB does not play a role in the early inflammation-associated neoplastic growth but, acting downstream of TNF, functions to suppress apoptosis and to allow cancer malignancy to progress.

Less direct but highly suggestive evidence for a role of NF-κB in inflammation-associated cancer has been presented. For example, NF-κB activation is suggested to promote neoplastic progression in Barrett's esophagus [134]. It is proposed that this is mediated through the ability of NF-κB to regulate COX-2 and IL-8 gene expression. Jung and colleagues [128] reported that IL-1β induces HIF-1α gene expression through a mechanism involving the induction of PGE_2 through the NF-κB-dependent upregulation of COX-2. These studies suggest a direct mechanism whereby cytokines such as IL-1β promote oncogenesis. Regarding the work of Sparmann and Bar-Sagi [70] described above, it is speculated that the ability of Ras to activate IL-8 gene expression and subsequent neutrophil recruitment is mediated through NF-κB. Additionally, it has been reported [135] that the p65 subunit of NF-κB interacts with and promotes the TNF-induced nuclear accumulation of human telomerase reverse transcriptase, potentially linking inflammation with extended cell proliferation capacity.

The evidence of a role for NF-κB in controlling inflammation-based oncogenesis suggests that inhibitors of this transcription factor are likely to serve as key components in the prevention of many cancers. Thus, studies demonstrating the ability of nonsteroidal antiinflammatory compounds to suppress the development of some cancers is consistent with this hypothesis. While several NSAIDs are described as inhibitors of COX-2, it is also possible that these compounds target NF-κB activity as well (see Chapter 10). In fact, it was recently shown *in vitro* that the COX-2 inhibitor celecoxib inhibits NF-κB activation induced by TNF through a mechanism that suppresses IKK and Akt activation [136]. Therefore, it is possible that a primary effect of NSAIDs in preventing cancer progression is through inhibition of NF-κB. Additionally, a number of established dietary chemopreventive compounds have been shown to inhibit NF-κB [103].

8.5 CONCLUSIONS

Extensive data collected during the past 15 years demonstrate that NF-κB plays a pivotal role in the establishment as well as the resolution of inflammation. Many of the roles that NF-κB plays in promoting inflammation — including suppression of apoptosis, promoting cellular invasion, and stimulating cell division — are also controlled by NF-κB during oncogenic conversion. Given the clear involvement of NF-κB in the major diseases, the challenge of the near future is to bring rational inhibitors of NF-κB or its upstream regulatory pathways to human disease therapy.

REFERENCES

[1]   Marshall, J.S., Mast cell responses to pathogens, *Nat. Rev. Immunol.*, 4, 787, 2004.

[2] Greaves, D. and Schall, T., Chemokines and myeloid cell recruitment, *Microbes Infect.*, 2, 331, 2000.

[3] Bosca, L. et al., Nitric oxide and cell viability in inflammatory cells: A role for NO in macrophage function and fate, *Toxicology*, 208, 249, 2005.

[4] Kluth, D., Erwig, L., and Rees, A., Multiple facets of macrophages in renal injury, *Kidney Int.*, 66, 542, 2004.

[5] Coleman, J.W., Nitric oxide in immunity and inflammation, *Int. Immunopharmacol.*, 1, 1397, 2001.

[6] Wilson, H.M. et al., Inhibition of macrophage NF-κB leads to a dominant anti-inflammatory phenotype that attenuates glomerular inflammation *in vivo, Am. J. Pathol.*, 167, 27, 2005.

[7] Luster, A.D., Chemokines: Chemotactic cytokines that mediate inflammation, *N. Engl. J. Med.*, 338, 436, 1998.

[8] Tangirala, R.K., Murao, K., and Quehenberger, O., Regulation of the expression of the human monocyte chemoattractant protein-1 receptor (hCCR2) by cytokines, *J. Biol. Chem.*, 272, 8050, 1997.

[9] Tokuyama, H. et al., The simultaneous blockade of chemokine receptors CCR2, CCR5, and CXCR3 by a non-peptide chemokine receptor antagonist protects mice from dextran sodium sulfate-mediated colitis, *Int. Immunol.*, E-pub ahead of print.

[10] Zisman, D.A. et al., MCP-1 protects mice in lethal endotoxemia, *J. Clin. Inv.*, 99, 2832, 1997.

[11] Warner, T.D. and Mitchell, J.A., Cyclooxygenases: New forms, new inhibitors, and lessons from the clinic, *FASEB J.*, 18, 790, 2004.

[12] Flavell, R.A., The relationship of inflammation and initiation of autoimmune disease: Role of TNF super family members, *Curr. Top. Microbiol. Immunol.*, 266, 1, 2001.

[13] Pfeffer, K. et al., Mice deficient for 55kD TNF receptor are resistant to endotoxic shock, yet succumb to L. monocytogenes infection, *Cell*, 73, 457, 1993.

[14] Leon, L.R., Cytokine regulation of fever: Studies using gene knockout mice, *J. Appl. Physiol.*, 92, 2648, 2002.

[15] Dinarello, C., Role of IL-1 receptor antagonist in blocking inflammation mediated by IL-1, *New Engl. J. Med.*, 343, 732, 2000.

[16] Peters-Golden, M. et al., Leukotrienes: Underappreciated mediators of innate immune responses, *J. Immunol.*, 174, 589, 2005.

[17] Goulding, N.J., The complexity of glucocorticoid actions in inflammation — A four ring circus, *Curr. Opin. Pharmacol.*, 4, 629, 2004.

[18] Reimold, A.M., TNFα as therapeutic target: New drugs, more applications, *Curr. Drug Targets Inflamm., Allergy* 1, 377, 2002.

[19] Barnes, P.J., Effects of β2-agonists on inflammatory cells, *J. Allergy Clin. Immunol.*, 104, S10, 1999.

[20] McCarey, D.W. et al., Trial of atorvastatin in rheumatoid arthritis: Double-blind, randomized placebo-controlled trial, *Lancet*, 363, 2015, 2004.

[21] Madonna, R. et al., Simvastatin attenuates expression of cytokine-inducible nitric oxide synthase in embryonic cardiac myoblasts, *J. Biol. Chem.*, 280, 13503, 2005.

[22] Gilroy, D.W. et al., Inducible cyclooxygenase may have anti-inflammatory properties, *Nature Med.*, 5, 698, 1999.

[23] Nencioni, A., Wesselburg, S., and Brossart, P., Role of peroxisome proliferator-activated receptor gamma and its ligands in the control of immune responses, *Crit. Rev. Immunol.*, 23, 1, 2003.

[24] Scher, J.U. and Pillinger, M.H., 15d-PGJ$_2$: The anti-inflammatory prostaglandin? *Clin. Immunol.*, 114, 100, 2005.

[25] Chadban, S.J. et al., Effect of IL-10 treatment on crescentic glomerulonephritis in rats, *Kidney Int.,* 51, 1809, 1997.
[26] Moore, K.M. et al., Interleukin-10 and the interleukin-10 receptor, *Ann. Rev. Immunol.,* 19, 683, 2001.
[27] Wilson, H.M. et al., Bone marrow derived macrophages genetically modified to produce IL-10 reduce injury in experimental glomerulonephritis, *Mol. Ther.,* 6, 710, 2002.
[28] Matsumoto, K. et al., Different roles for interleukin-10 in onset and resolution of asthmatic responses in allergen-challenged mice, *Respirology,* 10, 18, 2005.
[29] Teder, P. et al., Resolution of lung inflammation by CD44, *Science,* 296, 155, 2002.
[30] Nagaoka, T. et al., Delayed wound healing in the absence of ICAM-1 or L-selectin expression, *Am. J. Pathol.,* 15, 237, 2000.
[31] Savill, J., Apoptosis in resolution of inflammation, *J. Leukoc. Biol.,* 61, 375, 1997.
[32] Haller, D. et al., TGFβ-1 inhibits non-pathogenic Gram negative bacteria-induced NF-κB recruitment to the IL-6 promoter in intestinal epithelial cells through modulation of histone acetylation, *J. Biol. Chem.,* 278, 23851, 2003.
[33] De Winter, H. et al., Regulation of mucosal immune responses by IL-10 produced by intestinal epithelial cells in mice, *Gastroentrol.,* 122, 1829, 2002.
[34] Leung, D.W. and Bloom, J.W., Update on glucocorticoid action and resistance, *J. Allergy Clin. Immunol.,* 111, 3, 2003.
[35] Lawrence, T., Willoughby, D., and Gilroy, D., Anti-inflammatory lipid mediators and insights into the resolution of inflammation, *Nat. Rev. Immunol.,* 2, 787, 2002.
[36] Levy, B.D. et al., Lipid mediator class switching during acute inflammation: Signals in resolution, *Nature Immunol.,* 2, 612, 2001.
[37] Godson, C. et al., Lipoxins rapidly stimulate nonphlogistic phagocytosis of apoptotic neutrophils by monocyte-derived macrophages, *J. Immunol.,* 164, 1663, 2000.
[38] De Schoolmeester, M.L. et al., Reciprocal effects of IL-4 and interferon-γ on immunoglobulin E-mediated mast cell degranulation: A role for nitric oxide but not peroxynitrite or cyclic GMP, *Immunology,* 96, 138, 1999.
[39] Bingisser, R.M. et al., Macrophage-derived nitric oxide regulates T cell activation via reversible disruption of the Jak3/STAT5 signaling pathway, *J. Immunol.,* 160, 5729, 1998.
[40] Van der Veen, R.C. et al., Antigen presentation to Th1 but not Th2 cells by macrophages results in nitric oxide production and inhibition of T cell proliferation: Interferon-γ is essential but insufficient, *Cell. Immunol.,* 20, 125, 2000.
[41] Davis, B.J. et al., Nitric oxide inhibits IgE-dependent cytokine production and Fos and Jun activation in mast cells, *J. Immunol.,* 173, 6914, 2004.
[42] Nagase, H. and Woessner, J.F., Matrix metalloproteinases, *J. Biol. Chem.,* 274, 21491, 1999.
[43] Ramos, C.D. et al., Neutrophil migration induced by IL-8-activated mast cells is mediated by CINC-1, *Cytokine,* 21, 214, 2003.
[44] Fiocchi, C., Inflammatory bowel disease: Etiology and pathogenesis, *Gastroenterology,* 115, 182, 1998.
[45] Podolsky, D.K., Inflammatory bowel disease, *New Engl. J. Med.,* 347, 417, 2002.
[46] Kobayashi, K.S. et al., Nod2-dependent regulation of innate and adaptive immunity in the intestinal tract, *Science,* 307, 731, 2005.
[47] Hugot, J.P. et al., Association of Nod2 leucine-repeat variants with susceptibility to Crohn's disease, *Nature,* 411, 599, 2001.
[48] Suerbaum, S. and Michetti, P., *H. pylori* infection, *N. Engl. J. Med.,* 347, 1175, 2002.
[49] Keates, S. et al., H. pylori infection activates NF-κB gastric epithelial cells, *Gastroenterology,* 113, 1099, 1997.

[50] Munzenmaier, A. et al., A secreted/shed product of *H. pylori* activates transcription factor NF-κB, *J. Immunol.*, 159, 6140, 1997.

[51] Ishihara, S. et al., Essential role of MD2 in TLR4-dependent signaling during *H. pylori*-associated gastritis, *J. Immunol.*, 173, 1406, 2004.

[52] Viala, J. et al., Nod1 responds to peptidoglycan delivery by *H. pylori* cag pathogenicity island, *Nat. Immunol.*, 5, 1166, 2004.

[53] Mokdad, A.H. et al., Prevalence of obesity, diabetes, and obesity-related health risk factors, *JAMA*, 289, 76, 2003.

[54] Hotamisligil, G.S., Inflammatory pathways and insulin action, *Int. J. Obes. Relat. Metab. Disord.*, 27, S53, 2003.

[55] Hanahan, D. and Weinberg, R., The hallmarks of cancer, *Cell*, 100, 57, 2000.

[56] Campbell, P.M. and Der, C.J., Oncogenic Ras and its role in tumor cell invasion and metastasis, *Semin. Cancer Biol.*, 14, 105, 2004.

[57] Evan, G. and Vousden, K., Proliferation, cell cycle and apoptosis in cancer, *Nature*, 411, 342, 2001.

[58] Shay, J.W. and Bacchetti, S., A survey of telomerase activity in human cancer, *Eur. J. Cancer*, 33, 787, 1997.

[59] Johnsen, M. et al., Cancer invasion and tissue remodeling: Common themes in proteolytic matrix degradation, *Curr. Opin. Cell Biol.*, 10, 667, 1998.

[60] Hay, E.D., An overview of epithelio-mesenchymal transformation, *Acta Anat.*, 154, 8, 1995.

[61] Thiery, J.P., Epithelial-mesenchymal transitions in tumor progression, *Nat. Rev. Cancer*, 2, 442, 2003.

[62] Coussens, L. and Werb, Z., Inflammation and cancer, *Nature*, 420, 860, 2002.

[63] Turini, M.E. and Dubois, R.N., Cyclooxygenase-2: A therapeutic target, *Ann. Rev. Med.*, 53, 35, 2002.

[64] Greten, F.R. et al., IKKβ links inflammation and tumorigenesis in a mouse model of colitis-associated cancer, *Cell*, 118, 285, 2004.

[65] Pikarsky, E. et al., NF-κB functions as a tumor promoter in inflammation-associated cancer, *Nature*, 353, 180, 2004.

[66] Clevers, H., At the crossroads of inflammation and cancer, *Cell*, 118, 671, 2004.

[67] Rutter, M. et al., Severity of inflammation is a risk for colorectal neoplasia in ulcerative colitis, *Gastroenterology*, 126, 451, 2004.

[68] Bissell, M. and Radisky, D., Putting tumors in context, *Nat. Rev. Cancer*, 1, 46, 2001.

[69] Balkwill, F., Charles, K., and Mantovani, A., Smoldering and polarized inflammation in the initiation and promotion of malignant disease, *Cancer Cell*, 7, 211, 2005.

[70] Sparmann, A. and Bar-Sagi, D., Ras-induced IL-8 expression plays a critical role in tumor growth and angiogenesis, *Cancer Cell*, 6, 447, 2004.

[71] Alcamo et al., Targeted mutation of the TNF receptor 1 rescues the RelA-deficient mouse and reveals a critical role for NF-κB in leukocyte recruitment, *J. Immunol.*, 167, 1592, 2001.

[72] Grigoriadis, G. et al., The Rel subunit of NF-κB-like transcription factors is a positive and negative regulator of macrophage gene expression: Distinct roles for Rel in different macrophage populations, *Embo J.*, 15, 7099, 1996.

[73] Grossmann, M. et al., The combined absence of the transcription factor Rel and RelA leads to multiple hemopoietic cell defects, *Proc. Nat. Acad. Sci. USA*, 96, 11848, 1999.

[74] Lawrence, T. et al., Possible new role for NF-κB in the resolution of inflammation, *Nat. Med.*, 7, 1291, 2001.

[75]   Lai, W.C. et al., Differential regulation of LPS-induced MMP-1 and MMP-9 by p38 and ERK1/2 mitogen-activated protein kinases, *J. Immunol.*, 170, 6244, 2003.

[76]   Lu, Y. and Wahl, L., Production of MMP-9 by activated monocyes involves a PI3K/Akt/IKKα/NF-κB pathway, *J. of Leuk. Biol.*, 78, 1, 2005.

[77]   Poligone, B. and Baldwin, A.S., Positive and negative regulation of NF-κB by Cox-2: Roles of different prostaglandins, *J. Biol. Chem.*, 276, 38658, 2001.

[78]   Anthonsen, M.W., Solhaug, A., and Johansen, B., Functional coupling between secretory and cytosolic phospholipase A2 modulates TNFα- and IL-1β-induced NF-κB activation, *J. Biol. Chem.*, 276, 30527, 2001.

[79]   Cheah, F.C. et al., Detection of apoptosis by caspase-3 activation in tracheal aspirate neutrophils from premature infants: Relationship with NF-κB activation, *J. Leuk. Biol.*, 77, 432, 2005.

[80]   Taneja, R. et al., Delayed neutrophil apoptosis in sepsis is associated with maintenance of mitochondrial transmembrane potential and reduced caspase-9 activity, *Crit. Care Med.*, 32, 1460, 2004.

[81]   Ward, C. et al., Regulation of granulocyte apoptosis by NF-κB, *Biochem. Soc. Trans.*, 32, 465, 2004.

[82]   Gewirtz, A.T. et al., Bacterial flagellin activates basolaterally expressed TLR5 to induce epithelial proinflammatory gene expression, *J. Immunol.*, 15, 1882, 2001.

[83]   Neish, A.S. et al., Prokaryotic regulation of epithelial responses by inhibition of IκBα ubiquitination, *Science,* 289, 1560, 2000.

[84]   Kelly, D. et al., Commensal anaerobic gut bacteria attenuate inflammation by regulating nuclear-cytoplasmic shuttling of PPARγ and RelA, *Nat. Immunol.*, 5, 104, 2004.

[85]   Castrillo, A. et al., Inhibition of IKK and IκB phosphorylation by 15d-PGJ2 in activated murine macrophages, *Mol. Cell. Biol.,* 20, 1692, 2000.

[86]   Rossi, A. et al., Anti-inflammatory cyclopentonone prostaglandins are direct inhibitors of IκB kinase, *Nature,* 403, 103, 2000.

[87]   Schottelius, A. et al., Interleukin-10 signaling blocks IκB kinase activity and NF-κB DNA binding, *J. Biol. Chem.*, 274, 1868, 1999.

[88]   Reynaert, N.L. et al., Nitric oxide represses IκB kinase through S-nitrosylation, *Proc. Nat. Acad. Sci. USA*, 101, 1845, 2004.

[89]   Wilson, H.M. et al., Inhibition of macrophage NF-κB leads to a dominant anti-inflammatory phenotype that attenuates glomerular inflammation *in vivo, Am. J. Pathol.*, 167, 27, 2005.

[90]   Arkan, M. et al., IKKβ links inflammation to obesity-induced insulin resistance, *Nat. Med.*, 11, 191, 2005.

[91]   Cai, M. et al., Local and systemic insulin resistance resulting from hepatic activation of IKK-β and NF-κB, *Nat. Med.*, 11, 181, 2005.

[92]   Chen, L.W. et al., The two faces of IKK and NF-κB inhibition: Prevention of systemic inflammation but increased local injury following ischemia-reperfusion, *Nat. Med.*, 9, 575, 2002.

[93]   Maeda, S. et al., IKKβ couples hepatocyte death to cytokine-driven compensatory proliferation that promotes chemical hepatocarcinogenesis, *Cell,* 121, 977, 2005.

[94]   Firestein, G.S., NF-κB: Holy grail for rheumatoid arthritis?, *Arthr. Rheum.,* 50, 2381, 2004.

[95]   Miagkov, A.V. et al., NF-κB activation provides the potential link between inflammation and hyperplasia in the arthritic joint, *Proc. Nat. Acad. Sci. USA*, 95, 13859, 1998.

[96]   Jimi, E. et al., Selective inhibition of NF-κB blocks osteoclastogenesis and prevents inflammatory bone destruction *in vivo, Nat. Med.*, 10, 617, 2004.

[97]   Catley, M.C. et al., Validation of IKKβ as a therapeutic target in airway inflammation disease by adenoviral delivery of dominant negative IKKβ to pulmonary epithelial cells, *Br. J. Pharmacol.*, 145, 114, 2005.

[98]   Kopp, E. and Ghosh, S., Inhibition of NF-κB by sodium salicylate and aspirin, *Science*, 270, 2017, 1994.

[99]   Yin, M.J., Yamamoto, Y., and Gaynor, R.B., The anti-inflammatory agents aspirin and salicylate inhibit the activity of IKKβ, *Nature*, 396, 15, 1998.

[100]   Takada, Y. et al., Nonsteroidal anti-inflammatory agents differ in their ability to suppress NF-κB activation, inhibition of Cox-2 and cyclin D1, and abrogation of tumor cell proliferation, *Oncogene*, 23, 9247, 2004.

[101]   Farmer, P. and Pugin, J., β2-adrenergic agonists effect their anti-inflammatory effects on monocytic cells through the IκB/NF-κB pathway, *Am. J. Physiol. Lung Cell. Mol. Physiol.*, 279, L615, 2000.

[102]   Lawrence, T. et al., IKKα limits macrophage NF-κB activation and contributes to the resolution of inflammation, *Nature*, 434, 1238, 2005.

[103]   Baldwin, A.S., Control of oncogenesis and cancer therapy resistance by the transcription factor NF-κB, *J. Clin. Inv.*, 107, 241, 2001.

[104]   Karin, M. et al., NF-κB in cancer: From innocent bystander to major culprit, *Nat. Rev. Cancer*, 2, 301, 2002.

[105]   Pianetti, S. et al., Her-2/neu overexpression induces NF-κB via a PI3-kinase/Akt pathway involving calpain-mediated degradation of IκBα that can be inhibited by the tumor suppressor PTEN, *Oncogene*, 20, 1287, 2001.

[106]   Hanson, J.L. et al., NF-κB subunits RelA/p65 and c-Rel potentiate but are not required for Ras-induced cellular transformation, *Cancer Res.*, 64, 7248, 2004.

[107]   Reuther, J.Y. et al., A requirement for NF-κB in Bcr-Abl-mediated transformation, *Genes Dev.*, 12, 968, 1998.

[108]   Zhou, H., Du, M., and Dixit, V.M., Constitutive NF-κB activation by the t(11;18)(q21;q21) product in MALT lymphoma is linked to deregulated ubiquitin ligase activity, *Cancer Cell*, 7, 425, 2005.

[109]   Rocha, S. et al., Regulation of NF-κB and p53 through activation of ATR and Chk1 by the ARF tumor suppressor, *Embo J.*, 24, 1157, 2005.

[110]   Kim, J.H. et al., Transcriptional regulation of a metastasis suppressor gene by Tip60 and beta-catenin complexes, *Nature*, 434, 921, 2005.

[111]   Biswas, D.K., EGF-induced NF-κB activation: A major pathway of cell-cycle progression in estrogen receptor negative breast cancer cells, *Proc. Nat. Acad. Sci., USA*, 97, 8542, 2000.

[112]   Sitcheran, R. et al., Positive and negative regulation of EAAT2 by NF-κB: A role for N-myc in TNF-controlled gene repression, *Embo J.*, 24, 510, 2005.

[113]   Dajee, M. et al., NF-kB blockade and oncogenic Ras trigger invasive human epidermal neoplasia, *Nature*, 421, 639, 2003.

[114]   Zhang, J.Y. et al., NF-κB RelA opposes epidermal proliferation driven by TNFR1 and JNK, *Genes Dev.*, 18, 17, 2004.

[115]   Ryan, K.M. et al., Role of NF-κB in p53-mediated programmed cell death, *Nature*, 404, 892, 2000.

[116]   Deng, J. et al., β-catenin interacts with and inhibits NF-κB in human colon and breast cancer, *Cancer Cell*, 2, 323, 2002.

[117]   Perkins, N.D., NF-κB: Tumor promoter or suppressor, *Trends Cell Biol.*, 14, 64, 2004.

[118]   Lam, L.T. et al., Small molecule inhibitors of IKKs are selectively toxic for subgroups of diffuse large B-cell lymphoma defined by gene expression profiling, *Clin. Cancer Res.*, 11, 28, 2005.

[119] Mayo, M.W. et al., Requirement of NF-κB activation to suppress p53-independent apoptosis induced by oncogenic Ras, *Science,* 278, 1812, 1997.

[120] Bargou, R.C. et al., Constitutive nuclear NF-κB-RelA activation is required for proliferation and survival of Hodgkin's disease tumor cells, *J. Clin. Inv.*, 100, 2961, 1997.

[121] Hinz, M. et al., Constitutive NF-κB maintains high expression of a characteristic gene network, including CD40 and CD86, and a set of anti-apoptotic genes in Hodgkin/Reed-Sternberg cells, *Blood,* 97, 2798, 2001.

[122] Hu, M.C. et al., IKKβ promotes tumorigenesis through inhibition of forkhead Foxo3a, *Cell,* 117, 225, 2004.

[123] Nakanishi, C. and Toi, M., NF-κB inhibitors as sensitizers to anticancer drugs, *Nat. Revs. Cancer,* 5, 297, 2005.

[124] Wang, C.Y. et al., Control of inducible chemoresistance: Enhanced anti-tumor therapy through increased apoptosis by inhibition of NF-κB, *Nat. Med.*, 5, 412, 1999.

[125] Campbell, K., Rocha, S., and Perkins, N., Active repression of antiapoptotic gene expression by RelA/p65 NF-κB, *Mol. Cell,* 13, 853, 2004.

[126] Albanese, C. et al., IKKα regulates mitogenic signaling through transcriptional induction of cyclin D1 via TCF, *Mol. Cell. Biol.*, 14, 285, 2003.

[127] Cao, Y. et al., IKKα provides an essential link between RANK signaling and cyclin D1 expression during mammary gland development, *Cell,* 107, 763, 2001.

[128] Jung, Y.J. et al., IL-1β-mediated upregulation of HIF-1α via an NF-κB/COX-2 pathway identifies HIF-1α as a critical link between inflammation and oncogenesis, *FASEB J.,* 17, 2115, 2003.

[129] Lee, H. et al., Role of Rel-related factors in control of c-myc transcription in receptor-mediated apoptosis of the murine B-cell WEHI cell line, *J. Exp. Med.*, 181, 1169, 1995.

[130] Wu, C.Y. et al., H. pylori promote gastric cancer cell invasion through NF-κB and COX-2-mediated pathways, *World J. Gastroenterol.*, 11, 3197, 2005.

[131] Muerkoster, S. et al., Tumor stroma interactions and chemoresistance in pancreatic ductal carcinoma cells involving increased secretion and paracrine effects of nitric oxide and IL-1β, *Cancer Res.*, 64, 1331, 2004.

[132] Huber, M.A. et al., NF-κB is essential for epithelial-mesenchymal transition and metastasis in a model of breast cancer progression, *J. Clin. Inv.*, 114, 569, 2004.

[133] Li, Q., Withoff, S. and Verma, I.M., Inflammation-associated cancer: NF-κB is the lynchpin, *Trends in Immunol.*, 26, 318, 2005.

[134] Konturek, P.C. et al., Activation of NF-κB represents the central event in the neoplastic progression associated with Barrett's esophagus: A possible link to inflammation and overexpression of PPARγ and growth factors, *Dig. Dis. Sci.*, 49, 1075, 2004.

[135] Akiyama, M. et al., NF-κB mediates TNFα-induced nuclear translocation of telomerase reverse transcriptase protein, *Cancer Res.*, 63, 18, 2003.

[136] Shisodia, S. et al., Cox-2 inhibitor celecoxib abrogates TNF-induced NF-κB activation through inhibition of activation of IκB kinase and Akt in human non-small cell lung carcinoma: Correlation with suppression of COX-2 synthesis, *J. Immunol.*, 173, 2011, 2004.

9 NF-κB and Associated Human Genetic Pathologies

Alain Israël

CONTENTS

9.1 INTRODUCTION

NF-κB represents a family of transcription factors that are kept inactive in the cytoplasm through interaction with inhibitory molecules of the IκB family (Chapter 1). In response to multiple stimuli such as inflammatory cytokines, bacterial or viral products, or various types of stress, the IκB molecules become phosphorylated on two critical serine residues. This modification allows their polyubiquitination and destruction by the proteasome (Chapter 4). As a consequence, free NF-κB enters the nucleus and activates transcription of a variety of genes participating in immune and inflammatory response, cell adhesion, growth control, and protection against apoptosis (Chapters 6 through 8).

For many years, the kinase that phosphorylates IκB (inhibitor of κB kinase [IKK], or IκB kinase) remained elusive. Upon biochemical fractionation, it was finally identified as a high-molecular-weight complex migrating around 700 to 900 kD and containing two related catalytic subunits: IKKα and IKKβ. An additional component, nuclear factor κB essential modifier (NEMO)/IKKγ, has subsequently been identified through genetic complementation of an NF-κB activation-defective cell line [1] and sequencing of IKK-associated polypeptides [2,3]. Although NEMO does not have catalytic properties, cell lines defective for NEMO do not activate NF-κB in response to many stimuli, which demonstrates the key role of this protein in activating the NF-κB pathway. The human *nemo* gene is located on chromosome X at Xq28 [4] and encodes a 45 kD protein composed of several structural domains, among them two coiled-coil domains, a leucine zipper, and a C-terminal zinc finger. It is required for IKK activation in response to most NF-κB stimuli (classical or canonical pathway), as shown using mutant cell lines defective for this protein. In contrast, a subset of stimuli, including B cell activating factor (BAFF) and lympho-toxin-β (LTβ), does not require NEMO, but instead induces IKK activation through nuclear factor κB inducing kinase (NIK)-induced IKKα phosphorylation (alternative or noncanonical pathway) [5,6].

The exact mechanism of IKK activation and the identity of the molecules that regulate its activity remain poorly understood (Chapter 3). Very recently, it has been proposed that IKK activation by cytokines and lipopolysaccharide (LPS) may involve nondegradative ubiquitination events [7], a modification that might be negatively con-trolled by the cylindromatosis (CYLD) deubiquitinase (see Section 9.5 and Chapter 4).

9.2 INCONTINENTIA PIGMENTI AND NEMO LOSS-OF-FUNCTION MUTATIONS

The story starts with a consortium of geneticists looking for the gene whose mutation is responsible for an X-linked neurocutaneous genodermatosis called Incontinentia pigmenti (IP) [8]. In males, IP is lethal during early development. In females, the most characteristic feature of the disease is a dermatosis that usually begins after birth and evolves according to the following sequence: (1) an erythematous, vascular rash develops, accompanied by a massive eosinophilic granulocyte infiltration into the epidermis; (2) verrucous hyperkeratotic lesions evolve and disappear over time, leaving behind areas of hyperpigmentation; and (3) this hyperpigmentation fades and pale hairless patches or streaks remain on the skin.

In addition to these epidermal manifestations, IP patients also suffer from oph-talmologic (abnormalities of the developing retinal vessels), odontological (missing or deformed teeth) and, in rare cases, neurological (convulsive disorders, motor or mental retardation) problems. An important characteristic of IP is the skewed X-inac-tivation that occurs in several cell types or tissues of female patients. This skewing reflects a counter-selection of cells carrying the mutated copy of NEMO and can reach, in blood cells for instance, more than 95%.

The gene responsible for IP was localized to Xq28, similar to NEMO. This localization, as well as the high sensitivity to apoptosis of IP cells, suggested that it might be responsible for the observed pathology. Indeed, the analysis of a large number of patients showed that 85% of them carried the same complex rearrange-ment of the NEMO gene [9], resulting in the excision of the region between two repeated sequences located upstream of exon 4 and downstream of exon 10. The first three exons of NEMO, those that remain after the rearrangement, produce a truncated 133 amino acid molecule devoid of activity. Indeed, IP fetus-derived primary fibroblasts carrying this rearrangement exhibited a complete lack of NF-κB activation, no degradation of the IκB molecules when stimulated, and a high sensitivity to tumor necrosis factor (TNF)α-induced apoptosis. Most other mutations that have been described as a cause of IP result in more or less severe truncations of NEMO [9,10], but a few missense mutations have also been identified [11].

The phenotype of IP patients is difficult to interpret due to the complex interplay that exists in several tissues between cells expressing the normal copy of NEMO and those expressing the defective one; the latter seem to die from apoptosis or necrosis at different periods in the patient's life. At the skin level, elimination of mutated cells starts only after birth, and this process generates the characteristic dermatosis observed in IP patients. The signal that triggers this postbirth dermatosis is still unknown, but the cascade of events that occurs may be explained as follows: necrosis of some *nemo (−)* keratinocytes may occur following various types of stress, triggering a local inflammatory response and the secretion of cytokines, such as TNF. *nemo (−)* cells die as a result of TNF-induced apoptosis.

NEMO knockout mice have been engineered by several groups [12–14], and the phenotype of these mice is very similar to that of IP patients. Male mice die very early during embryogenesis (E12.5) from massive liver apoptosis. A similar phenotype has also been reported for p65 and IKKβ knockout mice, but death occurs later, at E15.5 and E14.5, respectively [15]. Importantly, liver apoptosis is triggered by TNF. As mentioned, double knockout mice (*p65$^{-/-}$/tnfα$^{-/-}$* and *ikkβ$^{-/-}$/tnf-r1$^{-/-}$*) can survive until birth, although they are affected by multiple immune dysfunctions. It remains to be determined whether liver apoptosis is also responsible for male lethality in IP. In contrast to males, NEMO knockout female mice develop normally but, soon after birth, exhibit patchy skin lesions with massive granulocyte infiltration. The lesions are associated with hyperproliferation and increased apoptosis of keratinocytes. This is strikingly similar to what occurs in IP patients. In addition, the extensive X-inactivation skewing in blood leukocytes, which is a major feature of IP patients, is also observed in NEMO knockout mice. The only difference between IP pathology and NEMO knockout mice is the high level of mortality that occurs in female mice within 6 to 10 days after birth, which is never observed in humans. Despite this difference, NEMO knockout mice should provide a very useful model for investigating IP pathology.

As an alternative to the analysis of heterozygous *nemo(−)* female mice, which is complicated by their mosaic state, mice with selective *ikkβ* ablation in the epidermis have been generated [16]. As NEMO and IKKβ are both required in the context of the IKK complex to mediate NF-κB activation, their inactivation is expected to produce similar outcomes. In these mice, epidermal development proceeds normally until birth, but at P4–P5 their skin starts becoming hard and by day P7–P8 a widespread scaling is observed that precedes death. At P7, mice exhibit a thickened epidermis, and this hyperproliferation is accompanied by cellular infiltration into the dermis and severe inflammation. Several inflammatory cytokines such as IL-1 and TNFα, as well as chemokines, accumulate in the epidermis or the dermis during progression of the disease. The role of TNFα in the development of the pathology appears essential since crossing these mice with *tnfr1$^{-/-}$* mice completely suppresses the skin symptoms. Importantly, purified keratinocytes do not exhibit hyperproliferation on their own, indicating that the defect is not cell-autonomous. Thus, *ikkβ* skin knockout mice demonstrate that hyperproliferation is rather a secondary event resulting from inflammation and suggest that the critical function of IKKβ in epidermal keratinocytes is to regulate mechanisms that maintain the immune homeostasis of the skin. In contrast to *nemo*[(-)], apoptosis of keratinocytes in *IKKβ* skin knockout mice does not contribute substantially to the pathology. Although the identity of the signals triggering the dermatosis exhibited by IP patients remains unclear, the data summarized suggest a plausible sequence of events and identify TNFα as a key participant in both the onset of inflammation and its resolution through clearance of NEMO mutated cells by apoptosis.

9.3 PHENOTYPES ASSOCIATED WITH HYPOMORPHIC NEMO MUTATIONS

9.3.1 EDA-ID

As indicated, whereease the complete loss of NF-κB activation is lethal for males during embryogenesis, females can survive, exhibiting a complex phenotype due to their mosaic character regarding X-inactivation. Because it is unclear to what extent the phenotype affecting female patients is directly due to NF-κB dysfunction and not to the interaction between normal and dying cells, it is difficult to draw a firm conclusion regarding the consequence of NF-κB deficiency in humans. Therefore, the question is, what would happen in the case of milder NEMO mutations allowing males to survive? The answer is provided by a series of recent data that report new syndromes exclusively affecting male patients. In this case, their single X chromosome carries the mutated gene, allowing the direct observation of the physiological consequences of NF-κB dysfunction in humans.

For years, a rare and complex syndrome exclusively affecting male patients and associating anhidrotic ectodermal dysplasia and immunodeficiency (EDA-ID; OMIM # 300291) had been described, but its genetic cause was unknown (see, for example, [17]). The fact that this syndrome only affected male patients, combined with the association with a perturbed immune response and some similarities with IP, prompted the analysis of the *nemo* gene. Most patients indeed carried mutations in *nemo,* but instead of leading to large truncations of the NEMO molecule as observed in IP, the mutations were missense mutations or small deletions only affecting the C-terminal zinc finger [18–21] (Figure 9.1). Interestingly, all of these mutations led to reduced but not abolished NF-κB activation, which explains why affected male patients survive.

Anhidrotic ectodermal dysplasia (EDA) is a well-described pathology characterized by the absence of sweat glands, sparse scalp hair, and missing teeth [22]. Three distinct loci have been shown to be responsible for EDA. The first, located on the X chromosome (the mutant mouse at this locus is called *tabby*), codes for two members of the TNF family generated by alternative splicing, EDA-A1, and EDA-A2 (ectodysplasin), which are produced in tissues of ectodermal origin, such

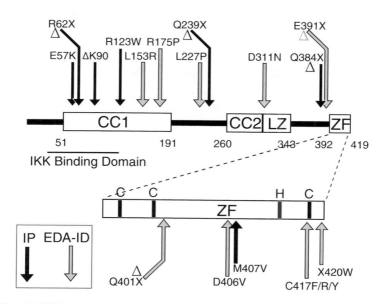

FIGURE 9.1 NEMO mutations in human pathologies. The structural domains of NEMO are indicated (CC1: coiled coil 1; CC2: coiled coil 2; LZ: leucine zipper; ZF: zinc finger). The zinc finger region is enlarged in the bottom part of the figure. The three Cys and the His predicted to coordinate the Zn atom are indicated. The region of interaction with the kinases is indicated. Mutations leading to IP (black) or to EDA-ID (grey) are indicated. Δ indicates that the mutation leads to a truncation. The rearrangement of the NEMO gene that is found in 85% of the IP families is not shown: it leads to the generation of a NEMO protein truncated after amino acid 133. Frameshift mutations that lead to truncations and C-terminal addition of a number of irrelevant amino acids are not shown: they lead to IP or EDA-ID, depending on the length of the C-terminal addition.

as keratinocytes, hair follicles, and sweat glands [23]. The second gene, *edar (down-less* or *dl* in the mouse), is located on chromosome 2 and codes for a member of the TNF receptor family [24,25]. Its expression is restricted to placodes, which are thickenings of epithelia where the formation of epidermal appendages begins. Recently, EDAR has been shown to be the receptor for EDA-A1 [26–29]. A second receptor, X-EDAR, encoded by a gene located on the X chromosome, binds EDA-A2. No pathology associated with mutations at this locus have been identified so far in humans, and mice lacking X-EDAR are indistinguishable from their wild-type littermates [30]. The third locus whose disruption leads to EDA has been identified recently in *crinkled* mice and was subsequently found to be mutated in a human family [31,32]. It encodes an adaptor protein, EDARADD, which binds EDAR and interacts with the adaptor molecule TRAF2. The fact that these mutations all result in an EDA phenotype can be explained by the alignment of these molecules (EDA, EDAR, EDARRAD, and NEMO) in a linear pathway that leads to NF-κB activation (Figure 9.2) [26,29,33]. Although the details of the ectodysplasin/EDAR signaling pathway are not fully characterized, it is clear from analyzing the mouse

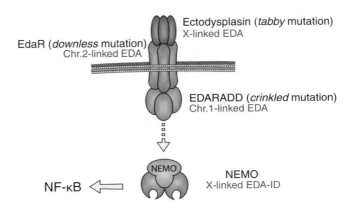

FIGURE 9.2 The EDA/EDAR/EDARADD/NEMO cascade. A simplified version of the signaling cascade is shown, emphasizing the molecules whose mutation leads to EDA. The names of the mouse mutations are indicated in parentheses. Mutations in the ligand, receptor, or adaptor lead to EDA, while mutations affecting NEMO lead to EDA-ID, as multiple other signaling pathways involved in the immune response target the IKK complex.

models of EDAR pathology that this pathway is involved very early during development of hair follicle morphogenesis [24], assigning a previously unrecognized role for NF-κB in this process.

The other symptom affecting EDA-ID patients — immunodeficiency — is less unexpected based on the well-known function played by NF-κB in both arms of the immune response (Chapters 6, 7, and 8). This immunodeficiency is characterized by unusually severe, life-threatening, and recurrent bacterial infections of the lower respiratory tract, skin, soft tissues, bones, and gastrointestinal tract, as well as meningitis and septicemia in early childhood. The causative pathogens are most often Gram-positive bacteria (*S. pneumoniae* and *S. aureus*), followed by Gram-negative bacteria (*Pseudomonas* spp. and *Haemophilus influenzae*) and mycobacteria.

A number of EDA-ID patients have elevated serum IgM levels — the so-called "hyper-IgM" phenotype (see, for example, [20]). In some EDA-ID patients, B cells have an impaired ability to switch in response to the cluster of differentiation 40 (CD40) ligand. In others, immunoglobulin switching is normal, but proliferation and differentiation are deficient, also resulting in a "hyper-IgM-like" syndrome. A more consistent feature of EDA-ID pathology is an impaired antibody response to polysaccharide antigens. In contrast to these B cell anomalies, patients with EDA-ID have normal T cell proliferation in response to both mitogens and antigens.

In conclusion, both symptoms exhibited by EDA-ID patients can be correlated with a defect in NF-κB activation. The immunological defect is due to the impairment of a series of important signaling pathways involved in both arms of the immune response; but the heterogeneity of the symptoms points to some previously unrecognized subtleties in the details of this signaling cascade, unless this heterogeneity can be assigned to the genetic background. On the other hand, the EDA symptom is specifically caused by an impairment of the downstream events of the EDA/EDAR/EDARADD signaling cascade.

9.3.2 EDA-ID with Osteopetrosis and Lymphedema (OL-EDA-ID)

Two EDA-ID patients exhibiting two additional defects, osteopetrosis and lymphoedema, have been described [9,33,34]. Osteopetrosis, which results from defective bone resorption by osteoclasts, has been previously linked to the NF-κB pathway since p50/p52 KO mice exhibit osteopetrosis [35,36]. In addition, the receptor activator of nuclear factor kappa B (RANK) pathway that plays an important role in osteoclast function is itself NF-κB-dependent [37]. Primary lymphedema is still poorly understood at the genetic/biochemical level, but the gene causing familial lymphedema, *vegfr-3*, has been identified, and overexpression of its product results in NF-κB activation [38].

Remarkably, the two reported OL-EDA-ID patients were found to carry the same genetic defect — the replacement of the NEMO stop codon with tryptophan, leading to the addition of 27 irrelevant amino acids at the C-terminus of the molecule [9,33]. This apparently benign mutation actually induces a strong destabilizing effect on the NEMO molecule, therefore explaining the severity of the symptoms.

9.3.3 Are IP and EDA-ID the Extreme Boundaries of a Genetic Continuum?

In addition to the EDA-ID cases described above, two male patients carrying different mutations in the NEMO gene have been described recently as suffering from immunodeficiency without EDA [39,40]. The first developed an atypical mycobacterial infection, while the second presented with a hyper-IgM phenotype.

Considering these multiple phenotypes, varying from lethal IP to immunodeficiency, one wonders whether there is a direct correlation between the level of impairment of the NF-κB pathway and the severity of the symptoms. While it is clear that IP results from a complete lack of NF-κB activity and leads to embryonic lethality, the variable phenotype observed in EDA-ID or ID patients is more difficult to reconcile with a quantitative variation in NF-κB activity. It is tempting to try to deduce from the severity of the symptoms the importance of a specific amino acid (or structural domain) in the activity of the NEMO molecule. However, one should be cautious and consider that the genetic background of the patients, as well as their history, might influence the symptoms they exhibit. For example, two patients carrying the C417R mutation (located in the C-terminal zinc finger) can exhibit slightly different immunological phenotypes (compare [18] with [20]). One additional problem is that it is difficult to establish an assay that gives an absolute measurement of NF-κB activity associated with a given mutation: the residual activity may vary a lot depending on the cell type (and cells derived from patients are not always available) or the stimulus analyzed.

Regarding NEMO truncations, those restricted to the zinc finger lead to EDA-ID, whereas larger ones lead to IP. However, the situation becomes more complicated when it comes to missense mutations: a single amino acid deletion (ΔK90) leads to IP, although it reconstitutes half of wt NF-κB activation in response to LPS when introduced into NEMO-deficient pre-B cells [11]. However, most other point mutations, scattered all along the molecule, lead to EDA-ID, and while they sometimes affect the C-terminal zinc finger (such as the C417R mutation described above, which probably interferes with Zn coordination), their effect on the structure of NEMO remains difficult to predict in most cases. Clearly, a crystal structure of the NEMO/IKK complex is needed before one can gain further understanding of these pathologies.

9.4 IκBα MUTATION IS ASSOCIATED WITH EDA AND T CELL IMMUNODEFICIENCY

Do mutations in other components of the NF-κB cascade lead to symptoms similar to those described above? Two recently described patients may help to answer this question but may, at the same time, point toward the complexity of this signaling pathway [41,42]. These patients exhibit an autosomal-dominant form of EDA-ID but show no mutation in the NEMO gene. The developmental, immunological, and infectious defects exhibited by these patients and by those carrying hypomorphic NEMO mutations largely overlap and include EDA, impaired cellular responses to ligands of Toll-like receptors (TLR) and tumor necrosis factor receptor (TNFR) superfamily members, and severe bacterial diseases. However, these new patients exhibit a unique T cell immunodeficiency; despite a marked blood lymphocytosis, no detectable memory T cells can be seen *in vivo,* and naïve T cells do not respond to CD3-TCR activation *ex vivo.* These patients have been demonstrated to carry a heterozygous missense mutation at serine 32 of IκBα. This residue, together with Ser36, is a target for IKK and is required to induce IκBα polyubiquitination and degradation. This mutation is a gain-of-function, as it enhances the inhibitory capacity of IκBα by preventing its degradation and results in impaired NF-κB activation. Currently, it is difficult to understand why these patients show this unique T cell defect. It must be concluded that partially inhibiting the degradation of the 3 IκB inhibitors (in the case of the hypomorphic NEMO mutations) and completely inhibiting the degradation of the most important of them, IκBα, leads to different outcomes (although the magnitude of the difference might vary depending on the cell type considered). These results highlight both the diversity of genotypes associated with EDA-ID and the diversity of immunological phenotypes associated with mutations in different components of the NF-κB signaling pathway.

9.5 CYLD DEUBIQUITINASE NEGATIVELY REGULATES NF-κB SIGNALING

In view of the importance of NEMO in human pathologies, interaction partners have been sought by numerous groups. Recently, three groups used different approaches to clone a cDNA, encoding a protein called CYLD, a member of a specific subclass of deubiquitinases [43–45]. CYLD had been shown to be a tumor suppressor, mutated in a pathology called cylindromatosis [46]. Familial cylindromatosis/Spiegler-Brooke syndrome (OMIM#132700) is a rare autosomal dominant inherited disease characterized by the development of adnexal tumors — mostly cylindromas but also trichoepitheliomas and spiradenomas — appearing during adulthood and exclusively derived from skin appendages such as eccrine and apocrine sweat glands. Malignant tumors may occur, usually with the features of a cylindrocarcinoma.

The interaction with NEMO suggests that CYLD is involved in the NF-κB signaling pathway. This suggestion is reinforced by the fact that CYLD has also been identified through an experimental approach involving a whole-genome search aimed at identifying members of the deubiquitinase family controling NF-κB activation by TNFα [45]. As discussed in Chapter 4, CYLD functions as a negative regulator of NF-κB signaling by acting as a specific deubiquitinase for K63-linked regulatory ubiquitination of NF-κB pathway components. Therefore, CYLD, whose expression was shown later to be induced by NF-κB [47], would be responsible for downregulating NF-κB activation by deubiquinating important regulatory molecules. More recently, it has been shown that CYLD is also a negative regulator of the jun N-terminal kinase (JNK) pathway [48].

As most mutations identified in cylindromatosis patients seem to abrogate the deubiquitinase activity of CYLD, it remains to be shown how upregulating NF-κB (and possibly the JNK pathway) may induce the specific symptoms observed, in particular the fact that cylindromas seem to affect mostly the head (and sometimes the trunk). This is a prerequisite before attempting to treat these patients with NF-κB inhibitors.

9.6 CONCLUSION

The study of NF-κB-related human pathologies during the last five years has led to a number of important conclusions regarding the physiological role of NF-κB. While the immunodeficiency that results from NF-κB dysfunction was not unexpected based on previous studies of genetically manipulated mice, the heterogeneity of the symptoms suggests some unexpected complexity in the mechanisms of regulation of NF-κB activation.

Possibly more interesting (at least from an academic point of view) is the demonstration and confirmation of the essential role that NF-κB plays in skin development and homeostasis. Although the symptoms observed in female IP patients are difficult to interpret, they show that dysfunction of this signaling pathway can trigger a complex dermatosis that combines inflammation, keratinocytes hyperproliferation, and apoptosis. On the other hand, the results obtained from the analysis of EDA-ID patients have allowed us to place the NF-κB pathway downstream of an important signaling cascade that specifically controls morphogenesis of skin appendages. It is unclear at the moment whether the phenotype observed in cylindromatosis patients can be explained along the same lines.

Finally, considering the wide range of dysfunctions generated by mutations of a single component of the NF-κB pathway, NEMO, it can be predicted that mutations affecting other molecules of the same pathway will be found in the future associated with genetic diseases showing discrete or general perturbations of the skin with or without the hematopoietic compartment.

REFERENCES

[1]   Yamaoka, S., Courtois, G., Bessia, C. et al., Complementation cloning of NEMO, a component of the IκB kinase complex essential for NF-κB activation, *Cell,* 93, 1231, 1998.

[2]   Mercurio, F., Murray, B.W., Shevchenko, A. et al., IκB kinase (IKK)-associated protein 1, a common component of the heterogeneous IKK complex, *Mol. Cell. Biol.,* 19, 1526, 1999.

[3]   Rothwarf, D.M., Zandi, E., Natoli, G. et al., IKK-γ is an essential regulatory subunit of the IκB kinase complex, *Nature,* 395, 297, 1998.

[4]   Jin, D.Y. and Jeang, K.T., Isolation of full-length cDNA and chromosomal localization of human NF-kappa B modulator NEMO to Xq28, *J. Biomed. Sci.,* 6, 115, 1999.

[5]   Pomerantz, J.L. and Baltimore, D., Two pathways to NF-kappaB, *Mol. Cell,* 10, 693, 2002.

[6]   Bonizzi, G. and Karin, M., The two NF-kappaB activation pathways and their role in innate and adaptive immunity, *Trends Immunol.,* 25, 280, 2004.

[7]   Sun, L. and Chen, Z.J., The novel functions of ubiquitination in signaling, *Curr. Opin. Cell. Biol.,* 16, 119, 2004.

[8]   Landy, S.J. and Donnai, D., Incontinentia pigmenti (Bloch Sulzberger syndrome), *J. Med. Genet.,* 30, 53, 1993.

[9]   Smahi, A., Courtois, G., Vabres, P. et al., Genomic rearrangement in NEMO impairs NF-κB activation and is a cause of incontinentia pigmenti, *Nature,* 405, 466, 2000.

[10] Aradhya, S., Woffendin, H., Jakins, T. et al., A recurrent deletion in the ubiquitously expressed NEMO (IKK-gamma) gene accounts for the vast majority of incontinentia pigmenti mutations, *Hum. Mol. Genet.,* 10, 2171, 2001.

[11] Fusco, F., Bardaro, T., Fimiani, G. et al., Molecular analysis of the genetic defect in a large cohort of IP patients and identification of novel NEMO mutations interfering with NF-kappaB activation, *Hum. Mol. Genet.,* 13, 1763, 2004.

[12] Makris, C., Godfrey, V.L., Krahn-Senftleben, G. et al., Female mice heterozygous for IKKγ/NEMO deficiencies develop a dermatopathy similar to the human X-linked disorder incontinentia pigmenti, *Mol. Cell,* 5, 969, 2000.

[13] Rudolph, D., Yeh, W.C., Wakeham, A. et al., Severe liver degeneration and lack of NF-κB activation in NEMO/IKK-γ deficient mice, *Genes Dev.,* 14, 854, 2000.

[14] Schmidt-Supprian, M., Bloch, W., Courtois, G. et al., NEMO/IKKγ-deficient mice model Incontinentia pigmenti, *Mol. Cell,* 5, 981, 2000.

[15] Gerondakis, S., Grossmann, M., Nakamura, Y. et al., Genetic approaches in mice to understand Rel/NF-κB and IκB function: Transgenics and knockouts, *Oncogene,* 18, 6888, 1999.

[16] Pasparakis, M., Courtois, G., Hafner, M. et al., TNF-mediated inflammatory skin disease in mice with epidermis-specific deletion of IKK2, *Nature,* 417, 861, 2002.

[17] Abinun, M., Spickett, G., Appleton, A.L. et al., Anhidrotic ectodermal dysplasia associated with specific antibody deficiency, *Eur. J. Pediatr.,* 155, 146, 1996.

[18] Döffinger, R., Smahi, A., Bessia, C. et al., X-linked anhidrotic ectodermal dysplasia with immunodeficiency is caused by impaired NF-kappaB signaling, *Nature Genet.,* 27, 277, 2001.

[19] Aradhya, S., Courtois, G., Rajkovic, A. et al., Atypical forms of Incontinentia pigmenti in males result from mutations of a cytosine tract in exon 10 of NEMO (IKKγ), *Am. J. Hum. Genet.,* 68, 765, 2001.

[20] Jain, A., Ma, C.A., Liu, S. et al., Specific missense mutations in NEMO result in hyper-IgM syndrome with hypohydrotic ectodermal dysplasia, *Nature Immunol.,* 2, 223, 2001.

[21] Zonana, J., Elder, M.E., Schneider, L.C. et al., A novel X-linked disorder of immune deficiency and hypohidrotic ectodermal dysplasia is allelic to incontinentia pigmenti and due to mutations in IKK-γ (NEMO), *Am. J. Hum. Genet.*, 67, 1555, 2000.

[22] Champlin, T.L. and Mallory, S.B., Hypohidrotic ectodermal dysplasia: a review, *J. Ark. Med. Soc.*, 86, 115, 1989.

[23] Kere, J., Srivastava, A.K., Montonen, O. et al., X-linked anhidrotic (hypohidrotic) ectodermal dysplasia is caused by mutation in a novel transmembrane protein, *Nature Genet.*, 13, 409, 1996.

[24] Headon, D.J. and Overbeek, P.A., Involvement of a novel TNF receptor homologue in hair follicle induction, *Nature Genet.*, 22, 370, 1999.

[25] Monreal, A.W., Ferguson, B.M., Headon, D.J. et al., Mutations in the human homologue of mouse dl cause autosomal recessive and dominant hypohidrotic ectodermal dysplasia, *Nature Genet.*, 22, 366, 1999.

[26] Kumar, A., Eby, M.T., Sinha, S. et al., The ectodermal dysplasia receptor activates the NF-κB, JNK and cell death pathways and binds to ectodysplasin A, *J. Biol. Chem.*, 276, 2668, 2001.

[27] Laurikkala, J., Mikkola, M., Mustonen, T. et al., TNF signaling via the ligand-receptor pair ectodysplasin and edar controls the function of epithelial signaling centers and is regulated by Wnt and activin during tooth organogenesis, *Dev. Biol.*, 229, 443, 2001.

[28] Tucker, A.S., Headon, D.J., Schneider, P. et al., EDAR/EDA interactions regulate enamel knot formation in tooth morphogenesis, *Dev.*, 127, 4691, 2000.

[29] Yan, M.H., Wang, L.C., Hymowitz, S.G. et al., Two-amino acid molecular switch in an epithelial morphogen that regulates binding to two distinct receptors, *Science*, 290, 523, 2000.

[30] Newton, K., French, D.M., Yan, M. et al., Myodegeneration in EDA-A2 transgenic mice is prevented by XEDAR deficiency, *Mol. Cell. Biol.*, 24, 1608, 2004.

[31] Headon, D.J., Emmal, S.A., Ferguson, B.M. et al., Gene defect in ectodermal dysplasia implicates a death domain adapter in development, *Nature*, 414, 913, 2001.

[32] Yan, M., Zhang, Z., Brady, J.R. et al., Identification of a novel death domain-containing adaptor molecule for ectodysplasin-A receptor that is mutated in crinkled mice, *Curr. Biol.*, 12, 409, 2002.

[33] Döffinger, R., Smahi, A., Bessia, C. et al., X-linked anhidrotic ectodermal dysplasia with immunodeficiency is caused by impaired NF-κB signaling, *Nature Genet.*, 27, 277, 2001.

[34] Mansour, S., Woffendin, H., Mitton, S. et al., Incontinentia pigmenti in a surviving male is accompanied by hypohidrotic ectodermal dysplasia and recurrent infection, *Am. J. Med. Genet.*, 99, 172, 2001.

[35] Iotsova, V., Caamano, J., Loy, J. et al., Osteopetrosis in mice lacking NF-kappaB1 and NF-kappaB2, *Nature Med.*, 3, 1285, 1997.

[36] Franzoso, G., Carlson, L., Xing, L. et al., Requirement for NF-kappaB in osteoclast and B-cell development, *Genes Dev.*, 11, 3482, 1997.

[37] Hsu, H., Lacey, D.L., Dunstan, C.R. et al., Tumor necrosis factor receptor family member RANK mediates osteoclast differentiation and activation induced by osteoprotegerin ligand, *Proc. Natl. Acad. Sci. USA*, 96, 3540, 1999.

[38] Karkkainen, M.J., Ferrell, R.E., Lawrence, E.C. et al., Missense mutations interfere with VEGFR-3 signalling in primary lymphoedema, *Nature Genet.*, 25, 153, 2000.

[39] Orange, J.S., Levy, O., Brodeur, S.R. et al., Human nuclear factor kappaB essential modulator mutation can result in immunodeficiency without ectodermal dysplasia, *J. Allergy Clin. Immunol.*, 114, 650, 2004.

[40] Niehues, T., Reichenbach, J., Neubert, J. et al., Nuclear factor kappaB essential modulator-deficient child with immunodeficiency yet without anhidrotic ectodermal dysplasia, *J. Allergy Clin. Immunol.,* 114, 1456, 2004.

[41] Courtois, G., Smahi, A., Reichenbach, J. et al., A hypermorphic IκBα mutation is associated with autosomal dominant anhidrotic ectodermal dysplasia and T cell immunodeficiency, *J. Clin. Invest.,* 112, 1108, 2003.

[42] Janssen, R., van Wengen, A., Hoeve, M.A. et al., The same IκBα mutation in two related individuals leads to completely different clinical syndromes, *J. Exp. Med.,* 200, 559, 2004.

[43] Trompouki, E., Hatzivassiliou, E., Tsichritzis, T. et al., CYLD is a deubiquitinating enzyme that negatively regulates NF-κB activation by TNFR family members, *Nature,* 424, 793, 2003.

[44] Kovalenko, A., Chable-Bessia, C., Cantarella, G. et al., The tumour suppressor CYLD negatively regulates NF-kappaB signalling by deubiquitination, *Nature,* 424, 801, 2003.

[45] Brummelkamp, T.R., Nijman, S.M., Dirac, A.M. et al., Loss of the cylindromatosis tumour suppressor inhibits apoptosis by activating NF-κB, *Nature,* 424, 797, 2003.

[46] Bignell, G.R., Warren, W., Seal, S. et al., Identification of the familial cylindromatosis tumour-suppressor gene, *Nature Genet.,* 25, 160, 2000.

[47] Jono, H., Lim, J.H., Chen, L.F. et al., NF-kappaB is essential for induction of CYLD, the negative regulator of NF-kappaB: evidence for a novel inducible autoregulatory feedback pathway, *J. Biol. Chem.,* 279, 36171, 2004.

[48] Reiley, W., Zhang, M., and Sun, S.C., Negative regulation of JNK signaling pathway by the tumor suppressor CYLD, *J. Biol. Chem.,* 279, 55161, 2004.

10 Regulating the Master Regulator NF-κB: From Natural Strategies to Rationally Designed Superdrugs

Sebo Withoff, Vinay Tergaonkar, and Inder M. Verma

CONTENTS

10.1 INTRODUCTION

Transcription factors are often the final arbiters of growth, differentiation, and development in response to intrinsic or extrinsic signals. Additionally, they coordinate specialized functions carried out by specific cell types. The NF-κB family of transcription factors is pivotal for the modulation of the innate and adaptive immune responses (Chapters 6 and 7). Additionally, they control cell numbers by regulating the apoptotic machinery of the cell. More importantly, as discussed in Chapter 8, there is ample data available showing that NF-κB hyperactivation contributes to inflammation-associated cancer and many severe chronic (inflammatory) diseases, such as rheumatoid arthritis, SLE, multiple sclerosis, insulin resistance, and Alzheimer's disease [1–3]. This realization has led to an explosion of interest in NF-κB signaling from basic researchers and pharmaceutical companies, with a significant part of their concerted effort aimed at the development of drugs that specifically inhibit NF-κB signaling. Considering the rapidly expanding array of strategies used to inhibit NF-κB signaling, here we can only give a bird's eye view of what we consider significant, rather than an exhaustive list of all the proposed strategies of inhibiting NF-κB activity. For a more complete list, we will direct the reader to excellent reviews where appropriate.

Out of necessity, nature has provided several ways to attenuate NF-κB activity, since continuous NF-κB signaling can be severely detrimental, leading to chronic inflammatory diseases, the development of autoimmunity, and even hyperplasia. To prevent sustained NF-κB activation, the cell has several endogenous molecules available that will downregulate the NF-κB pathway. In addition, investigators have found that viruses and bacteria encode proteins that inhibit NF-κB signaling as a means of preventing an immune response that would otherwise eradicate the viral or bacterial threat.

Many drugs (natural and synthetic) in use for decades to inhibit inflammation or malignancy have been shown recently to modulate NF-κB activity. Based on these findings, some of these drugs are being modified on the molecular level to yield more potent inhibitors. We will give examples of drugs (or drug classes) that are most interesting from a clinical point of view.

Instead of direct intervention in the NF-κB pathway, researchers are trying to modulate other pathways that mediate the adverse effects of constitutive NF-κB activation. Additionally, the application of antibodies against NF-κB-activating cytokines and soluble receptors for NF-κB-activating cytokines are now being applied clinically to decrease the induction of NF-κB signaling.

The lessons learned from nature and from clinical experience with drugs that are known to modulate NF-κB activity have been very helpful in rationally designing synthetic compounds, peptides, DNA oligonucleotides, and RNA molecules that can interact with and downregulate the NF-κB pathway.

10.2   NATURE'S WAY OF INHIBITING NF-κB

NF-κB plays a pivotal role in eliciting a strong proinflammatory, proliferative, and antiapoptotic response. After execution of the essential task, this powerful response has to be downregulated by endogenous proteins in order to prevent chronic inflammation, autoimmunity, and hyperproliferation. The best known endogenous NF-κB inhibitors are members of the IκB-family (IκBs) that can lead to cytosolic sequestering of NF-κB, thereby attenuating NF-κB signaling [4]. Here we will discuss endogenous NF-κB inhibitors other than IκB family members, as these are discussed at length in Chapters 2 and 4.

In addition to endogenous factors that modify NF-κB, many viruses and bacteria express proteins — exogenous NF-κB inhibitors — that inhibit NF-κB signaling to thwart the host immune response. Although these proteins may not be ideal for clinical development, they have provided new insights into NF-κB regulation, and small molecule inhibitors of NF-κB signaling have already been derived based on sequences present in these viral and bacterial proteins.

10.2.1 ENDOGENOUS INHIBITION OF NF-κB SIGNALING

Several tumor suppressor genes have been reported to regulate NF-κB activity. The tumor suppressor gene adenosine diphosphate (ADP) ribosylation factor 1 (ARF) has been reported to induce the phosphorylation of thr505 in the transactivation domain (TAD) of p65, resulting in recruitment of histone deacetylase-1 (HDAC1) and inhibition of NF-κB activity (Chapter 5) [5,6]. The tumor suppressor gene ING4 was shown to physically interact with p65 in the nucleus, forming a transcriptional complex in which NF-κB-induced gene expression is downregulated, leading to inhibition of angiogenesis in a brain tumor model [7]. Finally, it was suggested that p53 and NF-κB control each other's ability to express their target genes — a regulatory "transcriptional crosstalk." This effect was dependent on the relative level of p53 and NF-κB, and it was proposed that competition for a limiting pool of CBP/p300 complexes underlies this regulatory balance [8,9].

NF-κB signaling has become a standard example of how ubiquitination and deubiquitination play a role in the activation and deactivation of signaling pathways. As discussed in Chapter 4, it is now apparent that protein-ubiquitination does not always lead to degradation of the ubiquitinated protein but that certain proteins depend on ubiquitination for localization or activity. Several proteins involved in NF-κB signaling (e.g., tumor necrosis factor-receptor 1 [TNF-RI], tumor necrosis factor receptor associated factor 2 [TRAF2], receptor interacting protein [RIP], nuclear factor kappa B essential modifier [NEMO], and inhibitor of kappa B kinase β [IKKβ]) are ubiquitinated, although the functional implications of this are largely unclear. Cylindromatosis (CYLD) and A20, as discussed in Chapter 4, act as endogenous negative regulators of NF-κB, signaling by modifying the ubiquitination of these pathway components.

Sumoylation can also affect NF-κB activity at different levels [10]. Sumoylation of a small fraction of NEMO has been shown to govern its nuclear localization and prevent it from shuttling to the cytoplasm, as discussed in Chapters 2 and 4. More recently, it has been reported that the sumoylation of CBP/p300 enables the binding of sirtuin (silent mating type information regulation 2 homolog) 1 (SIRT1) to p300, thereby repressing p300. SIRT1, a member of the nicotinamide adenosinedinucleotide (NAD)-dependent deacetylases, represses several transcription factors, including NF-κB. As CBP/p300 is a limiting cofactor for the transcriptional activity of NF-κB, it was suggested that desumoylation of CBP/p300 leads to a relieving of p300-repression by SIRT1, which in turn could lead to enhanced transcription mediated by NF-κB (Chapter 5) [11].

Mutations in the cold-induced autoinflammatory syndrome 1 (CIAS1) gene have been described to be involved in three different chronic autoinflammatory diseases [12]. CIAS proteins have been reported to inhibit nuclear translocation of p65 in monocytes, which implies that p65-inhibition may contribute to the induction of the chronic inflammatory phenotype; however, it has to be kept in mind that CIAS-mutations can also lead to increased processing of IL-1β-precursor into mature IL-1β, a major proinflammatory cytokine [13].

Recently, a new family of NF-κB inhibitory proteins has been identified. The first family member, Murr1, was identified due to its involvement in copper metabolism [14,15]. Nine other family members have since been described, all of which are characterized by a copper metabolism gene Murr1 domain (COMMD) that functions as a homotypic protein–protein interaction domain [16]. COMMD1 through -10 form multimeric complexes, and several of them interact with and inhibit NF-κB. Murr1/COMMD1, the prototype of the family, does not inhibit nuclear translocation of NF-κB, but rather inhibits NF-κB signaling at the nuclear level by negatively regulating the association of p65 with chromatin. Co-IP experiments showed that Murr1/COMMD1 interacts with p65, c-Rel, RelB, p100, and p105. COMMD1 binds to the first 180 amino acids of the p65 Rel homology domain (RHD) containing the DNA binding domain and thus may directly interfere with binding to κB sites.

15-Deoxy-Δ-^{12,14}-prostaglandin J$_2$ (15d-PGJ$_2$), the ligand of the peroxisome proliferator-activated receptor-γ (PPARγ) transcription factor, is a recently discovered prostaglandin that has been suggested to elicit an antiinflammatory response [17]. PPARγ is thought to be complexed in the cytosol to corepressor proteins but is probably released from corepression by binding of 15d-PGJ$_2$ or other PPARγ-agonists such as thiazolidinediones (glitazones). It has been suggested that 15d-PGJ$_2$ inhibits NF-κB indirectly via PPARγ by antagonizing the transcriptional activity of NF-κB [18]. The interference of PPARγ (and other nuclear antagonists of NF-κB such as p53, ARF, Twists, and N-CoR) with NF-κB could be mediated by competition for histone acetyltransferase (HATs) or recruitment of HDACs to NF-κB [19,20]. Additionally, 15d-PGJ$_2$ has been described to inhibit NF-κB activity directly by binding to IKKβ via its cyclopentenone ring, thereby inhibiting IKKβ activity [21]. The same ring has also been reported to prevent NF-κB from binding to DNA by modifying p65 and p50 [22]. Finally, it has been reported that 15d-PGJ$_2$ inhibits the recruitment of CBP/p300, an essential cofactor for NF-κB's transcriptional activity, to p65 [23]. At present, 15d-PGJ$_2$ and other PPARγ-agonists have been shown to inhibit inflammation in murine models of arthritis, ischemia-reperfusion injury, inflammatory bowel disease, Alzheimer's disease, and lupus nephritis [17], which are all diseases in which NF-κB is known to be chronically activated.

The nuclear protein RelA (p65)-associated inhibitor (RAI), isolated from human placenta, has been found to inhibit tumor necrosis factor (TNF)-induced NF-κB signaling by binding to p65 and preventing its DNA-binding. This 351 amino-acid, 40 kDa protein contains multiple ankyrin repeats, just as the IκB family members, but it does not contain a proline, glutamic acid, serine, and threonine (PEST)-region [24]. Recently, it was shown that RAI is part of a larger protein (824 amino acids, 98 kDa) called NF-κB interacting protein1 (Nkip1), which is predominantly expressed in endothelial cells of the skin, testis, heart, and stomach [25]. Nkip1 was found to be mutated in *waved 3* (*wa3*) mutant mice that develop a rapidly progressive cardiomyopathy. Although *wa3/wa3* mice survive, they remain small in size, and 90% die within 4 weeks of birth. Histology revealed hearts of abnormal shape and size, containing areas with myofiber degeneration, mineralization, and necrosis. At the same time, macrophage infiltration, fibrosis, and neovascularization could be observed. In *wa3/wa3* mice, ICAM-1 expression was significantly increased, but other NF-κB target genes examined were not affected, suggesting that Nkip1 regulates a subset of NF-κB target genes. The authors suggested that loss of NF-κB repression through Nkip1 mutation induces constitutive activation of endothelial cells, leading to an environment that promotes inflammation, thrombosis, and ultimately cardiomyocyte death.

10.2.2 VIRAL AND BACTERIAL STRATEGIES TO INHIBIT OR PREVENT NF-κB SIGNALING

Several viruses and bacteria encode proteins that inhibit NF-κB activation and activity, and it has been suggested that they do so to prevent a functional immune response (immune evasion). Almost every step in the NF-κB pathway is fair game for the inhibitory approaches taken by microorganisms.

One strategy to inhibit or prevent NF-κB signaling is to express mimics of receptors of the TNF-R family, which scavenge TNF family members before they can reach their intended target receptor. Pox viruses encode viral orthologs of various receptors of the TNF-R family. Cow pox and mouse pox viruses encode a viral CD30 (vCD30) protein [26,27]. The ligand of CD30 (CD153) is involved in downregulation of autoimmune responses and in negative selection of T cells, provides costimulation to B and T cells, and stimulates cell proliferation [28]. vCD30 interferes in these processes by competing with endogenous CD30 for CD153-binding, preventing the antiviral response. The recent finding that CD30 activates the alternative NF-κB signaling pathway [29] implies that vCD30 is an (indirect) inhibitor of noncanonical NF-κB activation. Other pox viruses encode soluble or membrane-bound IL-1R, IL-18R, IFN-R, or TNF-Rs that bind their respective ligands, thereby preventing NF-κB activation [30–33]. This microbiological "decoy-receptor" approach is presently being explored clinically. For example, recombinant TNF-R is already applied clinically and recombinant BAFF-R is in preclinical development.

Adenoviruses are of special interest because they are being developed as vectors for gene delivery. Unfortunately, adenoviral particles induce a potent immune response, which is surprising since a variety of adenoviral proteins downregulate the immune response. The adenovirus early transcription region 3 (E3) encodes for seven proteins, of which five inhibit immune responses, some by inhibiting NF-κB activity [34]. For instance, the adenoviral proteins E3-10.4K and E3-14.5K form a complex in the membrane that inhibits the assembly of the TNF-R1 signaling complex at the membrane, leading to downregulation of TNF-R1 on the cell surface [35].

Pox viruses also encode proteins that inhibit NF-κB signaling downstream of the receptors. Two ORFs of the Vaccinia pox virus, A46R and A52R, share sequence homology with the Toll/IL-1 receptor (TIR) domain that is characteristic of the

IL-1R/Toll-like receptor (TLR) superfamily. A46R and A52R inhibit IL-1-, IL-18-, and TLR-induced NF-κB and MAPK signaling (and therefore host defense) by binding and sequestering adaptor proteins (MyD88, MyD88 adaptor-like, TRIF, and Toll/IL-1 resistance domain containing protein inducing interferon beta (TRIF)-related adaptor) that are involved in signaling downstream of ligand-receptor interaction and upstream of the IKK complex (see Chapter 3) [36–38]. A peptide derived from A52R was recently shown to inhibit cytokine production in bacterially induced inflammation [39]. The authors suggested that this peptide could be applied in the treatment of chronic inflammation caused by viral or bacterial infection.

Another virulence factor of Vaccinia virus, N1L, inhibits NF-κB signaling directly by binding to the IKK complex [38]. Though N1L inhibits host defense against Vaccinia infection by ten-thousand-fold, it does not affect adaptive immunity [40]. DiPerna and colleagues describe that N1L inhibits receptor-, adaptor-, TRAF-, and IKK-α, and IKKβ-dependent NF-κB signaling. Coimmunoprecipitation showed that N1L associates with several components of the IKK complex, while TANK-binding kinase 1 (TBK1) appears to be its principal target, suggesting that this protein may preferentially target the interferon pathway. Also, human papillomavirus (HPV) E7 has been reported to bind to the IKK complex, leading to inhibition of induced activation of IKKα and IKKβ [41] and, most recently, the Hepatitis C Virus (HCV) core protein has been reported to bind to IKKβ and inhibit NF-κB activation [42].

Perhaps the most important immunological cell type that is involved in the inflammatory response is the macrophage (see Chapter 8). Two viruses that exhibit tropism toward these cells and are able to inhibit NF-κB signaling are the Pichinde virus, which causes Lassa fever in humans, and the African swine fever virus (ASFV). Virulent strains of the Pichinde virus, but not attenuated strains, were shown to repress NF-κB activation [43]. The ASFV encodes a protein, A238Lp, that resembles a porcine IκBα and can bind to p65, resulting in cytoplasmic sequestration. Interestingly, A238Lp needs posttranslational modification and NF-κB signaling to be able to bind to p65, as the endogenous IκBα needs to be degraded first [44].

Two viral proteins have been described to interact with βTrCP-proteins — the components of the SCF-type ubiquitin ligase complex that plays such a pivotal role in NF-κB activation (see Chapter 4). The HIV1-protein Vpu sequesters βTrCP1 to the cytoplasm [45], preventing it from functioning in the degradation of IκB and thereby effecting NF-κB activation. The EBV protein LMP1 has been reported to interact with and inhibit βTrCP2 [46]. As discussed in Chapter 4, bacteria of the *Yersinia* species (*Yersinia pestis* causes the bubonic plague) also interfere with the ubiquitination of NF-κB signaling proteins through the YopJ protein. Another bacterial protein, AvrA of *Salmonella typhimurium,* which is homologous to YopJ, has been suggested to prevent the ubiquitination of IκBα, thus regulating NF-κB activity downstream of the IKK-complex [47]. Interestingly, *Salmonella typhi* and *Salmonella paratyphi,* which do cause severe systemic disease, do not contain the AvrA locus, suggesting that control of NF-κB signaling is an important quality of virulent *Salmonella* sp. [48].

Finally, viral proteins can also target NF-κB transcriptional activity. The E6 oncoprotein of HPV downregulates NF-κB-mediated transcription by binding to and inhibiting CBP/p300 [49]. Thus, scientists have come to recognize that there are a variety of means by which pathogens subvert the function of NF-κB in the immune response. Although this is in principle not a positive finding, it offers us the opportunity to learn from pathogen strategies, and uncover critical steps in NF-κB signaling that can be blocked to inhibit NF-κB activity.

10.3 SYNTHETIC AND NATURAL COMPOUNDS INHIBITING NF-κB

In recent years, great effort has been expended to find compounds in large drug screens that inhibit NF-κB. Various natural compounds and synthetic drugs have been identified by this approach, and upon their identification, some of these have been modified to enhance their specificity for NF-κB. Multiple antiinflammatory and anticancer agents have demonstrated NF-κB-inhibitory activity by targeting various steps in the NF-κB pathway, including IKK complex activity, IκB degradation, NF-κB nuclear localization, and NF-κB DNA binding capacity. In the following sections we describe the most promising, clinically relevant drug classes. The disadvantage of most of these drugs is that they are not very specific for NF-κB signaling and therefore potentially could display significant unwanted side effects when administered to patients.

10.3.1 PROTEASOME INHIBITORS

The ubiquitin-proteasome pathway regulates the turnover of proteins that are of major importance for the cell, such as cyclins (which have a role in cell cycle control), MDM2 and p53 (which have a role in cell cycle arrest, DNA repair, differentiation, senescence, and apoptosis), and NF-κB [50–52]. As discussed in Chapter 4, activation of the classical-, alternative-, and DNA-damage-induced NF-κB pathways depends on the action of the proteasome. Thus, it appears safe to argue that the application of proteasome inhibitors will result in NF-κB inhibition; however, it is also clear that other pathways will be affected as well.

Proteasome inhibitors can be divided into two subclasses: (1) inhibitors of the 20S proteasome core, such as peptide aldehydes (e.g., MG-132), PSI, lactacystins (e.g., PS-519), epoxomicin, and peptide boronic acids (e.g., bortezomib) and (2) inhibitors of substrate specific E3 ubiquitin ligases (the E3 ubiquitin ligase MDM2 can be inhibited by nutlins) [53]. The best-studied proteasome inhibitor at present is the small molecule inhibitor bortezomib (also known as Velcade or PS-341), which is used mostly in combination with other chemotherapeutic agents. Bortezomib inhibits the proteasome by binding to the chymotrypsin-like site of the 20S core of the proteasome [50–52]. It has been approved by the U.S. Food and Drug Administration (FDA) for the treatment of patients with advanced multiple myeloma [54] and has been described to display clinical activity in relapsed and refractory multiple myeloma, prostate cancer, non-Hodgkin's lymphoma, T cell leukemia, nonsmall-cell lung carcinoma, renal cell cancer, Lewis lung carcinoma, and pancreatic cancer [50–52,55,56].

As remarked previously, the efficacy of bortezomib is probably dependent on other mechanisms in addition to NF-κB inhibition. By directly comparing the efficacy of bortezomib and the IκB-kinase specific inhibitor PS-1145 in multiple myeloma cells, it was found that bortezomib inhibited cell proliferation completely in contrast to PS-1145, which inhibited only 20–50%, although both compounds inhibited NF-κB activity completely [57]. Kenneth C. Anderson and colleagues showed subsequently that in human multiple myeloma cells, bortezomib downregulates growth and survival pathways (e.g., IGF-1 receptor pathway; IGF-1 stimulates Akt-signaling) and antiapoptotic genes, and upregulates proapoptotic genes, resulting in apoptosis [55,58,59]. The authors suggest that bortezomib-induced NF-κB inhibition causes a decrease in the antiapoptotic protein BcL-2 and the BcL-2 family member A1/Bfl-1, leading to cytochrome C release and caspase-9 activation, in other words, activation of the intrinsic apoptotic pathway. Furthermore, they suggest that bortezomib causes p53-phosphorylation and subsequent jun N-terminal kinase (JNK)-activation, leading to Fas upregulation. Additionally, an increase in c-myc levels was observed, which correlated with an increase in FasL. The higher expression of both Fas and its ligand FasL would explain the observed activation of caspase-8, which is the initiator caspase of the extrinsic apoptotic pathway. The killing signal delivered by caspase-8 activation will be enhanced by the downregulation of NF-κB, as NF-κB inhibition will lead to a decrease in FLICE inhibitory protein (FLIP) levels, which is a physiological inhibitor of caspase-8 activation [60]. These observations show that inhibition of the proteasome in multiple myeloma cells leads to the concurrent modulation of multiple transcription factors that, together, tip the balance between life and death to the advantage of the latter. Presently, other proteasome inhibitors (with distinct structural features) are under development. For example, the lactacystin PS-519 has already entered clinical trials for treatment of patients with poststroke and myocardial infarction reperfusion injuries [51,52].

10.3.2 Nonsteroidal Antiinflammatory Drugs

At present, nonsteroidal antiinflammatory drugs (NSAIDs) are the most widely prescribed antiinflammatory drugs. It was estimated that in the United States, five to ten billion dollars are spent on NSAIDs, which are used in cardiovascular disease, for relief of pain symptoms in minor injuries and headaches, and to alleviate discomfort in serious inflammatory and joint diseases [61]. NSAIDs are known to inhibit the activity of cyclooxygenases (COXs), which are involved in prostaglandin (PG) synthesis [62,63]. As PGs play an important role in many processes other than inflammation, such as blood clotting, ovulation, bone metabolism, nerve function, wound healing, kidney function, and blood vessel tone, there is the accompanying risk of cytotoxic side effects associated with the application of COX-inhibitors [61]. Aspirin inhibits the activity of the constitutively expressed COX-1 by acetylating it; however, it has been shown that this is probably not the only mode of action of NSAIDs [64]. A second mode of action of aspirin is that it prevents the phosphorylation of IKKβ (and not IKKα) by binding to its adenosine triphosphate

(ATP)-binding site [65,174], thereby inhibiting NF-κB activation. The NSAID sulindac has also been shown to inhibit IKKβ activity [66]. As the expression of the inducible COX-2 enzyme is regulated by NF-κB, this implies that aspirin may inhibit PG synthesis by inactivating COX-1 and by downregulating COX-2 expression. Other studies have confirmed that NSAIDs inhibit NF-κB activity [67,68]. Takada and colleagues compared different NSAIDs, and the authors found that the tested compounds inhibited IKK activity and suppressed IκBα degradation and NF-κB reporter gene activation, although with different efficacy. Aspirin and ibuprofen were shown to be least potent in inhibiting NF-κB activity; naproxen, indomethacin, and diclofenac were inhibited intermediately; and resveratrol, curcumin, celecoxib, and tamoxifen were most potent [68]. Very recently, it was suggested that celecoxib suppresses NF-κB activation through inhibition of IKK and Akt [69].

There has been a considerable amount of excitement in the oncology field since reports found that long-term use of NSAIDs, in this case aspirin, is associated with the regression of intestinal polyps and a reduction in the incidence of colorectal cancer in arthritis and colorectal cancer patients [70,71]. Additionally, it was reported that women who had been taking aspirin every day for at least ten years had a 28% lower risk of developing breast cancer [72]. An unwanted side-effect of nonspecific COX-inhibitors (inhibiting both COX-1 and COX-2) is that these drugs can cause gastrointestinal irritation and ulceration, which was shown to be due to inhibition of COX-1. To avoid this, selective COX-2 inhibitors (celecoxib [Celebrex®] and rofecoxib [Vioxx®]) were developed. Initially, both these drugs were approved for the treatment of rheumatoid arthritis and osteoarthritis. Celecoxib was also shown to significantly reduce the incidence and multiplicity of colon tumors and the number of polyps in patients with familial adenomatous polyposis (FAP) [73]. Celecoxib has also been found to be effective against mammary, skin, and bladder cancer [74] and was tested for cancer prevention or treatment in nearly 50 clinical trials. However, in September 2004, rofecoxib and, shortly after that, celecoxib were withdrawn from the market because these drugs were found to increase the risk of heart attack and stroke. If these serious side effects can be prevented or overcome, NSAIDs still could be of major importance for treatment of various inflammation-associated cancers and chronic inflammatory diseases. It is interesting to note in this respect that the R-stereoisomer of the NSAID flurbiprofen (the latter being the S-isomer) has been shown to inhibit NF-κB (and also AP-1), while it does not inhibit COX activity [75]. Moreover, it does not cause gastrointestinal problems [76] or other cytotoxic effects, which have been associated with COX-1-inhibiting NSAIDs [61].

10.3.3 Glucocorticoids

The most widely prescribed antiinflammatory and immunosuppressive drugs are glucocorticoids (GCs), of which dexamethasone and hydrocortisone are the best known. The interaction of GCs with cytosolic GC-receptors (GCRs) induces the nuclear translocation of the GC-GCR-complex, in which the GRs interact with GC responsive elements (GREs), which are present, for instance, in promoters of cytokines. Initially, it was thought that the transcriptional effects of GCs were solely mediated by these GREs. However, later it was shown that GCs also inhibit the action of several transcription factors that are essential for immunity, such as AP-1, NF-AT, and NF-κB [77,78]. NF-κB inhibition was initially thought to be brought about by GC-induced upregulation of IκB [79–81], which could be caused by binding of GR to a GRE in the IκB promoter [82]. IκB upregulation is not a universal effect [83,84] and, although it may have a cell-type role with or without a stimulus specific role, several other mechanisms of GC/GR-induced NF-κB inhibition have been proposed [77,78]: steric hindrance of NF-κB binding to κB sites by GR bound to GREs; GR binding and sequestering NF-κB in the cytosol [83–85]; GR binding the TAD of DNA-bound p65 interfering with coactivator binding [86,87]; GR competing with NF-κB for limited nuclear coactivators (e.g., CBP/p300) [88,89]; or GC-induced deacetylation of histones [90,91]. Finally, it has been reported that GCs induce the GC-induced leucine zipper (GILZ) protein that inhibits NF-κB activation [92].

10.3.4 THALIDOMIDE

The glutamic acid derivative thalidomide was brought on the market in the late 1950s as a sedative and a treatment for morning sickness in pregnant women, although it was not approved by the FDA in the United States. In the early 1960s, it was realized that the intake of even one dose of thalidomide during gestation could lead to limb abnormalities and other congenital defects. Shortly thereafter, thalidomide was pulled from the market. The drug reemerged a few years later as a treatment modality for leprosy and the FDA approved its use in this disease in the late 1990s [93]. The devastating effects of the drug during pregnancy were thought to result from the drugs ability to inhibit vasculogenesis. It is now clear, however, that the drug also has antiinflammatory and immunomodulatory effects. Presently, studies are ongoing to investigate thalidomide's potential in treatment regimens for inflammatory and infectious diseases (dermatological, rheumatological, and gastrointestinal diseases, HIV-1, and congestive heart failure) and for treatment of a variety of malignancies (e.g., multiple myeloma and other hematological disorders, prostate cancer, renal-cell carcinoma, glioma, colorectal cancer, and more). In multiple myeloma patients, thalidomide induces a variety of effects, which include reduction in cell adhesion, reduction in drug resistance, induction of apoptosis, angiogenesis inhibition, and modulation of immunity. It is also known that thalidomide reduces the expression of TNFα, IL-6, IL-12, COX-2, and BcL-2 (inhibitor of the intrinsic apoptotic pathway) inhibitors of the extrinsic apoptotic pathway and NF-κB activity in general [93,94]. These clinical and biochemical features strongly suggest that NF-κB inhibition is central to the action of thalidomide, as NF-κB regulates the expression of proangiogenic factors (e.g., vascular endothelial growth factor [VEGF], $\alpha_v\beta_3$-integrins, IL-8, and PGs) [95,96], COX-2, antiapoptotic genes, and multiple proinflammatory and immunomodulatory cytokines (such as TNFα, IL-6, IL-12, and many more). Although there are conflicting reports about the effects of thalidomide on TNFα production in different cell types [97,98], there is evidence that thalidomide inhibits classical IKK activity, resulting in reduced levels of NF-κB-regulated genes downstream of cytokine receptor activation [99].

10.3.5 Antioxidants as NF-κB Inhibitors

The level of cellular free radicals (reactive oxygen species [ROS] and reactive nitrogen oxide species [RNOS]) can increase upon exposure to certain environmental stimuli (e.g., ozone and cigarette smoke) and upon NF-κB stimuli (e.g., phorbol myristate acetate [PMA], CD3-engagement, lipopolysaccharide [LPS], H_2O_2, TNF, IL-1, UV-light, and ionizing radiation) [100–102]. The production of free radicals seems to be cell-type specific [101]. In inflammation, ROS produced by inflammatory cells play an important role as a means to kill infected cells by oxidizing molecules essential for pathogen survival [103]. Free radicals such as NOS ($O_2^{\cdot-}$, $\cdot OH$) and RNOS ($NO^{\cdot}$, $ONOO^-$, N_2O_3) cause oxidation of proteins, DNA, and lipids that can directly or indirectly induce tissue damage with or without stress responses [104]. Products of lipid peroxidation have been shown to activate macrophages, to modulate the levels of prostaglandins and their receptors, and to induce multiple proinflammatory genes such as IL-8, COX-2, and monocyte chemoattractant protein-1 (MCP-1) [102,104]. Multiple studies have reported a direct link between increased ROS levels and the induction of NF-κB activity. Oxidative stress increases IκBα ubiquitination and degradation, and antioxidants have been reported to inhibit IκBα phosphorylation [105–107]. Moreover, it has been described that H_2O_2 can induce NF-κB through tyrosine-phosphorylation of IκBα and through activation of the IKK complex by inducing serine phosphorylation within the activation loops of IKKα and IKKβ [108,109]. Indirectly, ROS can also enhance NF-κB-mediated transcription by inhibiting HDAC activity [102]. The exact mechanism of ROS-induced NF-κB activation is largely unknown to date, and it is thought that various antioxidants inhibit at different steps in this pathway. However, it is clear that antioxidant application will inhibit NF-κB activity *in vitro* and *in vivo* [67,100], although other pathways will be affected as well. A number of antioxidants have been used to inhibit the NF-κB pathway (see Epinat and Gilmore [100] for an elaborate listing of antioxidants). An antioxidant [110] derived from the subtropical ginger *Languas galangas*, 1′-acetoxychavicol acetate (ACA), has recently been shown to inhibit the NF-κB pathway at the level of IKK activation, resulting ultimately in decreased expression of NF-κB-regulated proliferative genes, antiapoptotic genes, and genes involved in metastasis and inflammation [111]. Polyphenols curcumin, which can be isolated from the spice turmeric, and resveratrol, which is found in grapes and red wine [112], have been found to downregulate NIK and IKKα/β, resulting in inhibition of IKK activation, decreased IκBα-degradation, decreased p65 translocation, and a decrease in NF-κB activity [113–117]. Ascorbic acid (vitamin C) has been shown to display dual activity; first, it is known to quench ROS, and second, it has been described that, in the presence of ROS, ascorbic acid will be metabolized into dehydroascorbic acid, which has been shown to inhibit IKKα and IKKβ activity [118]. Other well-known antioxidants are thiols (e.g., glutathione, N-acetyl-L-cysteine [NAC], and N-Acystelyn [NAL]), vitamin E, NADPH, glutathione peroxidase, and MnSOD [102], all of which inhibit NF-κB activity to varying degrees.

10.3.6 POTENTIAL RISKS OF USING NF-κB INHIBITORS

In cancers with constitutive NF-κB activation and in chronic inflammation-associated cancers, NF-κB inhibition may prove to be a beneficial outcome [3]. However, it is important to point out that, in many cancers, unrestricted tumor outgrowth is facilitated by the fact that the cancer cells have found ways to evade apoptosis and immune surveillance. For this reason, immunotherapeutical anticancer therapies have been developed aiming at restoring and stimulating the proinflammatory functions of the patient's immune system as a palliative treatment option other than surgery, radiation, or chemotherapy [119]. The role of immune surveillance must be kept in mind when NF-κB inhibitors are applied as anticancer therapy, because such a treatment could inhibit whatever remains of the immunological antitumor response.

Another fact complicating the potential usefulness of NF-κB inhibitors in anticancer therapy is that NF-κB functions in a tissue-dependent manner. Remarkably, in keratinocytes, NF-κB inhibition can promote squamous cell carcinoma [120]. This implies that NF-κB-targeting anticancer therapy has to be tailored to specific tissues.

A concern for the use of NF-κB inhibitors to treat chronic inflammation is that NF-κB has been shown to be involved in inflammation resolution as well (Chapter 8). As expected, administration of broad-spectrum NF-κB inhibitors at the onset of inflammation leads to diminished inflammation; however, when these inhibitors are delivered later, they interfere with the ability to reduce inflammation [121]. It was hypothesized that during onset of inflammation, NF-κB stimulation leads to p65/p50 induced proinflammatory gene expression, while during resolution, translocation of repressive p50/p50 homodimers is induced. Alternatively, it is possible that p50/p50 homodimers, in collaboration with other cofactors, activate proapoptotic genes (e.g., BcL-3 could collaborate with p50-homodimers to activate transcription).

Another cause of concern is the occurrence of immunosuppression or other sequelae associated with a dysregulated immune system. For instance, several severe side effects have been related to anti-TNFα therapy, including infections, lymphoma development, congestive heart failure, demyelinating disease, lupus-like syndrome, and induction of autoantibodies [122]. These findings underscore the notion that modulators of the immune system have to be applied and dosed carefully.

Finally, it has to be realized that many of the drugs described in this section inhibit other pathways as well. This means that application of these drugs may induce unwanted side effects that have to be monitored carefully.

10.4 INDIRECT AND ALTERNATIVE STRATEGIES TO INHIBIT NF-κB

It is known that changes in other pathways can lead to the constitutive activation of NF-κB. Examples of such events are overexpression of HER2/neu, IGF-1R overexpression, BCR-ABL fusion, and activation of Ras/mitogen associated protein kinase (MAPK) or PI3K/Akt pathways [123]. If such changes are identified in diseases in which NF-κB activation plays an important role, then the proteins involved in these pathways represent an array of possible targets for treatment. As an alternative to inhibiting NF-κB signaling itself, a lot of attention is being focused on developing molecules that inhibit the triggering of NF-κB signaling (e.g., anticytokine antibodies and decoy receptors) and on molecules that stimulate pathways that counteract NF-κB signaling (e.g., antiinflammatory cytokines).

10.4.1 CYTOKINE AND ANTICYTOKINE THERAPY

A number of studies have associated cytokine expression with inflammation and carcinogenesis. For example, colony-stimulating factor-1 (CSF-1) is involved in breast cancer development [124], TNFα plays a role in skin tumorigenesis [125] and promotes metastasis of xenografted human gastric cancer [126], IL-1α is associated with an increased risk of *H. pylori*-induced gastric cancer [127], and IL-8 neutralization inhibits tumor-associated inflammation and Ras-induced tumor growth [128]. Most cytokines act through NF-κB signaling, which in turn induces the expression of a wide spectrum of cytokines. The result is a complex, intertwined network that makes it difficult to predict the clinical efficacy of cytokine inhibition. Nevertheless, the clinical successes obtained to date with anticytokine therapies shows much promise.

As described previously, the intervention with cytokines before they reach their target receptor is a strategy that has long been applied by microorganisms. Presently, TNFα-blocking agents such as infliximab (Remicade®, a chimeric monoclonal antibody specific for TNFα) and etanercept (Enbrel®, a soluble dimeric TNFα-receptor) have been applied successfully to treat autoimmune-inflammatory diseases such as rheumatoid arthritis, juvenile arthritis, psoriasis, psoriatic arthritis, and Crohn's disease [129]. The IL-1 receptor antagonist (IL-1Ra) anakinra (Kineret®) is based on a physiologically expressed membrane molecule that competes with the IL-R for IL-binding. Anakinra has recently been approved for treatment of rheumatoid arthritis [130–132]. Other cytokines and chemokines that are being targeted for anticytokine therapy include IFN-α, IFN-γ, IL-6, IL-8, IL-12, IL-15, IL-17, and IL-23 [131].

A relatively new player in the cytokine field is the TNF-family member B cell activating factor (BAFF). BAFF activates the alternative NF-κB pathway and is involved in the formation of secondary lymphoid organs and in lymphocyte development, thereby providing a link between the innate and the adaptive immune response (Chapter 7). BAFF is essential for B cell development (and also for T cell activation and development) and has been shown to be overexpressed in patients with autoimmune disorders (rheumatoid arthritis, SLE, and Sjögren's syndrome) and B cell lymphomas [133,134]. Patients with these diseases may benefit from treatment with neutralizing anti-BAFF antibodies or recombinant BAFF-R. The alternative pathway is triggered by a variety of TNF-R-family members [29], some of which are T cell costimulating molecules (HVEM, CD30, CD27, 4-1BB, GITR, OX40) [135]. If overexpression of the ligands for these receptors contributes to B or T cell malignancies or autoimmune disease, this would mean that a whole array of new antibody or decoy receptor targets have become accessible. The advantage of treatments that target the alternative NF-κB pathway instead of the classical NF-κB pathway is that they will leave the innate immune system intact. Finally, as an alternative to using inhibitors of proinflammatory cytokines, one could consider the use of antiinflammatory mediators such as IL-4, IL-10, IL-11, IFN-β, or TGFβ [136].

10.5 TAILOR-MADE NF-κB INHIBITION IN THE POSTGENOMIC ERA

The biotechnological revolution of the last decade has provided us with new types of drugs that now start to show their potential in the clinic, and some of them are now hailed as being the new "superdrugs." The proteasome inhibitors are good examples, which may have potential in the treatment of chronic inflammatory diseases, autoimmune diseases, and cancer. A disadvantage of proteasome inhibitors is that inhibiting the proteasome does not specifically inhibit NF-κB, and therefore these drugs will modulate the cell's physiology extensively. The lessons learned from nature's strategies to inhibit NF-κB and the results of clinical studies with drugs that have now been shown to inhibit NF-κB have provided us with new insights in the NF-κB pathway and have given us the opportunity to rationally design modulators of NF-κB signaling that do not interfere with other pathways. Additionally, the application of new molecular-biological techniques has provided us with a surge of information about how signals are transduced within the NF-κB pathway and how NF-κB proteins interact with each other and with other proteins. These new insights have led to the development of molecule-specific inhibitors of the NF-κB pathway, some of which are already in preclinical development. Improvements in other fields will also contribute to the development of more efficient drugs, such as improved delivery and targeting strategies, activation of the drugs specifically at the location where it is needed, and the coupling of the drugs to small peptide sequences that will facilitate the transport of the drug over the cell membrane [137,138].

One of the most potent inhibitors of the classical NF-κB pathway are dominant negative versions of IκB. One version of dominant negative IκBα (dn-IκBα) is a protein in which ser32 and ser36 phosphoacceptor sites have been replaced with alanines and multiple amino acids in the C-terminal domain have been replaced to confer stability to the mutant protein. The dominant negative IκBα variant (called IκBαM) described above cannot be phosphorylated and will therefore not be ubiquitinated and degraded, thus sequestering NF-κB to the cytosol [139]. Another nondegradable IκBα variant (cannot be phosphorylated because the first 36 amino acids have been deleted) has been coupled to a membrane translocating sequence (MTS) derived from the signal peptide of Kaposi Fibroblast Growth Factor. This protein was able to inhibit NF-κB activation *in vitro* and inhibited NF-κB activation *in vivo* in response to skin wounding in mice and in lungs after endotoxin treatment in sheep [140]. It is quite obvious that dominant negative variants of other NF-κB proteins (e.g., kinase-dead IKKs) will also inhibit NF-κB signaling. Moreover, using dominant negative p100 or IKKα, it may be possible to selectively downregulate the alternative NF-κB pathway without affecting the classical NF-κB pathway. In diseases in which the alternative pathway is dysregulated and needs to be inhibited, this approach will prevent adverse side effects associated with inhibition of the classical pathway. At present, the clinical value of such molecules does not seem to be very high; however, small peptide variants of such molecules, pathway specific peptides, or kinase-specific inhibitors may be more interesting in that respect.

IFN-β has been presented as one of the cytokines that can be used as an alternative to NF-κB inhibitors because of its antiinflammatory capacity. A very elegant approach has been to change this cytokine (or any other cytokine) into a latent form that will be activated at the site of inflammation. To achieve this, the latency associated protein (LAP) domain of TGF-β1 was fused to the IFN-β protein to prevent IFN-β from binding to its receptor. Between the LAP-domain and the IFN-β domain, a matrix-metalloprotease (MMP) cleavage site was introduced. Because MMPs are locally produced at sites where tissue damage or remodeling is taking place — for instance, at the site of inflammation or tumor expansion — IFN-β will be cleaved from the LAP domain and become active locally. LAP-MMP-IFN-β fusion-protein was shown to be cleaved by cerebrospinal or synovial fluid of patients with inflammatory diseases. Additionally, the latent protein was shown to be more efficient in a mouse model of arthritis than unmodified IFN-β [141].

As a converging point for multiple proinflammatory NF-κB stimuli [142], IKKβ is an ideal target for small-molecule kinase-inhibitors. Examples of IKKβ inhibitors under study are SPC-839, SC-514, PS-1145, and BMS-345541 (reviewed extensively in [143]). SPC839 (a quinazoline analogue) and SC-514 (an aminothiophenecarboxamide derivative) are reversible ATP-binding site-targeting ATP-competitors, with specificity for the ATP-binding site of IKKβ over that of IKKα and other kinases. To date, most IKK-inhibitors display a higher affinity for IKKβ than for IKKα. The realization that the alternative pathway may be hyperactivated in diseases in which lymphocytes go awry and the fact that this pathway depends on IKKα and not on IKKβ or NEMO, will undoubtedly spur the search for more specific IKKα inhibitors. Inhibitors of kinases upstream of the IKK complex (kinases that activate IKKα or IKKβ) could also be applied; however, great care will have to be taken regarding the side effects of these compounds when these kinases are not specific for the IKKs/NF-κB pathway. Finally, it has to be remarked that a third NF-κB pathway has been identified recently that is activated upon induction of DNA damage and is IKK-independent (Chapter 4) [144]. A variety of kinases have been found to play a role in this pathway [145–148]. Each of these kinases may represent a new target for small molecule inhibitors specifically inhibiting one of the three NF-κB pathways.

Another recent advance is the design of small peptides that interfere with protein–protein interactions that are important for NF-κB signaling. A short hexapeptide sequence (Leu-Asp-Trp-Ser-Trp-Leu) analogous to the NEMO-binding domain (NBD) of IKKα and IKKβ disrupts the association of NEMO with both IKK-α and IKK-β and inhibits cytokine induced NF-κB activation and inflammatory responses [149,150]. The peptide has already proven its potential by sensitizing TRAIL resistant cell lines to TRAIL, and by inhibiting mononuclear cell invasion and normalizing p65 expression in spinal cords of mice with experimental allergic encephalomyelitis (EAE, a model for multiple sclerosis) [151,152]. Interestingly, similar short NBD-peptides have been altered by adding moieties that facilitate delivery into the cell, such as HIV-TAT sequences, the homeodomain of antennapedia protein (a *Drosophila* transcription factor) and a cationic peptide transduction domain (PTD). A TAT-NBD fusion peptide has been shown to inhibit LPS-induced NF-κB activation in polymorphonuclear neutrophils (PMNs) [153] and to block osteoclastogenesis, to inhibit bone erosion, and to ameliorate inflammation in the joints of mice with arthritis [154,175]. Application of NBD fused to the antennapedia homeodomain improves lung edema and lung volume and reduces inflammation in a an acute airway distress syndrome model in piglets [155]. Finally, *in situ* delivery of PTD-5-NBD was shown to improve pancreatic islet function and viability prior to transplantation [156].

Other promising peptides that could be applied clinically are peptides blocking interactions at other levels in the pathway. For example, a peptide corresponding to amino acids 138–151 of mouse TIR domain-containing adaptor protein (TIRAP) has been coupled with the antennapedia protein to facilitate uptake. This TIRAP-fusion peptide blocked TIRAP's association with TLR4, resulting in inhibition of LPS-induced NF-κB activation in macrophages [157]. An alternative strategy to prevent IKK-activation has been described that is based on prevention of NEMO-oligomerization by delivery of peptides encompassing the minimal oligomerization domain of NEMO itself. These peptides were shown to inhibit NF-κB activation in pre-B lymphocytes [158]. Another elegant way of inhibiting NF-κB is by application of SN50, which is a fusion between a cell membrane-permeable motif (the hydrophobic domain of the signaling peptide) and a peptide corresponding to the nuclear localization sequence (NLS) of p50. Application of this peptide in T cells not only inhibits the nuclear translocation of NF-κB but also the translocation of AP-1, NFAT, and STAT1. The mechanism behind this inhibition is that SN50 interacts with and binds to an NLS-receptor complex present in the cytoplasm of these cells [159,160]. Takade and colleagues described the coupling of the antennapedia membrane-transport domain to a peptide encompassing a phosphorylation site of p65 whose phosphorylation is essential for NF-κB activity. This peptide was able to inhibit NF-κB signaling triggered by a range of stimuli [161].

The progress in DNA/RNA technology may provide new approaches to inhibit NF-κB activity, e.g., by application of antisense oligodeoxynucleotides (ODNs), ribozymes, decoy ODNs, and of course small interfering RNAs (siRNAs) [162,163]. Decoy ODNs are small DNA sequences carrying the consensus binding sequence for specific transcription factors, in this case the κB site to which NF-κB binds. In a mouse model for asthma (ova-induced airway allergy), κB-decoy oligos strongly inhibited NF-κB activity, lung inflammation, airway hyperresponsiveness, and production of mucus [164]. With the rapid advance in the small RNA field [163], its potential to modify NF-κB signaling is also beginning to be explored. Administration of siRNA directed against p65 (siP65) was shown to enhance sensitivity to chemotherapeutic agents, such as irinotecan, and modified NF-κB-mediated inflammation [165,166]. Various siRNA's against NF-κB mRNAs have already shown their inhibitory capacity *in vitro* [167,168]. Clearly, this will be an important therapeutic modality, but it will require safe and efficacious delivery methods.

10.6 OUTLOOK

At the dawn of twenty-first century, the advances in molecular biology and biochemistry have given investigators powerful new tools to study the molecular underpinnings of signaling pathways. The functions of IKKs (and several other kinases), so central in NF-κB signaling, can now be studied in whole organism models like *Drosophila* [169–171] and zebrafish [172,173], which will facilitate the application

TABLE 10.1
An Overview of the NF-κB Inhibitors That Are Discussed in Chapter 10 (See Text for Details and Abbreviations Where Appropriate)

		Endogenous inhibitors	Bacterial & viral inhibitors	Natural & synthetic inhibitors	Recombinant & small molecule inhibitors
TNF-family (ligands)			Pox-viral decoy receptors	Thalidomide	Anti-TNF (Remicade®), soluble TNF-R (Enbrel®), anti-BAFF, soluble BAFF-R
Binding to TNF-R-family (receptors)			Adenoviral E3-10.4K / E3-14.5K-complex		IL-1Ra (Kineret®)
Recruitment of adapter molecules & kinases		CYLD, A20	VV A46R & A52R	Antioxidants	A52R-peptide, Kinase inhibitors, TIRAP-peptide
Activation of IKK-complex		CYLD, SUMO, 15d-PGJ₂	VV N1L, HPV E7, HCV core protein, Yersinia YopJ	NSAIDs, Thalidomide, Antioxidants	IKKα- or β-inhibitors, NBD-peptides, NEMO oligomerisation peptides
Phosphorylation of IκB-family-members		SUMO	ASFV A238Lp, HIV1 Vpu, EBV LMP1, Salmonella AvrA	Proteasome inhibitors, Antioxidants	Dn-IκBα
NF-κB translocation		CIAS		Glucocorticoids	SN-50
NF-κB-DNA interaction & transcription		Arf1, P53, ING4, RAI / Nkip1, Murr1 / COMMD1, PPARγ, 15d-PGJ₂, Twist, N-CoR, SIRT1	HPV E6	Glucocorticoids, Antioxidants	P50-phosphorylation domain peptide, decoy-ODNs

of large-scale genetic screens to modulate NF-κB signaling. On the other hand, high-throughput, cell-based assays have been developed to screen large compound libraries, and the knowledge obtained has been applied to further develop lead compounds into drugs for various diseases. One area in which progress is expected in the next few years is the solving of the 3-dimensional structure of IKK complex. Nearly all the external signals leading to NF-κB stimulation are coordinated by the proteins present in this complex. Clearly, the kinases present and the associated proteins will make excellent drug targets. In the NF-κB field, the challenge will be to design drugs that specifically inhibit NF-κB signaling and, more specifically, each of the three NF-κB pathways (classical, alternative, and DNA-damage induced). At present, an impressive number of clinical trials are ongoing in which current drugs that inhibit NF-κB and new, rationally designed NF-κB inhibitors are being evaluated. Hopefully, this will lead to the identification of the new superdrug that will inhibit NF-κB without inducing severe side effects.

ACKNOWLEDGMENTS

Sebo Withoff is supported by National Institutes of Health (NIH) grants to IMV and by a fellowship of the Catharina Foundation. Vinay Tergaonkar is supported by a career development fellowship from the Leukemia and Lymphoma Society. Inder M. Verma is an American Cancer Society Professor of Molecular Biology supported in part by grants from NIH, the Larry L. Hillblom Foundation, Inc., the Lebensfeld Foundation, the Wayne and Gladys Valley Foundation, the H.N. and Frances C. Berger Foundation, Merck Research Laboratories, and the March of Dimes.

REFERENCES

[1] Shoelson, S.E., Lee, J., and Yuan, M., Inflammation and the IKKβ/IκB/NF-kappaB axis in obesity- and diet-induced insulin resistance, *Int. J. Obes. Relat. Metab. Disord.,* 27 Suppl 3, S49, 2003.
[2] Mattson, M.P. and Camandola, S., NF-kappaB in neuronal plasticity and neurode-generative disorders, *J. Clin. Invest.,* 107, 247, 2001.
[3] Li, Q., Withoff, S., and Verma, I.M., Inflammation-associated cancer: NF-kappaB is the lynchpin, *Trends Immunol.,* 26, 318, 2005.
[4] Prigent, M., Barlat, I., Langen, H. et al., IκBα and IκBα/NF-kappaB complexes are retained in the cytoplasm through interaction with a novel partner, RasGAP SH3-binding protein 2, *J. Biol. Chem.,* 275, 36441, 2000.
[5] Rocha, S., Campbell, K.J., and Perkins, N.D., p53- and MDM2-independent repression of NF-kappaB transactivation by the ARF tumor suppressor, *Mol. Cell,* 12, 15, 2003.
[6] Rocha, S., Garrett, M.D., Campbell, K.J. et al., Regulation of NF-kappaB and p53 through activation of ATR and Chk1 by the ARF tumour suppressor, *Embo J.,* 24, 1157, 2005.
[7] Garkavtsev, I., Kozin, S.V., Chernova, O. et al., The candidate tumour suppressor protein ING4 regulates brain tumour growth and angiogenesis, *Nature,* 428, 328, 2004.
[8] Webster, G.A. and Perkins, N.D., Transcriptional cross talk between NF-kappaB and p53, *Mol. Cell. Biol.,* 19, 3485, 1999.
[9] Tergaonkar, V., Pando, M., Vafa, O. et al., p53 stabilization is decreased upon NF-kappaB activation: A role for NF-kappaB in acquisition of resistance to chemotherapy, *Cancer Cell,* 1, 493, 2002.
[10] Gill, G., SUMO and ubiquitin in the nucleus: Different functions, similar mechanisms? *Genes Dev.,* 18, 2046, 2004.
[11] Bouras, T., Fu, M., Sauve, A.A. et al., SIRT1 deacetylation and repression of p300 involves lysine residues 1020/1024 within the cell cycle regulatory domain 1, *J. Biol. Chem.,* 280, 10264, 2005.
[12] O'Connor, W., Jr., Harton, J.A., Zhu, X. et al., Cutting edge: CIAS1/cryopy-rin/PYPAF1/NALP3/CATERPILLER 1.1 is an inducible inflammatory mediator with NF-kappaB suppressive properties, *J. Immunol.,* 171, 6329, 2003.

[13] Dinarello, C.A., Unraveling the NALP-3/IL-1beta inflammasome: A big lesson from a small mutation, *Immunity,* 20, 243, 2004.

[14] van De, S.B., Rothuizen, J., Pearson, P.L. et al., Identification of a new copper metabolism gene by positional cloning in a purebred dog population, *Hum. Mol. Genet.,* 11, 165, 2002.

[15] Tao, T.Y., Liu, F., Klomp, L. et al., The copper toxicosis gene product Murr1 directly interacts with the Wilson disease protein, *J. Biol. Chem.,* 278, 41593, 2003.

[16] Burstein, E., Hoberg, J.E., Wilkinson, A.S. et al., COMMD proteins, a novel family of structural and functional homologs of Murr1, *J. Biol. Chem.,* 280, 22222, 2005.

[17] Scher, J.U. Pillinger, M.H., 15d-PGJ2: The anti-inflammatory prostaglandin? *Clin. Immunol.,* 114, 100, 2005.

[18] Ricote, M., Li, A.C., Willson, T.M. et al., The peroxisome proliferator-activated receptor-gamma is a negative regulator of macrophage activation, *Nature,* 391, 79, 1998.

[19] Chen, F., Endogenous inhibitors of nuclear factor-kappaB, an opportunity for cancer control, *Cancer Res.,* 64, 8135, 2004.

[20] Dreyfus, D.H., Nagasawa, M., Gelfand, E.W. et al., Modulation of p53 activity by I-kappaBα: Evidence suggesting a common phylogeny between NF-kappa B and p53 transcription factors, *BMC Immunol.,* 6, 12, 2005.

[21] Rossi, A., Kapahi, P., Natoli, G. et al., Anti-inflammatory cyclopentenone prostaglandins are direct inhibitors of IκB kinase, *Nature,* 403, 103, 2000.

[22] Straus, D.S., Pascual, G., Li, M. et al., 15-deoxy-delta 12,14-prostaglandin J2 inhibits multiple steps in the NF-kappaB signaling pathway, *Proc. Natl. Acad. Sci. USA,* 97, 4844, 2000.

[23] Giri, S., Rattan, R., Singh, A.K. et al., The 15-deoxy-delta12,14-prostaglandin J2 inhibits the inflammatory response in primary rat astrocytes via down-regulating multiple steps in phosphatidylinositol 3-kinase-Akt-NF-kappaB-p300 pathway independent of peroxisome proliferator-activated receptor gamma, *J. Immunol.,* 173, 5196, 2004.

[24] Yang, J.P., Hori, M., Sanda, T. et al., Identification of a novel inhibitor of nuclear factor-kappaB, RelA-associated inhibitor, *J. Biol. Chem.,* 274, 15662, 1999.

[25] Herron, B.J., Rao, C., Liu, S. et al., A mutation in NFκB interacting protein 1 results in cardiomyopathy and abnormal skin development in *wa3* mice, *Hum. Mol. Genet.,* 14, 667, 2005.

[26] Panus, J.F., Smith, C.A., Ray, C.A. et al., Cowpox virus encodes a fifth member of the tumor necrosis factor receptor family: A soluble, secreted CD30 homologue, *Proc. Natl. Acad. Sci. USA,* 99, 8348, 2002.

[27] Saraiva, M., Smith, P., Fallon, P.G. et al., Inhibition of type 1 cytokine-mediated inflammation by a soluble CD30 homologue encoded by ectromelia (mousepox) virus, *J. Exp. Med.,* 196, 829, 2002.

[28] Benedict, C.A., Banks, T.A., and Ware, C.F., Death and survival: Viral regulation of TNF signaling pathways, *Curr. Opin. Immunol.,* 15, 59, 2003.

[29] Hauer, J., Puschner, S., Ramakrishnan, P. et al., TNF receptor (TNFR)-associated factor (TRAF) 3 serves as an inhibitor of TRAF2/5-mediated activation of the non-canonical NF-kappaB pathway by TRAF-binding TNFRs, *Proc. Natl. Acad. Sci. USA,* 102, 2874, 2005.

[30] Alcami, A., Symons, J.A., and Smith, G.L., The vaccinia virus soluble alpha/beta interferon (IFN) receptor binds to the cell surface and protects cells from the antiviral effects of IFN, *J. Virol.,* 74, 11230, 2000.

[31] Moss, B., *Poxviridae: The Viruses and Their Replication.* Philadelphia: Lippincott Williams and Wilkins, 2001.

[32] Reading, P.C., Khanna, A., and Smith, G.L., Vaccinia virus CrmE encodes a soluble and cell surface tumor necrosis factor receptor that contributes to virus virulence, *Virology*, 292, 285, 2002.

[33] Gracie, J.A., Robertson, S.E., and McInnes, I.B., Interleukin-18, *J. Leukoc. Biol.*, 73, 213, 2003.

[34] Horwitz, M.S., Function of adenovirus E3 proteins and their interactions with immunoregulatory cell proteins, *J. Gene Med.*, 6 Suppl 1, S172, 2004.

[35] Fessler, S.P., Chin, Y.R., and Horwitz, M.S., Inhibition of tumor necrosis factor (TNF) signal transduction by the adenovirus group C RID complex involves downregulation of surface levels of TNF receptor 1, *J. Virol.*, 78, 13113, 2004.

[36] Bowie, A., Kiss-Toth, E., Symons, J.A. et al., A46R and A52R from vaccinia virus are antagonists of host IL-1 and toll-like receptor signaling, *Proc. Natl. Acad. Sci. USA*, 97, 10162, 2000.

[37] Harte, M.T., Haga, I.R., Maloney, G. et al., The poxvirus protein A52R targets Toll-like receptor signaling complexes to suppress host defense, *J. Exp. Med.*, 197, 343, 2003.

[38] DiPerna, G., Stack, J., Bowie, A.G. et al., Poxvirus protein N1L targets the I-kappaB kinase complex, inhibits signaling to NF-kappaB by the tumor necrosis factor superfamily of receptors, and inhibits NF-kappaB and IRF3 signaling by toll-like receptors, *J. Biol. Chem.*, 279, 36570, 2004.

[39] McCoy, S.L., Kurtz, S.E., Macarthur, C.J. et al., Identification of a peptide derived from vaccinia virus A52R protein that inhibits cytokine secretion in response to TLR-dependent signaling and reduces *in vivo* bacterial-induced inflammation, *J. Immunol.*, 174, 3006, 2005.

[40] Kotwal, G.J., Hugin, A.W., and Moss, B., Mapping and insertional mutagenesis of a vaccinia virus gene encoding a 13,800-Da secreted protein, *Virology*, 171, 579, 1989.

[41] Spitkovsky, D., Hehner, S.P., Hofmann, T.G. et al., The human papillomavirus oncoprotein E7 attenuates NF-kappaB activation by targeting the IkappaB kinase complex, *J. Biol. Chem.*, 277, 25576, 2002.

[42] Joo, M., Hahn, Y.S., Kwon, M. et al., Hepatitis C virus core protein suppresses NF-kappaB activation and cyclooxygenase-2 expression by direct interaction with IkappaB kinase beta, *J. Virol.*, 79, 7648, 2005.

[43] Fennewald, S.M., Aronson, J.F., Zhang, L. et al., Alterations in NF-kappaB and RBP-Jkappa by arenavirus infection of macrophages *in vitro* and *in vivo*, *J. Virol.*, 76, 1154, 2002.

[44] Tait, S.W., Reid, E.B., Greaves, D.R. et al., Mechanism of inactivation of NF-kappaB by a viral homologue of IkappaBalpha. Signal-induced release of IkappaBalpha results in binding of the viral homologue to NF-kappaB, *J. Biol. Chem.*, 275, 34656, 2000.

[45] Besnard-Guerin, C., Belaidouni, N., Lassot, I. et al., HIV-1 Vpu sequesters beta-transducin repeat-containing protein (betaTrCP) in the cytoplasm and provokes the accumulation of beta-catenin and other SCFbetaTrCP substrates, *J. Biol. Chem.*, 279, 788, 2004.

[46] Tang, W., Pavlish, O.A., Spiegelman, V.S. et al., Interaction of Epstein-Barr virus latent membrane protein 1 with SCFHOS/beta-TrCP E3 ubiquitin ligase regulates extent of NF-kappaB activation, *J. Biol. Chem.*, 278, 48942, 2003.

[47] Collier-Hyams, L.S., Zeng, H., Sun, J. et al., Cutting edge: Salmonella AvrA effector inhibits the key proinflammatory, anti-apoptotic NF-kappaB pathway, *J. Immunol.*, 169, 2846, 2002.

[48] Prager, R., Mirold, S., Tietze, E. et al., Prevalence and polymorphism of genes encoding translocated effector proteins among clinical isolates of *Salmonella enterica*, *Int. J. Med. Microbiol.*, 290, 605, 2000.

[49] Patel, D., Huang, S.M., Baglia, L.A. et al., The E6 protein of human papillomavirus type 16 binds to and inhibits co-activation by CBP and p300, *Embo J.*, 18, 5061, 1999.

[50] Adams, J., The proteasome: A suitable antineoplastic target, *Nat. Rev. Cancer*, 4, 349, 2004.

[51] Adams, J., The development of proteasome inhibitors as anticancer drugs, *Cancer Cell*, 5, 417, 2004.

[52] Adams, J. and Kauffman, M., Development of the proteasome inhibitor Velcade (Bortezomib), *Cancer Invest.*, 22, 304, 2004.

[53] Burger, A.M. and Seth, A.K., The ubiquitin-mediated protein degradation pathway in cancer: Therapeutic implications, *Eur. J. Cancer*, 40, 2217, 2004.

[54] Richardson, P.G., Barlogie, B., Berenson, J. et al., A phase 2 study of bortezomib in relapsed, refractory myeloma, *N. Engl. J. Med.*, 348, 2609, 2003.

[55] Hideshima, T., Mitsiades, C., Akiyama, M. et al., Molecular mechanisms mediating antimyeloma activity of proteasome inhibitor PS-341, *Blood*, 101, 1530, 2003.

[56] Rajkumar, S.V., Richardson, P.G., Hideshima, T. et al., Proteasome inhibition as a novel therapeutic target in human cancer, *J. Clin. Oncol.*, 23, 630, 2005.

[57] Hideshima, T., Chauhan, D., Richardson, P. et al., NF-kappaB as a therapeutic target in multiple myeloma, *J. Biol. Chem.*, 277, 16639, 2002.

[58] Mitsiades, C.S., Mitsiades, N., Poulaki, V. et al., Activation of NF-kappaB and upregulation of intracellular anti-apoptotic proteins via the IGF-1/Akt signaling in human multiple myeloma cells: Therapeutic implications, *Oncogene*, 21, 5673, 2002.

[59] Mitsiades, N., Mitsiades, C.S., Poulaki, V. et al., Molecular sequelae of proteasome inhibition in human multiple myeloma cells, *Proc. Natl. Acad. Sci. USA*, 99, 14374, 2002.

[60] Irmler, M., Thome, M., Hahne, M. et al., Inhibition of death receptor signals by cellular FLIP, *Nature*, 388, 190, 1997.

[61] DuBois, R.N., Abramson, S.B., Crofford, L. et al., Cyclooxygenase in biology and disease, *FASEB J.*, 12, 1063, 1998.

[62] Vane, J.R., Inhibition of prostaglandin synthesis as a mechanism of action for aspirin-like drugs, *Nat. New Biol.*, 231, 232, 1971.

[63] Gupta, R.A. and DuBois, R.N., Colorectal cancer prevention and treatment by inhibition of cyclooxygenase-2, *Nat. Rev. Cancer*, 1, 11, 2001.

[64] Williams, C.S., Watson, A.J., Sheng, H. et al., Celecoxib prevents tumor growth in vivo without toxicity to normal gut: Lack of correlation between in vitro and in vivo models, *Cancer Res.*, 60, 6045, 2000.

[65] Yin, M.J., Yamamoto, Y., and Gaynor, R.B., The anti-inflammatory agents aspirin and salicylate inhibit the activity of I(kappa)B kinase-beta, *Nature*, 396, 77, 1998.

[66] Yamamoto, Y., Yin, M.J., Lin, K.M. et al., Sulindac inhibits activation of the NF-kappaB pathway, *J. Biol. Chem.*, 274, 27307, 1999.

[67] D'Acquisto, F., May, M.J., and Ghosh, S., Inhibition of nuclear factor kappaB (NF-κB): An emerging theme in anti-inflammatory therapies, *Mol. Interv.*, 2, 22, 2002.

[68] Takada, Y., Bhardwaj, A., Potdar, P. et al., Nonsteroidal anti-inflammatory agents differ in their ability to suppress NF-kappaB activation, inhibition of expression of cyclooxygenase-2 and cyclin D1, and abrogation of tumor cell proliferation, *Oncogene*, 23, 9247, 2004.

[69] Shishodia, S., Koul, D., and Aggarwal, B.B., Cyclooxygenase (COX)-2 inhibitor celecoxib abrogates TNF-induced NF-kappaB activation through inhibition of activation of IkappaBalpha kinase and Akt in human non-small cell lung carcinoma: Correlation with suppression of COX-2 synthesis, *J. Immunol.*, 173, 2011, 2004.

[70] Smalley, W.E. and DuBois, R.N., Colorectal cancer and nonsteroidal anti-inflammatory drugs, *Adv. Pharmacol.*, 39, 1, 1997.

[71] Thun, M.J., Henley, S.J., and Patrono, C., Nonsteroidal anti-inflammatory drugs as anticancer agents: Mechanistic, pharmacologic, and clinical issues, *J. Natl. Cancer Inst.*, 94, 252, 2002.

[72] Harris, R.E., Chlebowski, R.T., Jackson, R.D. et al., Breast cancer and nonsteroidal anti-inflammatory drugs: Prospective results from the Women's Health Initiative, *Cancer Res.*, 63, 6096, 2003.

[73] Steinbach, G., Lynch, P.M., Phillips, R.K. et al., The effect of celecoxib, a cyclooxygenase-2 inhibitor, in familial adenomatous polyposis, *N. Engl. J. Med.*, 342, 1946, 2000.

[74] Kismet, K., Akay, M.T., Abbasoglu, O. et al., Celecoxib: A potent cyclooxygenase-2 inhibitor in cancer prevention, *Cancer Detect. Prev.*, 28, 127, 2004.

[75] Tegeder, I., Niederberger, E., Israr, E. et al., Inhibition of NF-kappaB and AP-1 activation by R- and S-flurbiprofen, *FASEB J.*, 15, 2, 2001.

[76] Brune, K., Beck, W.S., Geisslinger, G. et al., Aspirin-like drugs may block pain independently of prostaglandin synthesis inhibition, *Experientia*, 47, 257, 1991.

[77] Almawi, W.Y. and Melemedjian, O.K., Molecular mechanisms of glucocorticoid antiproliferative effects: Antagonism of transcription factor activity by glucocorticoid receptor, *J. Leukoc. Biol.*, 71, 9, 2002.

[78] Smoak, K.A. and Cidlowski, J.A., Mechanisms of glucocorticoid receptor signaling during inflammation, *Mech. Ageing Dev.*, 125, 697, 2004.

[79] Scheinman, R.I., Cogswell, P.C., Lofquist, A.K. et al., Role of transcriptional activation of IkappaBalpha in mediation of immunosuppression by glucocorticoids, *Science*, 270, 283, 1995.

[80] Auphan, N., DiDonato, J.A., Rosette, C. et al., Immunosuppression by glucocorticoids: Inhibition of NF-kappaB activity through induction of IkappaB synthesis, *Science*, 270, 286, 1995.

[81] Aljada, A., Ghanim, H., Assian, E. et al., Increased IkappaB expression and diminished nuclear NF-kappaB in human mononuclear cells following hydrocortisone injection, *J. Clin. Endocrinol. Metab.*, 84, 3386, 1999.

[82] De Bosscher, K., Vanden Berghe, W., and Haegeman, G., The interplay between the glucocorticoid receptor and nuclear factor-kappaB or activator protein-1: Molecular mechanisms for gene repression, *Endocr. Rev.*, 24, 488, 2003.

[83] Kleinert, H., Euchenhofer, C., Ihrig-Biedert, I. et al., Glucocorticoids inhibit the induction of nitric oxide synthase II by down-regulating cytokine-induced activity of transcription factor nuclear factor-kappaB, *Mol. Pharmacol.*, 49, 15, 1996.

[84] De Bosscher, K., Schmitz, M.L., Vanden Berghe, W. et al., Glucocorticoid-mediated repression of nuclear factor-kappaB-dependent transcription involves direct interference with transactivation, *Proc. Natl. Acad. Sci. USA*, 94, 13504, 1997.

[85] Adcock, I.M., Brown, C.R., Gelder, C.M. et al., Effects of glucocorticoids on transcription factor activation in human peripheral blood mononuclear cells, *Am. J. Physiol.*, 268, C331, 1995.

[86] De Bosscher, K., Vanden Berghe, W., Vermeulen, L. et al., Glucocorticoids repress NF-kappaB-driven genes by disturbing the interaction of p65 with the basal transcription machinery, irrespective of coactivator levels in the cell, *Proc. Natl. Acad. Sci. USA,* 97, 3919, 2000.

[87] Nissen, R.M. and Yamamoto, K.R., The glucocorticoid receptor inhibits NFkappaB by interfering with serine-2 phosphorylation of the RNA polymerase II carboxyterminal domain, *Genes Dev.,* 14, 2314, 2000.

[88] Sheppard, K.A., Phelps, K.M., Williams, A.J. et al., Nuclear integration of glucocorticoid receptor and nuclear factor-kappaB signaling by CREB-binding protein and steroid receptor coactivator-1, *J. Biol. Chem.,* 273, 29291, 1998.

[89] McKay, L.I. and Cidlowski, J.A., CBP (CREB binding protein) integrates NF-kappaB (nuclear factor-kappaB) and glucocorticoid receptor physical interactions and antagonism, *Mol. Endocrinol.,* 14, 1222, 2000.

[90] Li, J., Lin, Q., Yoon, H.G. et al., Involvement of histone methylation and phosphorylation in regulation of transcription by thyroid hormone receptor, *Mol. Cell Biol.,* 22, 5688, 2002.

[91] Peterson, C.L. and Laniel, M.A., Histones and histone modifications, *Curr. Biol.,* 14, R546, 2004.

[92] Ayroldi, E., Migliorati, G., Bruscoli, S. et al., Modulation of T-cell activation by the glucocorticoid-induced leucine zipper factor via inhibition of nuclear factor-kappaB, *Blood,* 98, 743, 2001.

[93] Franks, M.E., Macpherson, G.R., and Figg, W.D., Thalidomide, *Lancet,* 363, 1802, 2004.

[94] Sleijfer, S., Kruit, W.H., and Stoter, G., Thalidomide in solid tumours: The resurrection of an old drug, *Eur. J. Cancer,* 40, 2377, 2004.

[95] Yoshida, S., Ono, M., Shono, T. et al., Involvement of interleukin-8, vascular endothelial growth factor, and basic fibroblast growth factor in tumor necrosis factor alpha-dependent angiogenesis, *Mol. Cell. Biol.,* 17, 4015, 1997.

[96] Stephens, T.D., Bunde, C.J., and Fillmore, B.J., Mechanism of action in thalidomide teratogenesis, *Biochem. Pharmacol.,* 59, 1489, 2000.

[97] Moreira, A.L., Sampaio, E.P., Zmuidzinas, A. et al., Thalidomide exerts its inhibitory action on tumor necrosis factor alpha by enhancing mRNA degradation, *J. Exp. Med.,* 177, 1675, 1993.

[98] Marriott, J.B., Clarke, I.A., Dredge, K. et al., Thalidomide and its analogues have distinct and opposing effects on TNF-alpha and TNFR2 during co-stimulation of both CD4(+) and CD8(+) T cells, *Clin. Exp. Immunol.,* 130, 75, 2002.

[99] Keifer, J.A., Guttridge, D.C., Ashburner, B.P. et al., Inhibition of NF-kappa B activity by thalidomide through suppression of IkappaB kinase activity, *J. Biol. Chem.,* 276, 22382, 2001.

[100] Epinat, J.C. and Gilmore, T.D., Diverse agents act at multiple levels to inhibit the Rel/NF-kappaB signal transduction pathway, *Oncogene,* 18, 6896, 1999.

[101] Bowie, A. and O'Neill, L.A., Oxidative stress and nuclear factor-kappaB activation: A reassessment of the evidence in the light of recent discoveries, *Biochem. Pharmacol.,* 59, 13, 2000.

[102] Rahman, I., Marwick, J., and Kirkham, P., Redox modulation of chromatin remodeling: Impact on histone acetylation and deacetylation, NF-kappaB and pro-inflammatory gene expression, *Biochem. Pharmacol.,* 68, 1255, 2004.

[103] Hensley, K., Robinson, K.A., Gabbita, S.P. et al., Reactive oxygen species, cell signaling, and cell injury, *Free Radic. Biol. Med.,* 28, 1456, 2000.

[104]   Hussain, S.P., Hofseth, L.J., and Harris, C.C., Radical causes of cancer, *Nat. Rev. Cancer,* 3, 276, 2003.

[105]   Bowie, A.G., Moynagh, P.N., and O'Neill, L.A., Lipid peroxidation is involved in the activation of NF-kappaB by tumor necrosis factor but not interleukin-1 in the human endothelial cell line ECV304. Lack of involvement of H2O2 in NF-kappaB activation by either cytokine in both primary and transformed endothelial cells, *J. Biol. Chem.,* 272, 25941, 1997.

[106]   Ginn-Pease, M.E. and Whisler, R.L., Optimal NF kappa B mediated transcriptional responses in Jurkat T cells exposed to oxidative stress are dependent on intracellular glutathione and costimulatory signals, *Biochem. Biophys. Res. Commun.,* 226, 695, 1996.

[107]   Cho, S., Urata, Y., Iida, T. et al., Glutathione downregulates the phosphorylation of I kappa B: Autoloop regulation of the NF-kappaB-mediated expression of NF-kappaB subunits by TNF-alpha in mouse vascular endothelial cells, *Biochem. Biophys. Res. Commun.,* 253, 104, 1998.

[108]   Janssen-Heininger, Y.M., Macara, I., and Mossman, B.T., Cooperativity between oxidants and tumor necrosis factor in the activation of nuclear factor (NF)-kappaB: Requirement of Ras/mitogen-activated protein kinases in the activation of NF-kappaB by oxidants, *Am. J. Respir. Cell Mol. Biol.,* 20, 942, 1999.

[109]   Kamata, H., Manabe, T., Oka, S. et al., Hydrogen peroxide activates IkappaB kinases through phosphorylation of serine residues in the activation loops, *FEBS Lett.,* 519, 231, 2002.

[110]   Nakamura, Y., Murakami, A., Ohto, Y. et al., Suppression of tumor promoter-induced oxidative stress and inflammatory responses in mouse skin by a superoxide generation inhibitor 1'-acetoxychavicol acetate, *Cancer Res.,* 58, 4832, 1998.

[111]   Ichikawa, H., Takada, Y., Murakami, A. et al., Identification of a novel blocker of I{kappa}B{alpha} kinase that enhances cellular apoptosis and inhibits cellular invasion through suppression of NF-{kappa}B-regulated gene products, *J. Immunol.,* 174, 7383, 2005.

[112]   Surh, Y.J., Cancer chemoprevention with dietary phytochemicals, *Nat. Rev. Cancer,* 3, 768, 2003.

[113]   Plummer, S.M., Holloway, K.A., Manson, M.M. et al., Inhibition of cyclo-oxygenase 2 expression in colon cells by the chemopreventive agent curcumin involves inhibition of NF-kappaB activation via the NIK/IKK signalling complex, *Oncogene,* 18, 6013, 1999.

[114]   Chun, K.S., Keum, Y.S., Han, S.S. et al., Curcumin inhibits phorbol ester-induced expression of cyclooxygenase-2 in mouse skin through suppression of extracellular signal-regulated kinase activity and NF-kappaB activation, *Carcinogenesis,* 24, 1515, 2003.

[115]   Philip, S. and Kundu, G.C., Osteopontin induces nuclear factor-kappaB-mediated promatrix metalloproteinase-2 activation through I kappa B alpha/IKK signaling pathways, and curcumin (diferulolylmethane) down-regulates these pathways, *J. Biol. Chem.,* 278, 14487, 2003.

[116]   Kundu, J.K. and Surh, Y.J., Molecular basis of chemoprevention by resveratrol: NF-kappaB and AP-1 as potential targets, *Mutat. Res.,* 555, 65, 2004.

[117]   Wessler, S., Muenzner, P., Meyer, T.F. et al., The anti-inflammatory compound curcumin inhibits Neisseria gonorrhoeae-induced NF-kappaB signaling, release of proinflammatory cytokines/chemokines and attenuates adhesion in late infection, *Biol. Chem.,* 386, 481, 2005.

[118] Carcamo, J.M., Pedraza, A., Borquez-Ojeda, O. et al., Vitamin C is a kinase inhibitor: Dehydroascorbic acid inhibits IkappaBalpha kinase beta, *Mol. Cell. Biol.,* 24, 6645, 2004.

[119] Smyth, M.J., Cretney, E., Kershaw, M.H. et al., Cytokines in cancer immunity and immunotherapy, *Immunol. Rev.,* 202, 275, 2004.

[120] Dajee, M., Lazarov, M., Zhang, J.Y. et al., NF-kappaB blockade and oncogenic Ras trigger invasive human epidermal neoplasia, *Nature,* 421, 639, 2003.

[121] Lawrence, T., Gilroy, D.W., Colville-Nash, P.R. et al., Possible new role for NF-kappaB in the resolution of inflammation, *Nat. Med.,* 7, 1291, 2001.

[122] Scheinfeld, N., A comprehensive review and evaluation of the side effects of the tumor necrosis factor alpha blockers etanercept, infliximab and adalimumab, *J. Dermatolog. Treat.,* 15, 280, 2004.

[123] Ravi, R. and Bedi, A., NF-kappaB in cancer — A friend turned foe, *Drug Resist. Updat.,* 7, 53, 2004.

[124] Lin, E.Y., Nguyen, A.V., Russell, R.G. et al., Colony-stimulating factor 1 promotes progression of mammary tumors to malignancy, *J. Exp. Med.,* 193, 727, 2001.

[125] Moore, R.J., Owens, D.M., Stamp, G. et al., Mice deficient in tumor necrosis factor-alpha are resistant to skin carcinogenesis, *Nat. Med.,* 5, 828, 1999.

[126] Mochizuki, Y., Nakanishi, H., Kodera, Y. et al., TNF-alpha promotes progression of peritoneal metastasis as demonstrated using a green fluorescence protein (GFP)-tagged human gastric cancer cell line, *Clin. Exp. Metastasis,* 21, 39, 2004.

[127] El Omar, E.M., Carrington, M., Chow, W.H. et al., Interleukin-1 polymorphisms associated with increased risk of gastric cancer, *Nature,* 404, 398, 2000.

[128] Sparmann, A. and Bar-Sagi, D., Ras-induced interleukin-8 expression plays a critical role in tumor growth and angiogenesis, *Cancer Cell,* 6, 447, 2004.

[129] Anderson, G.M., Nakada, M.T., and DeWitte, M., Tumor necrosis factor-alpha in the pathogenesis and treatment of cancer, *Curr. Opin. Pharmacol.,* 4, 314, 2004.

[130] Jiang, Y., Genant, H.K., Watt, I. et al., A multicenter, double-blind, dose-ranging, randomized, placebo-controlled study of recombinant human interleukin-1 receptor antagonist in patients with rheumatoid arthritis: Radiologic progression and correlation of Genant and Larsen scores, *Arthritis Rheum.,* 43, 1001, 2000.

[131] Vilcek, J. and Feldmann, M., Historical review: Cytokines as therapeutics and targets of therapeutics, *Trends Pharmacol. Sci.,* 25, 201, 2004.

[132] Furst, D.E., Anakinra: Review of recombinant human interleukin-I receptor antagonist in the treatment of rheumatoid arthritis, *Clin. Ther.,* 26, 1960, 2004.

[133] Stohl, W., B lymphocyte stimulator protein levels in systemic lupus erythematosus and other diseases, *Curr. Rheumatol. Rep.,* 4, 345, 2002.

[134] Mackay, F. and Tangye, S.G., The role of the BAFF/APRIL system in B cell homeostasis and lymphoid cancers, *Curr. Opin. Pharmacol.,* 4, 347, 2004.

[135] Watts, T.H., TNF/TNFR family members in costimulation of T cell responses, *Annu. Rev. Immunol.,* 23, 23, 2005.

[136] Andreakos, E.T., Foxwell, B.M., Brennan, F.M. et al., Cytokines and anti-cytokine biologicals in autoimmunity: Present and future, *Cytokine Growth Factor Rev.,* 13, 299, 2002.

[137] Tarner, I.H., Muller-Ladner, U., and Fathman, C.G., Targeted gene therapy: Frontiers in the development of "smart drugs," *Trends Biotechnol.,* 22, 304, 2004.

[138] Dietz, G.P. and Bahr, M., Delivery of bioactive molecules into the cell: The Trojan horse approach, *Mol. Cell Neurosci.,* 27, 85, 2004.

[139] Van Antwerp, D.J., Martin, S.J., Kafri, T. et al., Suppression of TNF-alpha-induced apoptosis by NF-kappaB, *Science,* 274, 787, 1996.

[140] Mora, A.L., Lavoy, J., McKean, M. et al., Prevention of NF-{kappa}B activation *in vivo* by a cell permeable NF-{kappa}B inhibitor peptide, *Am. J. Physiol. Lung Cell Mol. Physiol.*, 289, L536–544, 2005.

[141] Adams, G., Vessillier, S., Dreja, H. et al., Targeting cytokines to inflammation sites, *Nat. Biotechnol.*, 21, 1314, 2003.

[142] Li, Q. and Verma, I.M., NF-kappaB regulation in the immune system, *Nat. Rev. Immunol.*, 2, 725, 2002.

[143] Karin, M., Yamamoto, Y., and Wang, Q.M., The IKK NF-kappaB system: A treasure trove for drug development, *Nat. Rev. Drug Discov.*, 3, 17, 2004.

[144] Tergaonkar, V., Bottero, V., Ikawa, M. et al., IkappaB kinase-independent IkappaBalpha degradation pathway: Functional NF-kappaB activity and implications for cancer therapy, *Mol. Cell. Biol.*, 23, 8070, 2003.

[145] Ryan, K.M., Ernst, M.K., Rice, N.R. et al., Role of NF-kappaB in p53-mediated programmed cell death, *Nature*, 404, 892, 2000.

[146] Kato, T., Jr., Delhase, M., Hoffmann, A. et al., CK2 is a C-terminal IkappaB kinase responsible for NF-kappaB activation during the UV response, *Mol. Cell*, 12, 829, 2003.

[147] Panta, G.R., Kaur, S., Cavin, L.G. et al., ATM and the catalytic subunit of DNA-dependent protein kinase activate NF-kappaB through a common MEK/extracellular signal-regulated kinase/p90(rsk) signaling pathway in response to distinct forms of DNA damage, *Mol. Cell. Biol.*, 24, 1823, 2004.

[148] Bohuslav, J., Chen, L.F., Kwon, H. et al., p53 induces NF-kappaB activation by an IkappaB kinase-independent mechanism involving phosphorylation of p65 by ribosomal S6 kinase 1, *J. Biol. Chem.*, 279, 26115, 2004.

[149] May, M.J., Marienfeld, R.B., and Ghosh, S., Characterization of the IkappaB-kinase NEMO binding domain, *J. Biol. Chem.*, 277, 45992, 2002.

[150] May, M.J., D'Acquisto, F., Madge, L.A. et al., Selective inhibition of NF-kappaB activation by a peptide that blocks the interaction of NEMO with the IkappaB kinase complex, *Science*, 289, 1550, 2000.

[151] Thomas, R.P., Farrow, B.J., Kim, S. et al., Selective targeting of the nuclear factor-kappaB pathway enhances tumor necrosis factor-related apoptosis-inducing ligand-mediated pancreatic cancer cell death, *Surgery*, 132, 127, 2002.

[152] Dasgupta, S., Jana, M., Zhou, Y. et al., Antineuroinflammatory effect of NF-kappaB essential modifier-binding domain peptides in the adoptive transfer model of experimental allergic encephalomyelitis, *J. Immunol.*, 173, 1344, 2004.

[153] Choi, M., Rolle, S., Wellner, M. et al., Inhibition of NF-kappaB by a TAT-NEMO-binding domain peptide accelerates constitutive apoptosis and abrogates LPS-delayed neutrophil apoptosis, *Blood*, 102, 2259, 2003.

[154] Dai, S., Hirayama, T., Abbas, S. et al., The IkappaB kinase (IKK) inhibitor, NEMO-binding domain peptide, blocks osteoclastogenesis and bone erosion in inflammatory arthritis, *J. Biol. Chem.*, 279, 37219, 2004.

[155] Ankermann, T., Reisner, A., Wiemann, T. et al., Topical inhibition of nuclear factor-kappaB enhances reduction in lung edema by surfactant in a piglet model of airway lavage, *Crit. Care Med.*, 33, 1384, 2005.

[156] Rehman, K.K., Bertera, S., Bottino, R. et al., Protection of islets by in situ peptide-mediated transduction of the IkappaB kinase inhibitor NEMO-binding domain peptide, *J. Biol. Chem.*, 278, 9862, 2003.

[157] Horng, T., Barton, G.M., and Medzhitov, R., TIRAP: An adapter molecule in the Toll signaling pathway, *Nat. Immunol.*, 2, 835, 2001.

[158]   Agou, F., Courtois, G., Chiaravalli, J. et al., Inhibition of NF-kappaB activation by peptides targeting NF-kappaB essential modulator (*nemo*) oligomerization, *J. Biol. Chem.*, 279, 54248, 2004.

[159]   Lin, Y.Z., Yao, S.Y., Veach, R.A. et al., Inhibition of nuclear translocation of transcription factor NF-kappaB by a synthetic peptide containing a cell membrane-permeable motif and nuclear localization sequence, *J. Biol. Chem.*, 270, 14255, 1995.

[160]   Torgerson, T.R., Colosia, A.D., Donahue, J.P. et al., Regulation of NF-kappaB, AP-1, NFAT, and STAT1 nuclear import in T lymphocytes by noninvasive delivery of peptide carrying the nuclear localization sequence of NF-kappaB p50, *J. Immunol.*, 161, 6084, 1998.

[161]   Takada, Y., Singh, S., and Aggarwal, B.B., Identification of a p65 peptide that selectively inhibits NF-kappaB activation induced by various inflammatory stimuli and its role in down-regulation of NF-kappaB-mediated gene expression and up-regulation of apoptosis, *J. Biol. Chem.*, 279, 15096, 2004.

[162]   Morishita, R., Tomita, N., Kaneda, Y. et al., Molecular therapy to inhibit NF-kappaB activation by transcription factor decoy oligonucleotides, *Curr. Opin. Pharmacol.*, 4, 139, 2004.

[163]   Karagiannis, T.C. and El Osta, A., RNA interference and potential therapeutic applications of short interfering RNAs, *Cancer Gene Ther.*, 12, 787–795, 2005.

[164]   Desmet, C., Gosset, P., Pajak, B. et al., Selective blockade of NF-kappaB activity in airway immune cells inhibits the effector phase of experimental asthma, *J. Immunol.*, 173, 5766, 2004.

[165]   Guo, J., Verma, U.N., Gaynor, R.B. et al., Enhanced chemosensitivity to irinotecan by RNA interference-mediated down-regulation of the nuclear factor-kappaB p65 subunit, *Clin. Cancer Res.*, 10, 3333, 2004.

[166]   Pinkenburg, O., Platz, J., Beisswenger, C. et al., Inhibition of NF-kappaB mediated inflammation by siRNA expressed by recombinant adeno-associated virus, *J. Virol. Methods*, 120, 119, 2004.

[167]   Takaesu, G., Surabhi, R.M., Park, K.J. et al., TAK1 is critical for IkappaB kinase-mediated activation of the NF-kappaB pathway, *J. Mol. Biol.*, 326, 105, 2003.

[168]   Ducut Sigala, J.L., Bottero, V., Young, D.B. et al., Activation of transcription factor NF-kappaB requires ELKS, an IkappaB kinase regulatory subunit, *Science*, 304, 1963, 2004.

[169]   Silverman, N. and Maniatis, T., NF-kappaB signaling pathways in mammalian and insect innate immunity, *Genes Dev.*, 15, 2321, 2001.

[170]   Silverman, N., Zhou, R., Erlich, R.L. et al., Immune activation of NF-kappaB and JNK requires *Drosophila* TAK1, *J. Biol. Chem.*, 278, 48928, 2003.

[171]   Zhou, R., Silverman, N., Hong, M. et al., The role of ubiquitination in *Drosophila* innate immunity, *J. Biol. Chem.*, 280, 34048–55, 2005.

[172]   Correa, R.G., Tergaonkar, V., Ng, J.K. et al., Characterization of NF-kappaB/IkappaB proteins in zebra fish and their involvement in notochord development, *Mol. Cell Biol.*, 24, 5257, 2004.

[173]   Correa, R.G., Matsui, T., Tergaonkar, V. et al., Zebrafish IkappaB kinase 1 negatively regulates NF-kappaB activity, *Curr. Biol.*, 15, 1291, 2005.

[174]   Kapp, E. and Ghosh, S., Inhibition of NF-κB by sodium salicylate and aspirin, *Science*, 265, 956, 1994.

[175]   Jimi, E., Aoki, K., Sairo, H. et al., Selective inhibition of NF-κB blocks osteoclastogenesis and prevents inflammatory bone destruction, *in vivo, Nature Med.*, 10, 617, 2004.

Index